Algorithms for Intelligent Systems

This book series publishes research on the analysis and development of algorithms for intelligent systems with their applications to various real world problems. It covers research related to autonomous agents, multi-agent systems, behavioral modeling, reinforcement learning, game theory, mechanism design, machine learning, meta-heuristic search, optimization, planning and scheduling, artificial neural networks, evolutionary computation, swarm intelligence and other algorithms for intelligent systems.

The book series includes recent advancements, modification and applications of the artificial neural networks, evolutionary computation, swarm intelligence, artificial immune systems, fuzzy system, autonomous and multi agent systems, machine learning and other intelligent systems related areas. The material will be beneficial for the graduate students, post-graduate students as well as the researchers who want a broader view of advances in algorithms for intelligent systems. The contents will also be useful to the researchers from other fields who have no knowledge of the power of intelligent systems, e.g. the researchers in the field of bioinformatics, biochemists, mechanical and chemical engineers, economists, musicians and medical practitioners.

The series publishes monographs, edited volumes, advanced textbooks and selected proceedings.

More information about this series at http://www.springer.com/series/16171

Ibrahim Aljarah · Hossam Faris ·
Seyedali Mirjalili
Editors

Evolutionary Data Clustering: Algorithms and Applications

Springer

Editors
Ibrahim Aljarah
King Abdullah II School
for Information Technology
The University of Jordan
Amman, Jordan

Hossam Faris
King Abdullah II School
for Information Technology
The University of Jordan
Amman, Jordan

Seyedali Mirjalili
Center for Artificial Intelligence
Research and Optimization
Torrens University Australia
Brisbane, QLD, Australia

Griffith University
Brisbane, QLD, Australia

ISSN 2524-7565 ISSN 2524-7573 (electronic)
Algorithms for Intelligent Systems
ISBN 978-981-33-4190-6 ISBN 978-981-33-4191-3 (eBook)
https://doi.org/10.1007/978-981-33-4191-3

This Springer imprint is published by the registered company Springer Nature Singapore Pte Ltd.
The registered company address is: 152 Beach Road, #21-01/04 Gateway East, Singapore 189721, Singapore

Preface

This book provides an in-depth analysis of the current evolutionary clustering techniques, discussing the most highly regarded methods for data clustering. The book provides literature reviews about single objective and multi-objective evolutionary clustering algorithms. In addition, the book provides a comprehensive review of the fitness functions and evaluation measures that are used in most of the evolutionary clustering algorithms. Furthermore, the book provides a conceptual analysis including definition, validation and quality measures, applications, and implementations for data clustering using classical and modern nature-inspired techniques. It features a range of proven and recent nature-inspired algorithms used for data clustering, including particle swarm optimization, ant colony optimization, grey wolf optimizer, salp swarm algorithm, multi-verse optimizer, Harris hawks optimization, and beta-hill climbing optimization.

The book also covers applications of evolutionary data clustering in diverse fields such as image segmentation, security, medical applications, EEG-based person identification, and pavement infrastructure asset management.

Amman, Jordan — Dr. Ibrahim Aljarah
Amman, Jordan — Prof. Hossam Faris
Brisbane, Australia — Dr. Seyedali Mirjalili
August 2020

Contents

Editors and Contributors

About the Editors

Ibrahim Aljarah is an associate professor of BIG Data Mining and Computational Intelligence at the University of Jordan-Department of Information Technology, Jordan. Currently, he is the Director of the Open Educational Resources and Blended Learning Center at The University of Jordan. He obtained his PhD in computer science from the North Dakota State University, USA, in 2014. He also obtained the master degree in computer science and information systems from the Jordan University of Science and Technology – Jordan in 2006. He obtained the bachelor degree in Computer Science from Yarmouk University - Jordan, 2003. He participated in many conferences in the field of data mining, machine learning, and Big data such as CEC, GECCO, NTIT, CSIT, IEEE NABIC, CASON, and BIGDATA Congress. Furthermore, he contributed in many projects in USA such as Vehicle Class Detection System (VCDS), Pavement Analysis Via Vehicle Electronic Telemetry (PAVVET), and Farm Cloud Storage System(CSS) projects. He has published more than 60 papers in refereed international conferences and journals. His research focuses on Data Mining, Data Science, Machine Learning, Opinion Mining, Sentiment Analysis, Big Data, MapReduce, Hadoop, Swarm intelligence, Evolutionary Computation, and large-scale distributed algorithms.

Hossam Faris is a Professor in the Information Technology Department at King Abdullah II School for Information Technology at The University of Jordan, Jordan. Hossam Faris received his B.A. and M.Sc. degrees in computer science from the Yarmouk University and Al-Balqa' Applied University in 2004 and 2008, respectively, in Jordan. He was awarded a full-time competition-based scholarship from the Italian Ministry of Education and Research to peruse his Ph.D. degrees in e-Business at the University of Salento, Italy, where he obtained his Ph.D. degree in 2011. In 2016, he worked as a postdoctoral researcher with the GeNeura team at the Information and Communication Technologies Research Center (CITIC),

University of Granada, Spain. His research interests include applied computational intelligence, evolutionary computation, knowledge systems, data mining, semantic web, and ontologies.

Seyedali Mirjalili is an Associate Professor and the director of the Centre for Artificial Intelligence Research and Optimization at Torrens University Australia. He is internationally recognized for his advances in Swarm Intelligence and Optimization, including the first set of algorithms from a synthetic intelligence standpoint - a radical departure from how natural systems are typically understood - and a systematic design framework to reliably benchmark, evaluate, and propose computationally cheap robust optimization algorithms. He has published over 200 publications with over 20,000 citations and is in the list of 1% highly-cited researchers by Web of Science. Seyedali is a senior member of IEEE and an associate editor of several journals including Neurocomputing, Applied Soft Computing, Advances in Engineering Software, Applied Intelligence, and IEEE Access. His research interests include Robust Optimization, Engineering Optimization, Multi-objective Optimization, Swarm Intelligence, Evolutionary Algorithms, Machine Learning, and Artificial Neural Networks.

Contributors

Ammar Kamal Abasi School of Computer Sciences, Universiti Sains Malaysia, George Town, Pulau Pinang, Malaysia

Salwani Abdullah Faculty of Information Science and Technology, Center for Artificial Intelligence, Universiti Kebangsaan Malaysia, Bangi, Selangor, Malaysia

Ruba Abu Khurma King Abdullah II School for Information Technology, The University of Jordan, Amman, Jordan

Mohammed Azmi Al-Betar Department of Information Technology, MSAI, College of Engineering and Information Technology, Ajman University, Ajman, United Arab Emirates;
IT Department, Al-Huson University College, Al-Balqa Applied University, Irbid, Jordan

Hasnain Ali Formerly, Indian Institute of Technology Delhi, New Delhi, India

Ibrahim Aljarah King Abdullah II School for Information Technology, The University of Jordan, Amman, Jordan

Mohamad Al-laham Department of Management Information Systems, Amman University College, Al-Balqa Applied University (BAU), Amman, Jordan

Iman Almomani King Abdullah II School for Information Technology, The University of Jordan, Amman, Jordan;
Security Engineering Lab, Prince Sultan University, Riyadh, Saudi Arabia

Laila Al-Qaisi Information Systems and Networks Department, Faculty of Information Technology, The World Islamic Sciences and Education University, Amman, Jordan

Zaid Abdi Alkareem Alyasseri ECE Department, Faculty of Engineering, University of Kufa, Najaf, Iraq;
Faculty of Information Science and Technology, Center for Artificial Intelligence, Universiti Kebangsaan Malaysia, Bangi, Selangor, Malaysia

Ala' M. Al-Zoubi School of Science, Technology and Engineering, University of Granada, Granada, Spain

Mahmoud M. Al-Zoubi Department of Commutation and Information Technology, Yarmouk Water Company, Irbid, Jordan

Hossam Faris King Abdullah II School for Information Technology, The University of Jordan, Amman, Jordan

Saqib Gulzar Formerly, Indian Institute of Technology Delhi, New Delhi, India

Maria Habib King Abdullah II School for Information Technology, The University of Jordan, Amman, Jordan

Mohammad A. Hassonah King Abdullah II School for Information Technology, The University of Jordan, Amman, Jordan

Ahamad Tajudin Khader School of Computer Sciences, Universiti Sains Malaysia, George Town, Pulau Pinang, Malaysia

Majdi Mafarja Department of Computer Science, Birzeit University, Birzeit, West Bank, Palestine

Sharif Naser Makhadmeh School of Computer Sciences, Universiti Sains Malaysia, George Town, Pulau Pinang, Malaysia

Seyedali Mirjalili Center for Artificial Intelligence Research and Optimization, Torrens University Australia, Fortitude Valley, Brisbane, QLD, Australia

Syibrah Naim Technology Department, Endicott College of International Studies (ECIS), Woosong University, Daejeon, South Korea

Razan Nujoom King Abdullah II School for Information Technology, The University of Jordan, Amman, Jordan

João P. Papa Department of Computing, São Paulo State University - UNESP, Bauru, Brazil

Raneem Qaddoura Information Technology, Philadelphia University, Amman, Jordan

Dao Vu Truong Son International University, Vietnam National University - HCMC, Ho Chi Minh City, Vietnam

Pham Nhat Tan International University, Vietnam National University - HCMC, Ho Chi Minh City, Vietnam

Hamza Turabieh Information Technology Department, CIT College, Taif University, Taif, Saudi Arabia

Introduction to Evolutionary Data Clustering and Its Applications

Ibrahim Aljarah, Maria Habib, Hossam Faris, and Seyedali Mirjalili

Abstract Clustering is concerned with splitting a dataset into groups (clusters) that represent the natural homogeneous characteristics of the data. Remarkably, clustering has a crucial role in numerous types of applications. Essentially, the applications include social sciences, biological and medical applications, information retrieval and web search algorithms, pattern recognition, image processing, machine learning, and data mining. Even that clustering is ubiquitous over a variety of areas. However, clustering approaches suffer from several drawbacks. Mainly, they are highly susceptible to clusters' initial centroids which allows a particular dataset to easily fall within a local optimum. Handling clustering as an optimization problem is deemed an NP-hard optimization problem. However, metaheuristic algorithms are a dominant class of algorithms for solving tough and NP-hard optimization problems. This chapter anticipates the use of evolutionary algorithms for addressing the problem of clustering optimization. Therefore, it presents an introduction to clustering and evolutionary data clustering, reviews thoroughly the applications of evolutionary data clustering and its implementation approaches.

Keywords Data clustering · Optimization · Evolutionary computation · Nature-inspired algorithms · Swarm intelligence · Metaheuristics unsupervised learning · Machine learning · Data mining

---

I. Aljarah (✉) · M. Habib · H. Faris
King Abdullah II School for Information Technology, The University of Jordan, Amman, Jordan
e-mail: i.aljarah@ju.edu.jo

M. Habib
e-mail: mar8160671@fgs.ju.edu.jo; neutrinoscientist@gmail.com

H. Faris
e-mail: hossam.faris@ju.edu.jo

S. Mirjalili
Center for Artificial Intelligence Research and Optimization, Torrens University Australia, Fortitude Valley, Brisbane, QLD, Australia
e-mail: ali.mirjalili@laureate.edu.au

I. Aljarah et al. (eds.), *Evolutionary Data Clustering: Algorithms and Applications*, Algorithms for Intelligent Systems,
https://doi.org/10.1007/978-981-33-4191-3_1

1 Introduction

Clustering is a typical data mining technique with plentiful applications in diverse fields of science. For instance, it is popular in the case of grouping a cohort of patients based on similar symptoms or promoting the performance of a web search engine by finding documents of similar topics [2, 4, 5, 76]. Clustering is an unsupervised type of learning that groups unlabeled and m-dimensional data instances based on common characteristics. Clustering, in literature, is defined in several ways, but is generally identified as the process of recognizing the natural and homogeneous groups within data. While in which, the objects of every single cluster are all similar, but also distinct from other objects of other clusters. Quantifying the similarity or dissimilarity among data is achieved through the utilization of proximity measures such as distance measures. However, determining a quality measure of the obtained clusters leads to formulating the clustering problem as an optimization problem. Where a set of data that has (n) points needs to be clustered into a set of (k) groups by optimizing certain fitness or objective function. Such objective functions might be the sum squared error, homogeneity, purity, or others [38, 69, 70].

For a decade ago, clustering has been considered as a hard optimization problem [11]. For instance, searching for the optimal set of clusters requires calculating the minimum distances of all points from the initially selected centroids. Finding the minimum distances of all points is computationally expensive and infeasible to solve in a sensible amount of time. Primarily, optimization problems are divided into single-objective problems or multi-objective problems. In literature, however, the clustering problem is formulated as either a single-objective optimization problem or a multi-objective optimization problem. This chapter focuses on the problem of clustering as a single-objective optimization problem. Essentially, there are two popular ways for solving optimization problems; the deterministic algorithms or the stochastic algorithms. Deterministic algorithms always follow the same strategy to solve the problem, which results in having the same solutions each time. Even though the stochastic algorithms tend to produce approximate global optimum. Yet the stochastic algorithms have the ability to generate more efficient global optimal solutions, even though they are not always guaranteed. Primarily, stochastic algorithms adopt a randomization component that leads to having different solutions at each run of the algorithm [7, 9, 10, 67, 68].

A remarkable kind of stochastic algorithms is the metaheuristics. Metaheuristic algorithms can search at local-level and global-level, which is known as exploitation and exploration, respectively. A successful metaheuristic algorithm is the one that balances exploration and exploitation. Nature-inspired algorithms are well-established examples of metaheuristic algorithms, where they are designed in a way that mimics physical, biological, or chemical phenomena found in nature. Such examples of nature-inspired algorithms are the Genetic Algorithm (GA) [34], that are inspired by Darwinian principles of evolution, and the Particle Swarm Optimization (PSO) [29], which is inspired by the social behavior of birds flocking.

This chapter introduces the definition of clustering, its main approaches, as well as its limitations. Upon that, it presents an alternative outstanding approach that is the evolutionary data clustering. Evolutionary data clustering has shown substantial abilities in overcoming the drawbacks of classical data clustering methods, where it mainly integrates the evolutionary algorithms as an optimization algorithm. The aim is to address the clustering problem of the premature convergence by skipping the need to determine the initial centroids, or the prior setting of the number of clusters.

The rest of the chapter is organized as follows. Section 2 is an introduction to data clustering. Section 3 is an analytical discussion of common clustering approaches. Section 4 presents and analyzes an alternative approach of clustering, which is evolutionary clustering. Section 5 shows a thorough review of evolutionary clustering in a wide range of applications. Finally, we summarize our work and suggest possible future work in Sect. 6.

2 Clustering

Clustering or cluster analysis is a fundamental data mining technique and is the process of dividing a collection of data instances into groups. Each group is a cluster of similar objects which are dissimilar to other objects of other clusters. Each cluster is defined by a central point and a proximity metric that measures the level of similarity or dissimilarity of the candidate data points. Clustering analysis results in a set of clusters with the objective of having the most compact clusters of similar points and having the dissimilar clusters as the most separated from each other. Given a large amount of data, it is impractical to cluster the data manually by human's capacity, rather, there are specialized computational algorithms for clustering. However, different clustering algorithms result in different sets of clusters. Therefore, quantifying the quality of the produced clusters is an essential step throughout the clustering process [3, 6, 31, 32]. Figure 1 presents the potential output of clustering methods relying on the used algorithm and the application field.

As clustering categorizes the data into clusters, we can say that it performs a kind of implicit classification. Yet it is distinct from the classification process, where mainly the data is unlabeled. As a result, it is known as an unsupervised type of learning. A typical clustering algorithm is described as follows; assuming a dataset D which has a set of n data instances $\{x_1, x_2, ..., x_n\}$, where each data instance has d dimensions. The clustering analysis results in a set of clusters C, with K number of clusters; so that $C = \{C_1, C_2, ..., C_K\}$. However, a set of constrains should be satisfied

1. The combination of all clusters $\{C_1 \cup C_2 \cup ... \cup C_K\}$ equals to D.
2. The intersection of any two clusters is an empty set $\{C_1 \cap C_2 = \phi\}$.
3. The cardinality of any cluster is a non-empty set $\{|C_j| \neq \phi\}$.

Based on the previous definition of clustering; each data instance must be attached to one cluster that is known as an exclusive type of clustering. However, a broader

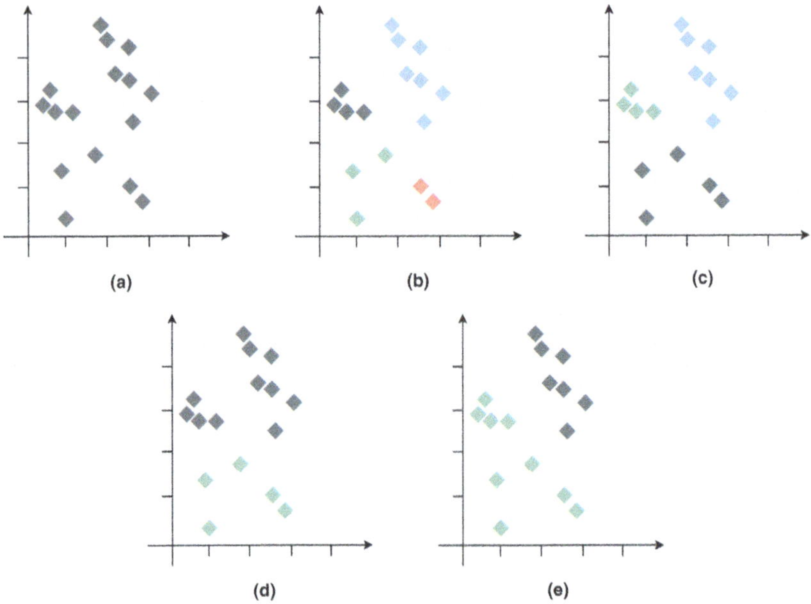

Fig. 1 A representation of different clustering outcomes depending on either the clustering algorithm or the application area. **a** shows the data before clustering. In **b** and **c** diverse groups of data may be obtained. While, **d** and **e** exhibit how different algorithms result in different clusters

type of clustering is the fuzzy clustering. Taking an example, the task of classifying a library book of bioinformatics to one class of materials, where it belongs to the biology section, as well as to the computer science section. So, in fuzzy clustering, each data instance might belong to several clusters to some extent based on a predefined membership function. An exceptional case of fuzzy clustering is when the membership function has either 1 (belong) or 0 (does not belong), which results in having each data point belonging to one cluster [78].

Since the eminent advantage of clustering is to group the similar objects; clustering has numerous applications in various fields. For instance, in marketing we may be interested in finding customers with similar shopping behaviors, or in academic institutions, we may analyze the students' academic performance by grouping the students who have similar studying behaviors. In addition, clustering has vast applications areas such as image segmentation and outlier detection, and the detection of tumor or fraudulent activities. Furthermore, clustering is considered a significant technique for revealing hidden and unknown patterns within data. Although clustering is adopted in various fields, it still faces several challenges [78].

In order to achieve the objective of clustering; this requires clustering to handle large and different types of data such as data with complex graphs and networks. Further, the data itself might be high-dimensional data which perhaps be highly dispersed and skewed; hence increases the computational cost. Besides that, the clusters might be in a complex, random form that is a non-convex shape, which makes the process of finding the optimal set of clusters more challenging and time-consuming [38].

3 Approaches of Clustering

This section presents an overview of basic clustering methodologies that are broadly categorized into partitioning and hierarchical approaches.

3.1 *Partitioning*

In partitioning approaches, the data is divided into several nonoverlapping groups, where each group is a cluster with at least one element. Mainly, partitioning algorithms perform an exclusive clustering, which means that each object in the dataset belongs only to one cluster. Partitioning algorithms calculate the proximity scores among the objects by the utilization of distance measurements. In literature, there are several common partitioning algorithms; including K-means [55], K-medoids [64], PAM [47], CLARA [47], CLARANS [59]. However, among all, k-Means is the most common partitioning-based clustering method. Hence, the following subsection describes the k-means algorithm.

3.1.1 k-Means Clustering

The k-Means clustering is a heuristic approach for grouping a dataset into a set of clusters. The k-means algorithm starts by determining the number of clusters k that is preferred for dividing the data. Assuming a dataset D with n objects that have m dimensions and k number of clusters; then the k-means will result in a set of clusters $C = \{C_1, C_2, ..., C_k\}$, where each cluster is a unique set of objects, while any cluster C_j must belong to the dataset D. The aim of clustering generally and k-means particularly is to gather the most similar objects in one cluster while, simultaneously, assures that they are dissimilar to any other object resides in other clusters. In other words, the objective is to increase the intra-cluster (within the cluster) similarity and decrease the inter-cluster (outside the cluster) similarity.

Handling k-means as a single-objective optimization problem requires determining an objective function to be optimized. One popular objective function of k-means is the sum squared error. Hence, treating k-means as an optimization problem means to best minimize the sum squared error.

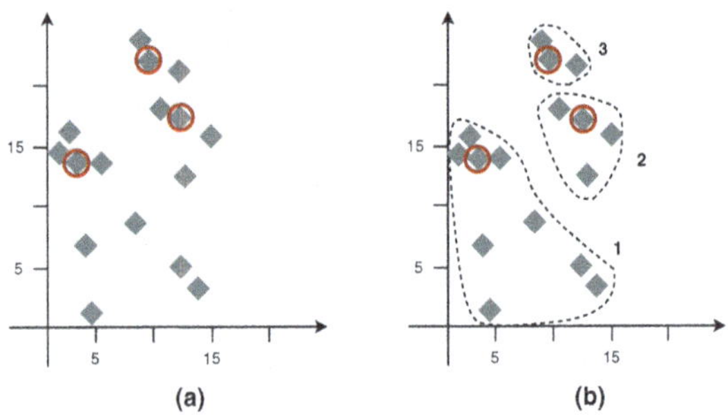

Fig. 2 An illustration of k-means procedure, which includes selecting the initial centroids, forming the clusters, and iterating until the centroids are stable

Primarily, each cluster C_j is represented by a center point c_j. Thus, assessing the quality of a cluster is performed by calculating the sum squared error of the distances of all points (n) that belong to C_j from the centroid point c_j, as defined in Eq. 1.

$$SSE = \sum_{j=1}^{k} \sum_{i=1}^{|C_j|} dist(n_i, c_j)^2 \tag{1}$$

In which, SSE is the sum squared error of all points in the dataset, dist is the Euclidean distance between the ith point n of xth dimension and the centroid c_j. The center c_j is defined by Eq. 2, in case of m-dimensional space.

$$dist(n_i, c_j) = \sqrt{\sum_{x=1}^{m} (n_{ix} - c_{jx})^2} \tag{2}$$

The procedure of k-means algorithm starts by determining the number of clusters k and by initializing the clusters to random centroid values. For instance, at subfigure (a) in Fig. 2, a three random centroids are selected for categorizing the data into three clusters; hence, c_1 is representing cluster (1), c_2 represents cluster (2), and c_3 substitutes cluster (3).

By calculating the Euclidean distance for each point from the three centroids; each candidate point is assigned to the cluster that has the minimum distance to it. As shown in subfigure (b), where at this iteration, the centroid of each cluster is calculated by finding the mean of all points in the corresponding cluster. Hence, at each iteration ever since, k-means calculates the distances of all points and find the means for centroids until the centroids are stable, then the algorithm stops. Subfigure (c) shows the results of k-means at the final iteration, where the centroids are stable. Algorithm 1 presents a pseudo-code of the k-means clustering algorithm.

Algorithm 1 K-means clustering algorithm pseudo-code

1: **procedure** K- MEANS (DATASET D, CLUSTERS- NUMBER K)
2: Select randomly initial clusters centers c_k
3: **for** each instance $n_i \in D$ **do**
4: Search for the nearest cluster center c_k
5: Assign each n_i instance to the nearest cluster S_k
6: Recompute the clusters' centers based on the allocated instances
7: Repeat steps (4–6) until centers are stable
8: **end for**
9: **end procedure**

Even that k-means is very popular and a well-regarded algorithm, but it experiences different challenges. Hence, one of the obvious drawbacks of k-means is that it can easily get stuck in a local optimum during the search for optimal clustering. Since it is highly susceptible to the initial, random centroids of the initial clusters. Furthermore, k-means is sensitive to the presence of outliers, which might reflect false centroids that leads to improperly build the clusters. Further, it is particularly applied for detecting spherical-like clusters, as well as it is best suited with relatively small-sized datasets. Addressing clustering problems as optimization problems and reaching global optimality is very costly and exhaustive using partitioning-based approaches. Thus, optimizing the minimum sum square error of clustering for $k \geq 2$ is considered an NP-hard problem [11, 38].

3.2 Hierarchical

Essentially, hierarchical clustering approaches group the data into different levels in a hierarchical way. The output of hierarchical clustering are clusters that consist of sub-clusters visualized by dendrogram or binary tree. Hierarchical clustering is achieved depending on proximity metrics. An example of hierarchical partitioning is the grouping of the employees of a company into managers, officers, and trainee. Mainly, hierarchical approaches are divided into agglomerative and divisive. The agglomerative methods are bottom-up processes that start by handling each data point as a standalone cluster, then repeatedly combine them into larger clusters. Whereas the divisive methods are top-down processes. They start with one cluster of all data points and then iteratively separates them into smaller clusters. The divisive algorithms proceed until each point represents a single cluster, which is computationally very expensive. The most used hierarchical methods are BIRCH [85], CURE [35], ROCK [36], and Chameleon [45].

The procedure of agglomerative clustering algorithms follows a set of consecutive steps. A general agglomerative clustering method starts by determining the number of clusters k, calculate the proximity matrix of all clusters, then find the minimum distances between clusters. Upon that, the closest clusters are merged together. Thereafter, iteratively, the proximity matrix is calculated again, the distances are calculated again, and the clusters are merged until there are one cluster [38].

Despite that, hierarchical approaches overcome some drawbacks of partitioning methods involving the initialization of the first initial centroids. But hierarchical methods still encounter different obstacles. The main challenges of hierarchical approaches are the high execution time; due to performing many merging and splitting points. In addition to the requirement of selecting the optimal merging and splitting points; which if not chosen correctly, then the quality of the obtained clusters will deteriorate.

Generally, there are several further clustering methods based on density, grid, fuzzy theory, graph theory, and many others [28]. However, all these methods are facing difficulties either in generalizing to cope with the big data of various densities, or dealing with certain kinds of clusters, or handling outliers, or even the requirement of the prior defination of some parameters as the number of clusters. Hence, diverse alternative approaches were proposed for clustering. Evolutionary clustering are salient alternative methodologies for clustering in compare with the partitioning and hierarchical approaches. Evolutionary algorithms perform higher-level heuristic for reaching the optimal clustering.

4 Evolutionary Clustering

Typically, finding high-quality solutions (clusters) is considered one of hardest optimization problems, where reaching global optimality is not trivial using classical clustering methods. Traditional clustering algorithms experience problems of falling in local optimum due to the improper selection of the initial centroids or the expected number of clusters. A more resilient strategy is desired for performing a local search, as well as going further and search at a global scale. Stochastic algorithms are an intriguing type of solutions for tackling optimization problems in a reasonable amount of time. Stochastic search algorithms are heuristic or metaheuristic types of algorithms, where the heuristic algorithms can generate very good solutions, but does not necessarily ensure that the obtained solutions are the optimal solutions. Whereas, metaheuristic algorithms are higher-level search algorithms that search at local and global scales resulting in high-quality solutions in a reasonable amount of time [83].

Nature-inspired algorithms are involved under the realm of metaheuristic algorithms that are best described as the pioneering algorithms for hard optimization problems. Roughly speaking, evolutionary algorithms are classified under the umbrella of nature-inspired algorithms, where the nature-inspired algorithms mainly consist of evolutionary algorithms, swarm intelligence algorithms, and other algorithms that are inspired by physical, chemical, or social phenomena [83]. The integration of evolutionary algorithms into the traditional clustering methods avoids the main drawbacks of the stagnation into local optima or the premature convergence toward global optimality.

Evolutionary algorithms are stochastic and inspired by the natural evolution principles in nature and Darwinian theories. As evolutionary algorithms compromise between diversification (search at global scale) and intensification (search at local scale); this hinders them from the getting stuck in the local optimum. Conventionally, evolutionary algorithms depend on an initial population of solutions that incorporate iteratively essential components to share information during the search process. The evolutionary components include the selection, mutation, and crossover (combination), which stand on the principle of the survival of the fittest. A popular well-regarded evolutionary algorithm is the GA algorithm.

On the contrary, swarm intelligence algorithms are inspired by the collective social behavior of birds, insects, or animals. Where they imitate, for instance, the birds flocking, fish schooling, or the foraging behavior of bacteria. Both evolutionary algorithms and the swarm intelligence algorithm adopt a fitness or objective function, which aims for evaluating the quality of solutions until a stopping criterion is satisfied.

Evolutionary clustering as a term implies the application of evolutionary algorithms for performing clustering. Recently, evolutionary clustering has been considered as one of the most popular kinds of clustering approaches in various fields of science and industry. Evolutionary clustering is involved in various clustering tasks such as to find the optimal number of clusters, to best optimize the quality of the potential clusters, or for performing feature selection. One of the early proposed evolutionary clustering approaches is the genetic algorithm-based clustering, by Maulik and Bandyopadhyay in 2000 [56], where its objective was to find the most appropriate number of clusters by taking advantage of the searchability of the GA algorithm. The next subsection demonstrates the basic principle of clustering using the GA algorithm.

4.1 GA Algorithm for Clustering

The main idea of using the GA algorithm for clustering is to exploit the searchability of GA in finding the optimal set of centers. GA based clustering algorithm follows the same steps of the original GA algorithm. GA algorithm starts by initializing a population of individuals, then enter a loop for performing the selection, combination, mutation, and then the evaluation of individuals. GA algorithm stops whenever a stopping criterion is met. The pseudo-code steps of GA algorithm are represented in Algorithm 2.

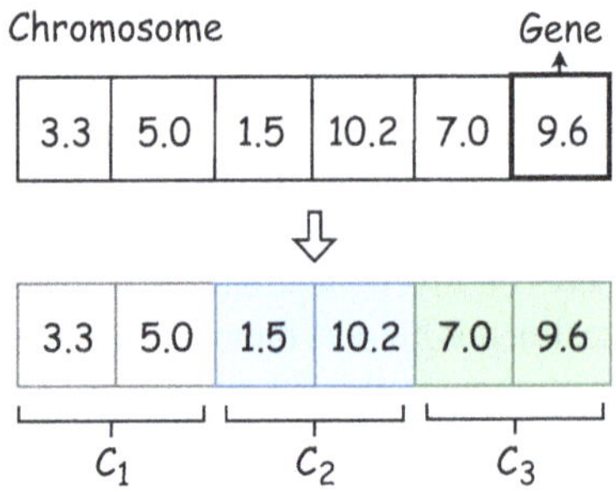

Fig. 3 A representation of GA individual for clustering. The data has 2-dimensions, in which, the centroid of the first cluster is $c_1 = (3.3, 5.0)$, the second center is $c_2 = (1.5, 10.5)$ and the third center is $c_3 = (7.0, 9.6)$

Algorithm 2 Conventional GA algorithm pseudo-code

```
procedure GA (POP-SIZE, CROSSOVER-RATE, MUTATION-RATE)
    Initialize the population of random individuals
    Evaluate each individual
    while the stopping criteria is not reached do
        Select the best individuals for the next generation
        Crossover (recombine) the best parent individuals
        Mutate genes of offspring
        Evaluate each new individual
        Increment the loop counter
    end while
end procedure
```

In order to prepare the GA for performing clustering, the individuals should be encoded properly, and a fitness criterion is defined. Hence, as the problem is to find the best clusters, the individuals represent the potential centroids. Taking into consideration the dimensionality of the problem m, each individual represents the candidate centroids of all clusters. Therefore, if the number of clusters is k, then the length of the individual (chromosome) is $m \times k$. Each gene in a chromosome represents the value of one dimension of a center point. Thus, each chromosome is a vector of continuous values. Figure 3 shows a representation for an individual.

Initially, all individuals in the population are assigned the random center points from the dataset. In order to evaluate an individual, each unlabeled point is categorized into one of the clusters based on a fitness value. Herein, the fitness function is the Euclidean distance between the data instance and the respective centroid. The smaller the Euclidean distance, the higher the similarity value; therefore, GA algorithm seeks for the minimum distance value for each point from the corresponding centers. After the assignment of the data instances, the values of the individuals were replaced by the mean values of the new points. In other words, if cluster 1 has two data instances; $n_1 = (1.0, 2.5)$ and $n_2 = (3.0, 2.0)$; then the new centroid is $c_1 = ((3.3 + 1 + 3)/3, (5.0 + 2.5 + 2)/3)$, which is $c_1 = (2.4, 3.2)$.

In the GA-clustering algorithm, all the genetic operators including the selection, crossover, and mutation are performed similarly as in GA algorithm. The individual that results in the minimum sum of distances is preserved throughout all generations as the optimal solution.

5 Applications of Evolutionary Clustering

As clustering is applied to various fields of science. This section presents a review of the clustering applications in data mining and machine learning, web search, bioinformatics, and business intelligence.

5.1 Data Mining and Machine Learning

The implementation of evolutionary algorithms for solving the clustering problem had accomplished several decades ago. An early implementation of evolutionary algorithms for clustering was originally mostly dependent on the GA algorithm. In [57], the authors used the GA algorithm for optimizing the sum of squared Euclidean distances for k-means, in an attempt for reaching the optimal set of clusters. In which, the proposed algorithm remarkably escaped the problem of premature convergence throughout all the conducted experiments. In [74], a different evolutionary clustering method is developed, which is based on the evolutionary programming algorithm. The developed algorithm aimed to search for the optimal number of clusters, besides identifying the best representative centroids. However, it is designed to handle spherical or crisp-like types of clusters.

In [79], the authors offered an evolutionary clustering method using the particle swarm optimization algorithm, which was implemented mainly to search for the initial clustering centroids, as well as to promote the clustering process toward optimality. Where the proposed algorithm outperformed k-means in terms of inter-cluster and intra-cluster distances, in addition to the quantization error. Whereas in [13], four evolutionary variants of k-means proposed that adopted diverse genetics operators for the objective of having the best grouping of data. In which, the authors used the Rand Index and the Simplified Silhouette as fitness functions. Nevertheless, as k-means faces a premature convergence; authors in [60] design a hybrid evolutionary algorithm for clustering, which consists of particle swarm optimization and simulated annealing for searching for the optimal clustering, where it accomplished better convergence ability than the k-means algorithm.

Further, in [75], a novel clustering algorithm is designed that is relied on Firefly Algorithm (FA). Whereas, [44] presented a new evolutionary clustering algorithm that is based on Artificial Bee Colony optimization algorithm, which was utilized for seeking the superior clustering set. Even that the authors used the proposed ABC-Clustering for classification purposes, but it showed efficient performance results.

In addition, [43] proposed an ABC algorithm for optimizing the fuzzy clustering algorithms by minimizing the objective function of the sum squared error. [4] developed a parallel PSO algorithm for clustering that showed effective scalingability with large datasets. Whereas, [14] proposed a parallel Bat algorithm for clustering large-scale data. [6] introduced a clustering approach based on the Glowworm swarm algorithm for determining the optimal centers of clusters, which showed superior results in terms of purity and entropy. Nonetheless, [27] presented the integration of an improved adaptive genetic algorithm into fuzzy c-means, in order to optimize the initial cluster centers by optimizing the sum squared error based on the Euclidean distance.

Authors in [76] implemented static and dynamic clustering approaches based on a recent evolutionary algorithm which is the Multi-verse Optimizer (MVO). In which, MVO was adopted for searching for the optimal grouping of data that is named as the static clustering method. As well as, MVO has been used as a dynamic clustering approach for finding the optimal number of clusters. The proposed evolutionary clustering algorithms outperformed the Clustering Genetic Algorithm (CGA) and the Clustering Particle Swarm Optimization (CPSO), in regards to purity and entropy evaluation measures.

Interestingly, authors in [82] utilized a multi-objective evolutionary algorithm for tackling the problem of the optimal number of clusters. Hence, the Non-dominated Sorting Genetic Algorithm (NSGA-II) is used to perform a multi-clustering that takes advantage of the parallelism of NSGA-II. In which, the number of clusters and the sum squared distances were used as objective functions. Further, in [8], a novel evolutionary clustering method is designed for approaching the problem of stagnation in local optimal, which showed superior performance results. The novel method is a combination of Grey Wolf Optimizer (GWO) and Tabu search, where the fitness function is the sum of squared errors. Whereas, [23] utilized a many-objective ABC algorithm for clustering software modules, which achieved robust results against different multi-objective evolutionary algorithms as the NSGA-III and MOEA/D.

Moreover, [10] investigated the effectiveness of a Multi-verse optimizer for addressing the problem of clustering. Where it surpassed the efficiency of PSO, GA, and Dragonfly algorithms regarding purity, homogeneity, and completeness. [58] proposed a hybrid of memetic algorithm and adaptive differential evolution (DE) mutation strategy. The objective of the proposed algorithm is overcoming the premature convergence of traditional clustering. Apparently, it exhibited remarkable results over the classical clustering algorithms. While [12], designed a modified differential evolution algorithm for data clustering. The designed algorithm attempted to address the slow convergence, increase solutions diversity, and balance between the exploration and exploitation abilities. Where it obtained high compactness of the clustering and improved the convergence speed as well.

5.2 Image Processing and Pattern Recognition

The authors in [61], used the evolutionary algorithms in the field of image clustering, which applied a PSO particularly on synthetic, Magnetic Resonance Imaging (MRI) and satellite images. The objective of the particle swarm optimization algorithm is to search for the optimal set of centroids for a predefined number of clusters. The proposed algorithm outperformed k-means, fuzzy c-means clustering, k-harmonic means, and GA-based clustering algorithms in terms of quantization error. Further, [39] implemented a fuzzy clustering approach based on Ant Colony Optimization (ACO) for image segmentation.

As image segmentation plays a vital role in image processing and computer vision applications; [84] developed a combination of biogeography-based optimization algorithm and fuzzy c-means for image segmentation. The designed approach exhibited very good performance over PSO and ABC based clustering algorithms. In [86], a multi-objective evolutionary fuzzy clustering method is utilized for image segmentation, where the compactness and separation were the objective functions. Additionally, [51] proposed an improved GWO based on DE and fuzzy c-means for segmenting synthetic aperture radar (SAR) images. In which, DE-GWO is employed for searching the initial number of clusters and the optimal centroids. In the field of computer vision, [20] proposed a deep clustering approach of k-means and convolutional neural networks for learning and grouping visual features, where the authors used the ImageNet and YFCC100M datasets for testing.

Nonetheless, [53] the authors deployed a clustering approach for driving pattern investigation for an electric vehicle. Whereas, [21] applied a clustering approach for railway delay pattern recognition. Furthermore, [33] implemented a hierarchical clustering approach for recognizing clinically useful patterns of patient-generated data. And [26] anticipated the application of GA-clustering approach for pattern identification of errors or outages of smart power grids.

5.3 Web Search and Information Retrieval

Recently, the number of web pages has increased rapidly. Therefore, clustering documents (text) is important for boosting the performance of web search engines. [1] presented an evolutionary clustering algorithm-based on Krill herd optimization algorithm for web text clustering. In which, the proposed algorithm outperformed k-means in terms of purity and entropy. Remarkably, [42] developed a method to solve the problem of abstract clustering. Which elevates the information retrieval process of a biomedical literature database (MEDLINE). In which, the authors utilized a hybrid of GA, vector space model that represents the text and agglomerative clustering algorithm. Where the agglomerative clustering is used to generate the initial population and to find the similar texts based on similarity metrics. [62] designed an algorithm called (EBIC) that concerns with finding a complex pattern of information within complex data. Particularly, in gene expression datasets. Mainly, the proposed algorithm stands on the implementation of evolutionary parallel biclustering algorithms.

Further, [72] designed a clustering-based genetic algorithm for community detection over social networks. The detection of modular communities is essential for investigating the behavior of complex environments. As well as, aids in the development of recommender systems. The authors utilized a modularity metric for quantifying the quality of the candidate clusters. While, [22] demonstrated the application of a heterogeneous evolutionary clustering method for the prediction of the rating of a collaborative filtering approach. More interestingly, [19] introduced a novel algorithm for language and text identification, which is called genetic algorithms image clustering for document analysis (GA-ICDA). In [50], the authors implemented the binary PSO for large-scale text clustering, in which, the binary PSO is used for performing feature selection. Furthermore, [54] designed a bio-inspired fuzzy clustering ensemble approach for better personalized recommendations through collaborative filtering.

Aspect-based summaries have a significant role in opinion mining. [66] proposed an evolutionary clustering algorithm for aspect-summarization. [73] examined the usage of multi-objective based DE algorithm for clustering, in the context of scientific documents clustering. In which, the Self-Organizing Map (SOM) is implemented with DE for searching the optimal set of clusters, while the Pakhira-Bandyopadhyay-Maulik index and Silhouette index were used as a fitness functions for optimization.

5.4 Bioinformatics

Digital image processing is an integral part of several scientific areas. In particular, the case of the detection of cancer. Where [40] proposed a hybrid of firefly algorithm and k-means clustering for the detection of brain tumors. In which, the algorithm objective is to perform brain image segmentation with Otsu's method as a fitness function for FA and the inter-cluster distance for k-means. The aim of k-means is to search for the optimal centroids. Hence, it obtained very good results in terms of different image segmentation quality measures. Halder et al. [37] proposed a genetic algorithm-based spatial fuzzy c-means for medical image segmentation. Further, [41] anticipated the integration of multi-objective evolutionary algorithms for clustering complex networks. Thus, the authors proposed a novel multi-objective evolutionary algorithm-based on decomposition and membrane structure for community detection. Evidently, the authors stated that the proposed algorithm is convenient for clustering disease-gene networks and DNA binding protein networks identification. Moreover, [52] interpreted the implementation of multi-objective evolutionary clustering in case of RNA sequencing, where the high-dimensionality and sparseness of data are the major challenges. Nonetheless, [65] deployed a multi-objective algorithm that is hybridized with intensification and diversification components for gene clustering. In which, quantifying genetic similarities has benefits in advancing the science of diseases and metabolic disorders. The cluster-quality index, compactness, and separation were all utilized for assessing the fitness function. In [15], the authors presented the potential of stochastic biclustering algorithms in the field of microarray data analysis, which has significant importance in finding the similar genes of similar gene expression data.

Bara'a et al. [17] applied a novel multi-objective evolutionary algorithm, for detecting the functional modules of protein-protein interaction networks. In which, the authors adopted a heuristic component called protein complex attraction and repulsion for assimilating the topological features of the networks. Additionally, [16] proposed a hybrid of cuckoo search with Nelder–Mead algorithm for biclustering microarray, gene expression data. Where the cuckoo search algorithm seeks for the optimal clusters. The proposed algorithm showed reasonable improvements over other swarm intelligence and biclustering algorithms. While [71], designed two approaches of multi-objective optimization, which are based on PSO and differential evolution for clustering gene expression data in the context of cancer classification. Regarding the proposed algorithms, the Xie–Beni index and FSym-index were used as objective functions.

5.5 Business Intelligence and Security

Chou et al. [24] proposed a hybrid of GA and fuzzy c-means for bankruptcy prediction. In which, the fuzzy c-means is integrated as a fitness function in order to seek for the best set of features that improve the prediction accuracy of GA algorithm. Additionally, [49] developed a two-stage clustering method for order clustering. It aims for minimizing the production time and the machine idle time. The proposed approach depends on neural networks and a hybrid of GA and PSO, by utilizing the sum of Euclidean distances as a fitness function. In which, the designed approach outperformed the GA, PSO, GA-PSO, and other variants. Further, Vehicular Ad-hoc Networks (VANETs) is a sort of intelligent transportation systems. Whereas, developing a routing protocol is fundamental for VANETs since its stochastic topology. In [30], the authors designed a GWO-clustering approach for best controlling the routing and stability of such scalable networks. Recommender systems play a significant role in e-commerce and social networks. Since they seek for customizing the recommendations for each user based on their preferences. Berbague et al. [18] provided an evolutionary clustering approach for enhancing the procedure of recommender systems. The proposed approach is a combination of GA and k-means. In which, the fitness function is the summation of both the group-precision and centers-diversity. Whereas, [46] designed a hybrid of cuckoo search and k-means for an efficient collaborative filtering approach for a movie recommender system. Mashup services is a technology to enhance the application of the Internet of things. In [63], the authors presented a GA-based clustering approach which aims to cluster mashup services over a cloud of Internet of things. In which, a structural similarity method (SimRank) is used, while the GA is utilized to find the optimal number of clusters. Furthermore, [25] designed an evolutionary approach with fuzzy c-means for clustering solar energy; in order to establish a solar power plant. The proposed approach compares each of PSO, DE, and GA for the best clustering, in terms of Calinski-Harabasz, Davies-Bouldin, and Silhouette indices. Hence, the performance results were superior over the fuzzy c-means.

Enhancing the security of computational services is of critical importance. Recently, Intrusion Detection Systems (IDSs) have attracted crucial awareness worldwide. Numerous research studies were conducted to promote the effectiveness of IDS and reduce its false alarm rate. In [80], the authors developed a hybrid approach of artificial neural networks and fuzzy clustering in order to improve the efficiency of such IDS. [48] presented a hybrid of FA and k-means for intrusion detection. In which, the FA algorithm is utilized in order to improve the slow convergence of k-means. Whereas, [81] designed an improved clustering approach based on adaptive GA and fuzzy c-means algorithms for the detection of public security events. In which, the GA algorithm searches for the optimal centroids. Additionally, [77] proposed a GA based on k-means for network intrusion detection. Where the GA algorithm is utilized to search for the superior clusters regarding a combination of inner-cluster distance and inter-cluster distance.

6 Conclusion

This chapter introduced a common unsupervised data mining technique; which is data clustering. The chapter includes a definition of clustering or cluster analysis. The main approaches for performing clustering (partitioning and hierarchical), alongside the most used algorithms such as the k-means clustering. In addition, presented a description of their major challenges and limitations. Nonetheless, the chapter investigated the formulation of clustering as NP-hard optimization problem. Where seeking for the optimal clustering with at least 2 clusters, is deemed as a tough optimization problem.

Therefore, the chapter suggested the integration of an alternative approach, which is the evolutionary algorithms for clustering. Evolutionary algorithms showed great efficiency in finding global optimal solutions. Evolutionary clustering uses evolutionary algorithms such as the GA algorithm for searching the optimal clustering. The optimal clustering involves the search of the best centroids or the optimal number of clusters by optimizing certain fitness function. Evolutionary clustering has a wide range of applications in various fields. So, in this chapter, the implementation of evolutionary clustering is covered in the area of data mining and machine learning. In image processing, web search, bioinformatics, and business intelligence applications.

References

1. Abualigah, Laith Mohammad, Ahamad Tajudin, Khader, Mohammed Azmi, Al-Betar, and Mohammed A. Awadallah. 2016. A krill herd algorithm for efficient text documents clustering. In *2016 IEEE symposium on computer applications and industrial electronics (ISCAIE)*, pp. 67–72. IEEE.
2. Al-Madi, Nailah, Ibrahim, Aljarah, and Simone A. Ludwig. 2014. Parallel glowworm swarm optimization clustering algorithm based on mapreduce. In *2014 IEEE Symposium on Swarm Intelligence*, pp. 1–8. IEEE.
3. Al Shorman, Amaal, Hossam, Faris, and Ibrahim, Aljarah. 2020. Unsupervised intelligent system based on one class support vector machine and grey wolf optimization for iot botnet detection. *Journal of Ambient Intelligence and Humanized Computing*, 11(7):2809–2825.
4. Aljarah, Ibrahim, and Simone A. Ludwig. 2012. Parallel particle swarm optimization clustering algorithm based on mapreduce methodology. In *2012 Fourth World Congress on Nature and Biologically Inspired Computing (NaBIC)*, pp. 104–111. IEEE.
5. Aljarah, Ibrahim, and Simone A. Ludwig. 2013. Mapreduce intrusion detection system based on a particle swarm optimization clustering algorithm. In *2013 IEEE congress on evolutionary computation*, pp. 955–962. IEEE.
6. Aljarah, Ibrahim, and Simone A. Ludwig. 2013. A new clustering approach based on glowworm swarm optimization. In *2013 IEEE congress on evolutionary computation*, pp. 2642–2649. IEEE.
7. Aljarah, Ibrahim, and Simone A. Ludwig. 2013. Towards a scalable intrusion detection system based on parallel pso clustering using mapreduce. In *Proceedings of the 15th annual conference companion on Genetic and evolutionary computation*, pp. 169–170.
8. Aljarah, Ibrahim, Majdi, Mafarja, Ali Asghar, Heidari, Hossam, Faris, and Seyedali, Mirjalili. 2019. Clustering analysis using a novel locality-informed grey wolf-inspired clustering approach. *Knowledge and Information Systems* 1–33.
9. Aljarah, Ibrahim, Majdi, Mafarja, Ali Asghar, Heidari, Hossam, Faris, and Seyedali, Mirjalili. 2020. Clustering analysis using a novel locality-informed grey wolf-inspired clustering approach. *Knowledge and Information Systems* 62(2):507–539.
10. Aljarah, Ibrahim, Majdi, Mafarja, Ali Asghar, Heidari, Hossam, Faris, and Seyedali, Mirjalili. 2020. Multi-verse optimizer: theory, literature review, and application in data clustering. In *Nature-Inspired Optimizers*, pp. 123–141. Berlin: Springer
11. Aloise, Daniel, Amit Deshpande, Pierre Hansen, and Preyas Popat. 2009. Np-hardness of euclidean sum-of-squares clustering. *Machine Learning* 75 (2): 245–248.
12. Alswaitti, Mohammed, Mohanad Albughdadi, and Nor Ashidi Mat Isa. 2019. Variance-based differential evolution algorithm with an optional crossover for data clustering. *Applied Soft Computing* 80: 1–17.
13. Alves, Vinícius S., Ricardo J.G.B. Campello, and Eduardo R. Hruschka. 2006. Towards a fast evolutionary algorithm for clustering. In *2006 IEEE international conference on evolutionary computation*, pp. 1776–1783. IEEE.
14. Ashish, Tripathi, Sharma, Kapil, and Bala, Manju. 2018. Parallel bat algorithm-based clustering using mapreduce. In *Networking communication and data knowledge engineering*, pp. 73–82. Berlin: Springer.
15. Ayadi, Wassim, Mourad, Elloumi, and Jin-Kao, Hao. 2018. 14 systematic and stochastic biclustering algorithms for microarray data analysis. *Microarray Image and Data Analysis: Theory and Practice* 369.
16. Balamurugan, R., A.M. Natarajan, and K. Premalatha. 2018. A new hybrid cuckoo search algorithm for biclustering of microarray gene-expression data. *Applied Artificial Intelligence* 32 (7–8): 644–659.
17. Bara'a, A. Attea, and Qusay Z. Abdullah. 2018. Improving the performance of evolutionary-based complex detection models in protein—protein interaction networks. *Soft Computing* 22(11):3721–3744.

18. Berbague, Chems Eddine, Nour, El Islem Karabadji, and Hassina, Seridi. 2018. An evolutionary scheme for improving recommender system using clustering. In *IFIP International Conference on Computational Intelligence and Its Applications*, pp. 290–301. Berlin: Springer.
19. Brodić, Darko, Alessia, Amelio, and Zoran N. Milivojević. 2018. Language discrimination by texture analysis of the image corresponding to the text. *Neural Computing and Applications* 29(6):151–172.
20. Caron, Mathilde, Piotr, Bojanowski, Armand, Joulin, and Matthijs, Douze. 2018. Deep clustering for unsupervised learning of visual features. In *Proceedings of the European conference on computer vision (ECCV)*, pp. 132–149.
21. Cerreto, Fabrizio, Bo Friis, Nielsen, Otto Anker, Nielsen, and Steven S. Harrod. 2018. Application of data clustering to railway delay pattern recognition. *Journal of Advanced Transportation* 2018.
22. Chen, Jianrui, Hua Wang, Zaizai Yan, et al. 2018. Evolutionary heterogeneous clustering for rating prediction based on user collaborative filtering. *Swarm and Evolutionary Computation* 38: 35–41.
23. Chhabra, Jitender Kumar et al. 2018. Many-objective artificial bee colony algorithm for large-scale software module clustering problem. *Soft Computing*, 22(19):6341–6361.
24. Chou, Chih-Hsun, Su-Chen Hsieh, and Chui-Jie Qiu. 2017. Hybrid genetic algorithm and fuzzy clustering for bankruptcy prediction. *Applied Soft Computing* 56: 298–316.
25. de Barros Franco, David Gabriel, and Maria Teresinha Arns, Steiner. 2018. Clustering of solar energy facilities using a hybrid fuzzy c-means algorithm initialized by metaheuristics. *Journal of Cleaner Production* 191:445–457.
26. De Santis, Enrico, Antonello Rizzi, and Alireza Sadeghian. 2018. A cluster-based dissimilarity learning approach for localized fault classification in smart grids. *Swarm and Evolutionary Computation* 39: 267–278.
27. Ding, Yi, and Fu Xian. 2016. Kernel-based fuzzy c-means clustering algorithm based on genetic algorithm. *Neurocomputing* 188: 233–238.
28. Dongkuan, Xu, and Yingjie Tian. 2015. A comprehensive survey of clustering algorithms. *Annals of Data Science* 2: 165–193.
29. Eberhart, Russell, and James, Kennedy. 1995. A new optimizer using particle swarm theory. In *MHS'95. Proceedings of the sixth international symposium on micro machine and human science*, pp. 39–43. IEEE.
30. Fahad, Muhammad, Farhan, Aadil, Salabat, Khan, Peer Azmat, Shah, Khan, Muhammad, Jaime, Lloret, Haoxiang, Wang, Jong Weon, Lee, Irfan, Mehmood, et al. 2018. Grey wolf optimization based clustering algorithm for vehicular ad-hoc networks. *Computers & Electrical Engineering* 70:853–870.
31. Hossam Faris, Ibrahim Aljarah, and Ja'far Alqatawna. 2015. Optimizing feedforward neural networks using krill herd algorithm for e-mail spam detection. In *2015 IEEE Jordan conference on applied electrical engineering and computing technologies (AEECT)*, pp. 1–5. IEEE.
32. Faris, Hossam, Ibrahim, Aljarah, Seyedali, Mirjalili, Pedro A. Castillo, and Juan Julián Merelo, Guervós. 2016. Evolopy: An open-source nature-inspired optimization framework in python. In *IJCCI (ECTA)*, pp. 171–177.
33. Feller, Daniel J., Marissa, Burgermaster, Matthew E. Levine, Arlene, Smaldone, Patricia G. Davidson, David J. Albers, and Lena, Mamykina. 2018. A visual analytics approach for pattern-recognition in patient-generated data. *Journal of the American Medical Informatics Association*, 25(10):1366–1374.
34. Goldberg, David E., and John H. Holland. 1988. Genetic algorithms and machine learning. *Machine Learning*, 3(2):95–99.
35. Guha, Sudipto, Rajeev, Rastogi, and Kyuseok, Shim. 1998. Cure: an efficient clustering algorithm for large databases. In *ACM Sigmod Record*, vol. 27, pp. 73–84. ACM.
36. Guha, Sudipto, Rajeev Rastogi, and Kyuseok Shim. 2000. Rock: A robust clustering algorithm for categorical attributes. *Information Systems* 25: 345–366.
37. Halder, Amiya, Avranil, Maity, and Ananya, Das. 2019. Medical image segmentation using ga-based modified spatial fcm clustering. In *Integrated Intelligent Computing, Communication and Security*, pp. 595–601. Berlin: Springer.

38. Han, Jiawei, Jian Pei, and Micheline Kamber. 2011. *Data mining: concepts and techniques*. Elsevier.
39. Han, Yanfang, and Pengfei Shi. 2007. An improved ant colony algorithm for fuzzy clustering in image segmentation. *Neurocomputing* 70 (4–6): 665–671.
40. Hrosik, Romana CAPOR, Eva, Tuba, Edin, Dolicanin, Raka, Jovanovic, and Milan, Tuba. 2019. Brain image segmentation based on firefly algorithm combined with k-means clustering. *Stud. Inform. Control*, 28:167–176.
41. Ying, Ju, Songming Zhang, Ningxiang Ding, Xiangxiang Zeng, and Xingyi Zhang. 2016. Complex network clustering by a multi-objective evolutionary algorithm based on decomposition and membrane structure. *Scientific Reports* 6: 33870.
42. Karaa, Wahiba Ben Abdessalem, Amira S. Ashour, Dhekra Ben, Sassi, Payel, Roy, Noreen, Kausar, and Nilanjan, Dey. 2016. Medline text mining: an enhancement genetic algorithm based approach for document clustering. In *Applications of intelligent optimization in biology and medicine*, pp. 267–287. Berlin: Springer.
43. Karaboga, Dervis, and Celal Ozturk. 2010. Fuzzy clustering with artificial bee colony algorithm. *Scientific Research and Essays* 5 (14): 1899–1902.
44. Karaboga, Dervis, and Celal Ozturk. 2011. A novel clustering approach: Artificial bee colony (abc) algorithm. *Applied Soft Computing* 11 (1): 652–657.
45. Karypis, George, Eui-Hong Sam, Han, and Vipin, Kumar. 1999. Chameleon: Hierarchical clustering using dynamic modeling. *Computer*, (8):68–75.
46. Rahul Katarya and Om Prakash Verma. 2017. An effective collaborative movie recommender system with cuckoo search. *Egyptian Informatics Journal* 18 (2): 105–112.
47. Kaufman, Leonard, and Peter J. Rousseeuw. 2009. *Finding groups in data: an introduction to cluster analysis*, vol. 344. New York: Wiley.
48. Kaur, Arvinder, Saibal K. Pal, and Amrit Pal, Singh. 2018. Hybridization of k-means and firefly algorithm for intrusion detection system. *International Journal of System Assurance Engineering and Management* 9(4):901–910.
49. Kuo, R.J., and L.M. Lin. 2010. Application of a hybrid of genetic algorithm and particle swarm optimization algorithm for order clustering. *Decision Support Systems* 49 (4): 451–462.
50. Kushwaha, Neetu, and Millie Pant. 2018. Link based bpso for feature selection in big data text clustering. *Future Generation Computer Systems* 82: 190–199.
51. Li, M.Q., L.P. Xu, Na, Xu, Tao, Huang, and Bo, Yan. 2018. Sar image segmentation based on improved grey wolf optimization algorithm and fuzzy c-means. *Mathematical Problems in Engineering* 2018.
52. Li, Xiangtao, and Ka-Chun, Wong. 2019. Single-cell rna sequencing data interpretation by evolutionary multiobjective clustering. *IEEE/ACM transactions on computational biology and bioinformatics*.
53. Li, Xuefang, Qiang Zhang, Zhanglin Peng, Anning Wang, and Wanying Wang. 2019. A data-driven two-level clustering model for driving pattern analysis of electric vehicles and a case study. *Journal of Cleaner Production* 206: 827–837.
54. Logesh, R., V. Subramaniyaswamy, D. Malathi, N. Sivaramakrishnan, and V. Vijayakumar. 2019. Enhancing recommendation stability of collaborative filtering recommender system through bio-inspired clustering ensemble method. *Neural Computing and Applications*, pp. 1–24.
55. MacQueen, James, et al. 1967. Some methods for classification and analysis of multivariate observations. In *Proceedings of the fifth Berkeley symposium on mathematical statistics and probability*, vol. 1, pp. 281–297. Oakland, CA, USA.
56. Maulik, Ujjwal, and Sanghamitra Bandyopadhyay. 2000. Genetic algorithm-based clustering technique. *Pattern Recognition* 33 (9): 1455–1465.
57. Murthy, Chivukula A., and Nirmalya, Chowdhury. 1996. In search of optimal clusters using genetic algorithms. *Pattern Recognition Letters*, 17(8):825–832.
58. Mustafa, Hossam M.J., Masri, Ayob, Mohd Zakree Ahmad, Nazri, and Graham, Kendall. 2019. An improved adaptive memetic differential evolution optimization algorithms for data clustering problems. *PloS One* 14(5):e0216906.

59. Ng, Raymond T., and Jiawei, Han. 2002. Clarans: A method for clustering objects for spatial data mining. *IEEE Transactions on Knowledge & Data Engineering* (5):1003–1016.
60. Niknam, Taher, Babak Amiri, Javad Olamaei, and Ali Arefi. 2009. An efficient hybrid evolutionary optimization algorithm based on pso and sa for clustering. *Journal of Zhejiang University-SCIENCE A* 10 (4): 512–519.
61. Omran, Mahamed, Andries Petrus, Engelbrecht, and Ayed, Salman. 2005. Particle swarm optimization method for image clustering. *International Journal of Pattern Recognition and Artificial Intelligence* 19(03):297–321.
62. Orzechowski, Patryk, Moshe, Sipper, Xiuzhen, Huang, and Jason H, Moore. 2018. Ebic: an evolutionary-based parallel biclustering algorithm for pattern discovery. *Bioinformatics* 34(21):3719–3726.
63. Pan, Weifeng, and Chunlai Chai. 2018. Structure-aware mashup service clustering for cloud-based internet of things using genetic algorithm based clustering algorithm. *Future Generation Computer Systems* 87: 267–277.
64. Park, Hae-Sang, and Chi-Hyuck Jun. 2009. A simple and fast algorithm for k-medoids clustering. *Expert systems with Applications* 36: 3336–3341.
65. Parraga-Alava, Jorge, Marcio Dorn, and Mario Inostroza-Ponta. 2018. A multi-objective gene clustering algorithm guided by apriori biological knowledge with intensification and diversification strategies. *BioData Mining* 11 (1): 16.
66. Priya, V., and K, Umamaheswari. 2019. Aspect-based summarisation using distributed clustering and single-objective optimisation. *Journal of Information Science* 0165551519827896.
67. Qaddoura, Raneem, Hossam Faris, and Ibrahim Aljarah. 2020. An efficient clustering algorithm based on the k-nearest neighbors with an indexing ratio. *International Journal of Machine Learning and Cybernetics* 11 (3): 675–714.
68. Qaddoura, Raneem, Hossam, Faris, Ibrahim, Aljarah, and Pedro A., Castillo. 2020. Evocluster: An open-source nature-inspired optimization clustering framework in python. In *International Conference on the Applications of Evolutionary Computation (Part of EvoStar)*, pp. 20–36. Berlin: Springer.
69. Qaddoura, R., H. Faris, and I. Aljarah. 2020. An efficient evolutionary algorithm with a nearest neighbor search technique for clustering analysis. *Journal of Ambient Intelligence and Humanized Computing*, 1–26.
70. Qaddoura, R., H. Faris, I. Aljarah, J. Merelo, and P. Castillo. 2020. Empirical evaluation of distance measures for nearest point with indexing ratio clustering algorithm. In *Proceedings of the 12th International Joint Conference on Computational Intelligence - Volume 1: NCTA*, ISBN 978-989-758-475-6, pp. 430–438. https://doi.org/10.5220/0010121504300438.
71. Saha, Sriparna, Ranjita Das, and Partha Pakray. 2018. Aggregation of multi-objective fuzzy symmetry-based clustering techniques for improving gene and cancer classification. *Soft Computing* 22 (18): 5935–5954.
72. Said, Anwar, Rabeeh Ayaz, Abbasi, Onaiza, Maqbool, Ali, Daud, and Naif Radi, Aljohani. 2018. Cc-ga: A clustering coefficient based genetic algorithm for detecting communities in social networks. *Applied Soft Computing*, 63:59–70.
73. Saini, Naveen, Sriparna Saha, and Pushpak Bhattacharyya. 2019. Automatic scientific document clustering using self-organized multi-objective differential evolution. *Cognitive Computation* 11 (2): 271–293.
74. Sarkar, Manish, B. Yegnanarayana, and Deepak, Khemani. 1997. A clustering algorithm using an evolutionary programming-based approach. *Pattern Recognition Letters*, 18(10):975–986.
75. Senthilnath, J., S.N. Omkar, and V. Mani. 2011. Clustering using firefly algorithm: performance study. *Swarm and Evolutionary Computation* 1 (3): 164–171.
76. Shukri, Sarah, Hossam Faris, Ibrahim Aljarah, Seyedali Mirjalili, and Ajith Abraham. 2018. Evolutionary static and dynamic clustering algorithms based on multi-verse optimizer. *Engineering Applications of Artificial Intelligence* 72: 54–66.
77. Sukumar, J.V. Anand, I. Pranav, M.M. Neetish, and Jayasree, Narayanan. 2018. Network intrusion detection using improved genetic k-means algorithm. In *2018 international conference on advances in computing, communications and informatics (ICACCI)*, pp. 2441–2446. IEEE.

78. Theodoridis, Sergios, and Konstantinos, Koutroumbas. 2006. Clustering: basic concepts. *Pattern Recognition*, 483–516.
79. Van der Merwe, D.W., and Andries Petrus, Engelbrecht. 2003. Data clustering using particle swarm optimization. In *The 2003 Congress on Evolutionary Computation, 2003. CEC'03.*, vol. 1, pp. 215–220. IEEE.
80. Wang, Gang, Jinxing Hao, Jian Ma, and Lihua Huang. 2010. A new approach to intrusion detection using artificial neural networks and fuzzy clustering. *Expert Systems with Applications* 37 (9): 6225–6232.
81. Wang, Heng, Zhenzhen, Zhao, Zhiwei, Guo, Zhenfeng, Wang, and Xu, Guangyin. 2017. An improved clustering method for detection system of public security events based on genetic algorithm and semisupervised learning. *Complexity* 2017.
82. Wang, Rui, Shiming Lai, Wu Guohua, Lining Xing, Ling Wang, and Hisao Ishibuchi. 2018. Multi-clustering via evolutionary multi-objective optimization. *Information Sciences* 450: 128–140.
83. Yang, Xin-She. 2010. *Nature-inspired metaheuristic algorithms*. Luniver Press. 2010.
84. Zhang, Minxia, Weixuan, Jiang, Yu, Xiaohan Zhou, Xue, and Shengyong, Chen. 2019. A hybrid biogeography-based optimization and fuzzy c-means algorithm for image segmentation. *Soft Computing* 23(6):2033–2046.
85. Zhang, Tian, Raghu, Ramakrishnan, and Miron, Livny. 1996. Birch: an efficient data clustering method for very large databases. In *ACM Sigmod Record*, vol. 25, pp. 103–114. ACM.
86. Zhao, Feng, Hanqiang, Liu, Jiulun, Fan, Chang Wen, Chen, Rong, Lan, and Na, Li. 2018. Intuitionistic fuzzy set approach to multi-objective evolutionary clustering with multiple spatial information for image segmentation. *Neurocomputing*, 312:296–309.

A Comprehensive Review of Evaluation and Fitness Measures for Evolutionary Data Clustering

Ibrahim Aljarah, Maria Habib, Razan Nujoom, Hossam Faris, and Seyedali Mirjalili

Abstract Data clustering is among the commonly investigated types of unsupervised learning; owing to its ability for capturing the underlying information. Accordingly, data clustering has an increasing interest in various applications involving health, humanities, and industry. Assessing the goodness of clustering has been widely debated across the history of clustering analysis, which led to the emergence of abundant clustering evaluation measures. The aim of clustering evaluation is to quantify the quality of the potential clusters which is often referred to as clustering validation. There are two broad categories of clustering validations; the external and the internal measures. Mainly, they differ by relying on external true-labels of the data or not. This chapter considers the role of evolutionary and swarm intelligence algorithms for data clustering, which showed extreme advantages over the classical clustering algorithms. The main idea of this chapter is to present thoroughly the clustering validation indices that are found in literature, indicating when they were utilized with evolutionary clustering and when used as an objective function.

Keywords Data clustering · Optimization · Evolutionary computation · Evaluation measures · Swarm intelligence · Metaheuristics · Fitness functions · Validation indices

---

I. Aljarah (✉) · M. Habib · R. Nujoom · H. Faris
King Abdullah II School for Information Technology, The University of Jordan, Amman, Jordan
e-mail: i.aljarah@ju.edu.jo

M. Habib
e-mail: neutrinoscientist@gmail.com; mar8160671@fgs.ju.edu.jo

R. Nujoom
e-mail: razan.nujoom@gmail.com

H. Faris
e-mail: hossam.faris@ju.edu.jo

S. Mirjalili
Center for Artificial Intelligence Research and Optimization, Torrens University Australia, Fortitude Valley, Brisbane, Australia
e-mail: ali.mirjalili@laureate.edu.au

I. Aljarah et al. (eds.), *Evolutionary Data Clustering: Algorithms and Applications*, Algorithms for Intelligent Systems,
https://doi.org/10.1007/978-981-33-4191-3_2

1 Introduction

Clustering is a well-regarded unsupervised learning process in data mining [3, 8, 54, 55]. Even that clustering is endlessly utilized in various applications, but it still faces different challenges. For instance, with the plentiful clustering algorithms, selecting the appropriate algorithm is not trivial, choosing the number of clusters, assessing the clustering tendency, quantifying the optimal similarity among the data points with the respective problem, and avoiding the randomly-created clusters are all such example obstacles. Apart from that is how to evaluate the goodness of obtained clusters and compare them to the results of other clustering algorithms. Essentially, assessing the quality of clusters encompasses three aspects; the clustering evaluation, the clustering tendency, and the clustering stability. Clustering evaluation measures the goodness of clusters based on validation metrics. The clustering tendency concentrates on determining if the respective dataset has nonarbitrary structures which result in ambiguous clusters. While the clustering stability denotes results susceptibility when varying the algorithm parameters. Hence, this chapter focuses on the clustering evaluation with deeper coverage within the context of evolutionary clustering [9, 11, 12, 140, 141].

Fundamentally, there are two essential concepts relating to the clustering problem. First is the clustering validation which is also known as the evaluation using evaluation metrics or quality measures. Second is the clustering objective function or the fitness function that is often used in the context of clustering optimization. Primarily, the validation of clustering involves assuring that the clustering algorithm does not create arbitrary clusters. In addition, it is used for setting the number of clusters and for comparing the obtained results with other implemented clustering algorithms. Clustering Validation Indices (CVI) are measures for evaluating the results of clustering analysis. Generally, CVI are categorized into external measures, internal measures, and relative measures. The external measures depend on data that is not used through the clustering process as the known class labels. They measure how close the clustering is to the class labels; so, these measures are useful when the ground truth is available [2, 6, 7, 172].

There are numerous indices were used as external measures, such as Fowlkes-Mallows Scores [58], Rand Index [146], Adjusted Rand Index [78], Mutual Information based scores [180], V-measure [154], Jaccard Index [80], and Czekanowski-Dice Index [42]. Whereas, the internal measures are used to quantify the compactness and separation of formed clusters. Where the compactness measures how much the data points in a single cluster are close to each other, which usually refers to within-cluster variance metrics. The separation finds how much the formed clusters are well-disconnected from each other, which can be measured by finding the distances between clusters' centers. Figure 1 shows an illustration of compactness and separation concepts. In which, for example, C_1 reach the optimal compactness by having the minimum distance between all points and the center as between (a, b_1) points. While the clusters (C_1, C_2, C_3) have the best separation by having the furthest dis-

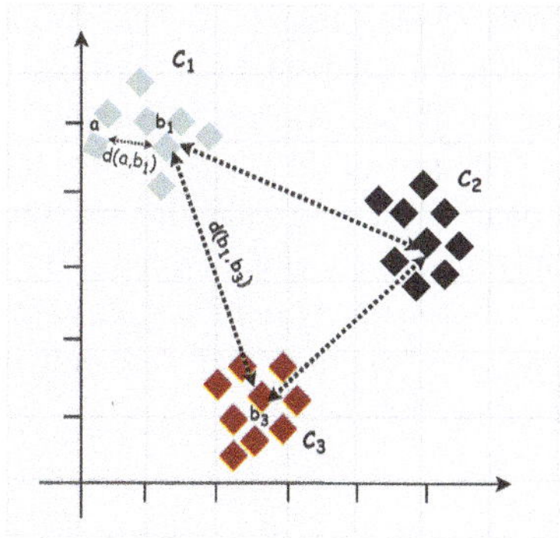

Fig. 1 An Illustration of compactness and separation of three imaginary clusters in a two-dimensional Cartesian coordinate system. All the clusters (C_1, C_2, C_3) are well-separated from each other

tances, which could be the representative distance between centroids. As the distance between (C_1, C_3).

In literature, there are numerous numbers of internal measures, such measures are including Sum of Squared Errors (SSE) [72], Davies-Bouldin Index [47], Calinski-Harabasz Index [26], Silhouette Coefficient [156], the Partition Coefficient [23], Xie-Beni Index [194], I Index [114], Average Between Group Sum of Squares (ABGSS) [90], Intra-cluster Entropy [151], Overall Cluster Deviation [74], Cluster Separation [151], Dunn Index [53], and Edge Index [170]. While the relative CVIs are a hybrid of external and internal indices.

On the other hand, the aim of objective (fitness) functions is to assess the quality of the clustering results during the clustering process [142, 143]. As clustering can be formulated as an optimization problem, the objective function is the optimization function. Traditional clustering algorithms as the k-means algorithm uses the sum of squared errors as an objective function. Usually, the sum of squared errors function uses the Euclidean distance to find the distances between each data point and the corresponding centroid. In which, the algorithm forms clusters and stops whenever it reaches the minimum sum of squared errors value. In the context of evolutionary clustering, the evolutionary algorithms optimize the clustering process by optimizing a fitness function aiming for searching the optimal solutions. Broadly speaking, evolutionary objective functions for clustering can be classified into distance measures and similarity or dissimilarity measures. There are a variety of utilized distance measures such as the Euclidean, Mahalanobis, Minkowski, and cosine distances. Also, a popular similarity measures are the Jaccard and cosine similarity indices.

Over the decades, wide-range of clustering validation measures have been developed. Few of them have been utilized for evolutionary data clustering. Milligan and Cooper [119] presented a comprehensive survey of CVI accounting for 30 measures. Figure 2 shows a classification of internal and external clustering validation measures indicating whether they used within the area of evolutionary computation. In the figure, the red circles denote that the respective index is utilized with evolutionary clustering. The black and green circles indicate the internal and external indices, respectively. The blue one refers to newly proposed indices. The light orange indicates a widely used index, and the grey circle presents a measure for fitness functions. Noticeably, the red circles with an overline dash symbol represents an index rarely used with evolutionary clustering.

The rest of the chapter is organized as follows. Section 2 represents the evolutionary clustering validation measures, encompassing external and internal measures only integrated within evolutionary clustering approaches, and presents other indices not used with evolutionary clustering, as described in Fig. 2. Section 3 presents the objective functions that were only used as fitness measures. While Sect. 4 is a summary of the chapter.

2 Evolutionary Clustering Validation Indices (CVI)

This section presents the clustering validation measures used with evolutionary clustering including both the external and the internal. In addition, it points out their implementation in further applications.

2.1 *External Measures*

External validation measures are used when the ground truth labels are known. They measure the extent to which cluster labels match externally supplied class labels. Hence, it targets maximizing cluster homogeneity and completeness. Maximizing the homogeneity means a maximized purity, while higher completeness indicates better ability to assign points to their true clusters. The following subsections show the external validation measures for evolutionary clustering.

2.1.1 Rand Index

Rand Index (RI) measures the similarity and consistency between the resultant groups of two random clusterings of a dataset [146]. In which, the comparison is conducted for all data points in each group (cluster) of each clustering (partition). In other words, it examines if two data points are in the same cluster at all partitions or even if they are in different clusters between partitions. Given two data points (y, y'), the two points (y, y') are paired points if they exist in the same cluster of a partition. Equation 1 defines the formula of Rand index given two partitions (P_1, P_2) and n data points [146].

$$RI = \frac{a+d}{a+b+c+d} = \frac{a+d}{\binom{n}{2}} \tag{1}$$

where a is the number of paired points that located in the two partitions (P_1, P_2). Parameter d is the number of points that are not paired in any of the partitions

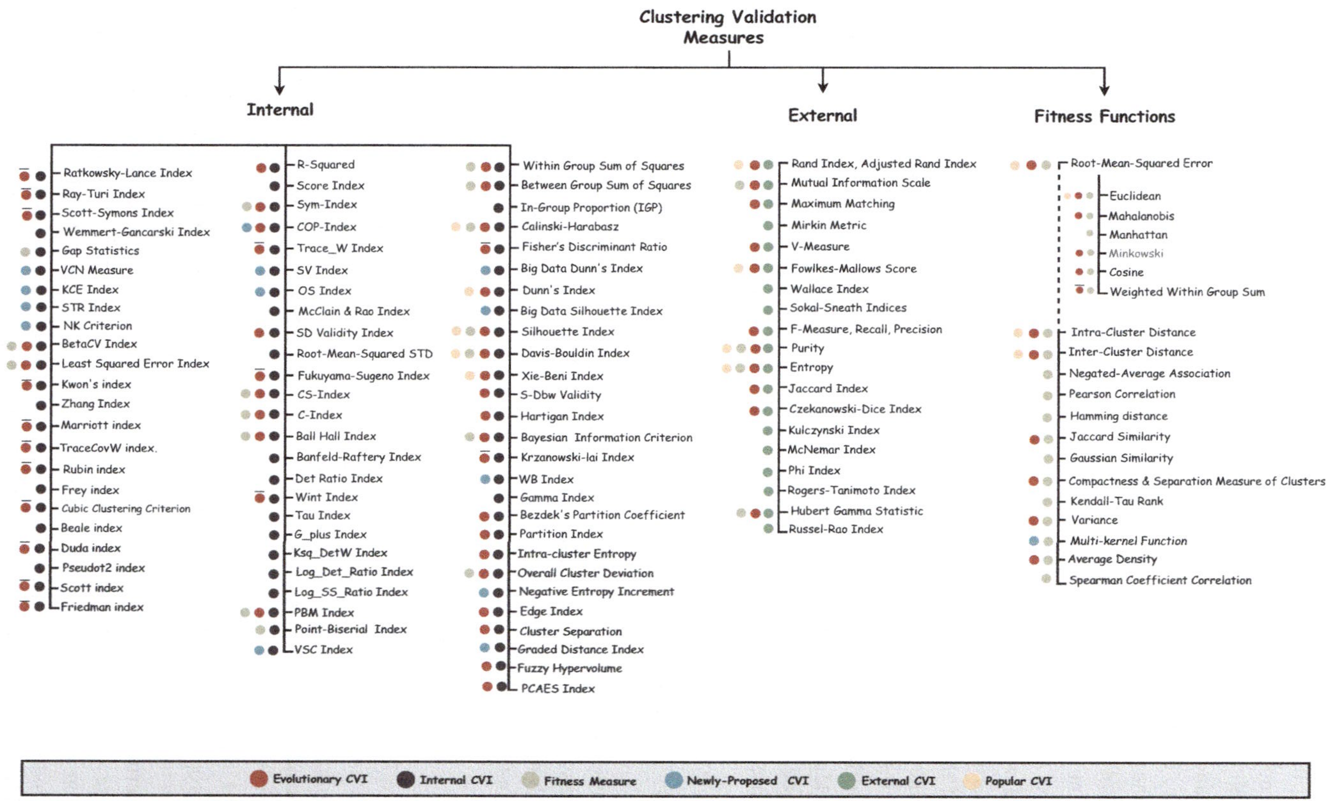

Fig. 2 A tree-like illustration of internal and external clustering validation measures, as well as the used fitness functions with evolutionary clustering

(P_1, P_2). While b is the number of points that are paired in one partition P_1 and c is the number of points that are paired just in the second partition P_2.

RI has a value in the range [0, 1]. Where 0 means the output of the clusterings does not have any matches of the clustered data points. While RI with value 1 means that all clusterings have the same output.

Rand index was used to evaluate a new evolutionary automatic clustering technique. The proposed algorithm is implemented by combining the k-means algorithm and the teaching-learning-based optimization algorithm (TLBO) [95]. A major drawback of RI is the sensitivity to the number of clusters in the partitions. An alternative variant of RI was proposed to overcome its weakness which is the Adjusted Rand Index (ARI) [78].

RI was also used in a novel combinatorial merge-split approach for automatic clustering using imperialist competitive algorithm [5]. In order to evaluate the quality of the clustering method, the new approach was compared with various methods including basic Imperialist Competitive Algorithm (ICA), Hybrid K Modify Imperialist Competitive Algorithm (K-MICA), Improved Imperialist Competitive Algorithm with mutation operator (IICA-G), and Explorative Imperialist Competitive Algorithm (EXPLICA) on 10 datasets. The results indicated that the proposed method was better than other algorithms. Another application of this index is that it was used in [196] to evaluate the performance of an evolutionary clustering method.

2.1.2 Adjusted Rand Index

Adjusted Rand (AR) index [78] is an external validity index that measures the similarity between two clusterings. It is the corrected for chance version of the Rand index; which means that a random result has 0 score. Equation 2 shows the formulation of AR Index.

$$AR = \frac{RI - E[RI]}{max(RI) - E[RI]} \tag{2}$$

where RI is the Rand Index, $E[RI]$ is the Expected Rand index, and max$[RI]$ is the Maximum Rand index.

Adjusted Rand index was used in clustering stability-based evolutionary K-Means to measure the clustering accuracy [77]. To evaluate the evolutionary K-Means, its results were compared to two consensus clustering algorithms; a clustering stability-based algorithm and a multi-index clustering approach. The results indicated that the proposed method is more robust to noise. The strength of this index is that it is corrected for chance. Adjusted Rand index was also used in evolutionary spectral clustering with adaptive forgetting factor as in [195], in order to evaluate the performance when using different values of the forgetting factor. Experiments indicated that the proposed method has results better than the usage of fixed forgetting factors. AR index was also used in a merge-split approach for automatic clustering using imperialist competitive algorithm [5].

2.1.3 Mutual Information Based Scores

Mutual Information (MI) based scores [180] are external validity indexes that evaluate the similarity of two clusterings. They compute the amount of information collected about a random variable by studying the other random variable. There are two versions; the Normalized Mutual Information (NMI) and the Adjusted Mutual Information (AMI). Equations (3–5) define MI score, NMI, and AMI, respectively [189].

$$MI(X,Y) = \sum_{i=1}^{R}\sum_{j=1}^{C} \frac{n_{ij}}{N} \log \frac{n_{ij}/N}{a_i b_j/N^2}. \tag{3}$$

where R is the number of clusters in clustering X, C the number of clusters in clustering Y, n_{ij} the number of objects that are common to clusters X_i and Y_j, and N the number of data items.

$$NMI(X,Y) = \frac{I(X,Y)}{\sqrt{H(X)H(Y)}}. \tag{4}$$

where X and Y are two clusterings, $H(X)$ the entropy of X, $H(Y)$ the entropy of Y, and $I(X,Y)$ is the Mutual Information between them.

$$AMI_{max}(X,Y) = \frac{I(X,Y) - E\{I(X,Y)\}}{\max\{H(X),H(Y)\} - E\{I(X,Y)\}} \tag{5}$$

where $E\{I(X,Y)\}$ is the expected mutual information of X and Y, (Eq. 6) is the expected mutual information.

$$E\{I(X,Y)\} = \sum_{i=1}^{R}\sum_{j=1}^{C} \sum_{n_{ij}=\max(a_i+b_j-N,0)}^{\min(a_i,b_j)} \frac{n_{ij}}{N} \log(\frac{N.n_{ij}}{a_i b_j}) \times Q \tag{6}$$

where

$$Q = \frac{a_i!b_j!(N-a_i)!(N-b_j)!}{N!n_{ij}!(a_i-n_{ij})!(b_j-n_{ij})!(N-a_i-b_j+n_{ij})!} \tag{7}$$

Normalized Mutual Information index is an important entropy measure in information theory. It calculates the similarity between two clusterings. It was used to evaluate a multi-objective and evolutionary clustering method for dynamic networks. It measured the performance of the proposed evolutionary clustering method on a dynamic network of a fixed number of communities, and then on a dynamic network of a variable number of communities [56]. AMI was also used to evaluate a clustering stability-based evolutionary K-Means [77]. The weakness of NMI and AMI is that they are not adjusted against chance.

2.1.4 V-measure

V-measure [154] is an entropy-based external index that measures how homogeneity and completeness are fulfilled. Homogeneity means that each cluster involves only the data points that are members of a single class, and completeness means that all data points of a given class are assigned to the same cluster. Homogeneity is defined by Eq. 8.

$$h = \begin{cases} 1 & \text{if } H(C, K) = 0 \\ 1 - \frac{H(C|K)}{H(C)} & \text{otherwise} \end{cases} \tag{8}$$

where $H(C|K)$ is the conditional entropy of the class distribution given the proposed clustering, and $H(C)$ is the entropy of the classes. Equation 9 is the conditional entropy, and Eq. 10 is the entropy of the classes.

$$H(C|K) = -\sum_{k=1}^{|K|}\sum_{c=1}^{|C|} \frac{a_{ck}}{N} \log \frac{a_{ck}}{\sum_{c=1}^{|C|} a_{a_{ck}}} \tag{9}$$

$$H(C) = -\sum_{c=1}^{|C|} \frac{\sum_{k=1}^{|K|} a_{ck}}{n} \log \frac{\sum_{k=1}^{|K|} a_{ck}}{n} \tag{10}$$

where N is the number of data points, C the set of classes, K the set of clusters, a_{ij} is the number of data points that are members of class c_i, and elements of cluster k_j.

Completeness is symmetrical to homogeneity, it is described in Eq. 11.

$$c = \begin{cases} 1 & \text{if } H(K, C) = 0 \\ 1 - \frac{H(K|C)}{H(K)} & \text{otherwise} \end{cases} \tag{11}$$

where $H(K|C)$ is the conditional entropy of the proposed cluster distribution given the class of the component data points, and $H(K)$ is the entropy of clusters. Equation 12 is the conditional entropy and Eq. 13 is the entropy of clusters.

$$H(K|C) = -\sum_{c=1}^{|C|}\sum_{k=1}^{|K|} \frac{a_{ck}}{N} \log \frac{a_{ck}}{\sum_{k=1}^{|K|} a_{a_{ck}}} \tag{12}$$

$$H(K) = -\sum_{k=1}^{|K|} \frac{\sum_{c=1}^{|C|} a_{ck}}{n} \log \frac{\sum_{c=1}^{|C|} a_{ck}}{n} \tag{13}$$

V-measure considers homogeneity and completeness, a higher value of this index means a better performance. Hence, Eq. 14 is the formulation of V-measure.

$$V_\beta = \frac{(1+\beta) * h * c}{(\beta * h) + c} \tag{14}$$

V-measure was used in the evaluation of various simulation scenarios, one is the usage of a probabilistic method for inferring intra-tumor evolutionary lineage trees from somatic single nucleotide variants of single cells. Where V-measure was used to evaluate the clustering performance [155]. This index provides accurate evaluations and solves problems such as the dependence on clustering algorithm or dataset.

2.1.5 Fowlkes-Mallows Index

Fowlkes-Mallows (FM) index [58] is an external validity measure that evaluates the similarity between two clusterings. The range of this index is between 0 and 1, while a higher index value means a better clustering. It is defined as the geometric mean of precision and recall. Equation 15 is Fowlkes-Mallows index.

$$FM = \frac{TP}{\sqrt{(TP + FP)(TP + FN)}} \tag{15}$$

where TP is the number of True Positive, FP is the number of False Positive, and FN is the number of False Negative.

In [73], the authors suggested a multi-objective differential evolution approach for clustering and feature selection. Further, [166] applied an evolutionary clustering approach for breast cancer prediction, which utilized various internal and external clustering validations including FM index. FM index was used to evaluate a swarm intelligence-based clustering method in [197]. Several indices were used as fitness functions, where the results stated that using silhouette statistic achieved the best results in most of the datasets.

2.1.6 Jaccard Index

Jaccard Index [80] (also known as Jaccard Similarity Coefficient) is an external validation index that measures similarity between two sets of data. Jaccard distance is complementary to this index and it measures dissimilarity. Equation 16 represents the formula of Jaccard Index [48].

$$JI = \frac{TP}{TP + FN + FP} \tag{16}$$

where TP is the number of True Positive, FP is the number of False Positive, and FN is the number of False Negative. Jaccard distance is calculated by subtracting Jaccard coefficient from 1.

Jaccard index was used for overlapping correlation clustering. In [14], the authors implemented a Biased Random-Key Genetic Algorithm (GA) method for overlapping correlation clustering. In which, Jaccard similarity coefficient and set-intersection indicator utilized as performance metrics. Furthermore, [33] suggested a local update strategy-based multi-objective evolutionary algorithm for community (cluster) detection. where the Jaccard index is implemented for measuring the similarity of nodes.

2.1.7 Czekanowski-Dice Index

Czekanowski-Dice index is an external validity measure that takes into consideration both precision and recall. It has several names; Ochiai index, Sørensen-Dice index, F1 Score, Dice's coefficient, and Dice similarity coefficient. This index was independently developed by Czekanowski [42], Dice [49], and Sorensen [176]. Equation 17 defines Czekanowski-Dice index [48]. Where TP is the number of True Positive, FP is the number of False Positive, and FN is the number of False Negative.

$$C - Dice = \frac{2 \times TP}{2 \times TP + FN + FP} \tag{17}$$

Czekanowski-Dice index was used in image segmentation using enhanced particle swarm optimization. It evaluated the performance of the proposed method and has been compared with related studies using different datasets. The mean of Dice coefficients over 10 runs was used as the criteria for performance comparison [181]. The main strength of this measure is that it reflects a realistic performance; since it considers both Precision and Recall. This index was also used to evaluate the performance of a new evolutionary clustering algorithm that is an improvement of the multi-objective clustering with automatic determination of the number of clusters (MOCK) algorithm [111].

2.1.8 Purity

Purity [204] is an external validation index that calculates to what degree a cluster contains data points of the same class. So, it finds the majority class in the cluster and computes the proportion of data points that belong to that class. The larger value of this index means better clustering. Equation 18 defines the formula of Purity [8].

$$Purity = \frac{1}{n} \sum_{j=1}^{k} \max_{i}(|L_i \cap C_j|) \tag{18}$$

where C_j contains all data instances assigned to cluster j, n is the number of data instances, k is the number of generated clusters, L_i is the true assignments of the data instances in cluster i.

Purity index was also used to evaluate the performance of a new clustering approach that is based on glowworm swarm optimization [8], and to assess a clustering algorithm using a novel grey wolf-inspired clustering approach [10]. Furthermore, authors in [57] designed a clustering strategy that integrates background information. In the proposed approach, the purity was used for evaluating the objective of the clustering algorithm, whereas the Davies-Bouldin index was used to measure the quality of the final obtained clusters.

2.1.9 Entropy

Entropy [204] is an external validation index that evaluates the distribution of the classes within each cluster. The smaller value of this index means less diversity of classes in the same cluster which indicates that the clustering is better. Equation 19 shows the definition of Entropy [8]. Where C_j contains all data instances assigned to cluster j, n is the number of data instances, k is the number of generated clusters, and $E(C_j)$ is the individual entropy of cluster C_j which is defined by Eq. 20. In which, L_i is the true assignments of the data instances in cluster i and q is the number of actual clusters.

$$Entropy = \sum_{j=1}^{k} \frac{|C_j|}{n} E(C_j) \tag{19}$$

$$E(C_j) = -\frac{1}{\log q} \sum_{i=1}^{q} \frac{|C_j \cap L_i|}{|C_j|} \log(\frac{|C_j \cap L_i|}{|C_j|}) \tag{20}$$

Entropy index was used to evaluate the clustering quality of a new clustering approach that is based on glowworm swarm optimization [8]. Three different fitness functions were compared with different datasets. Where the results indicated that the new clustering approach is efficient in comparison to well-known clustering algorithms. Moreover, the entropy was utilized in a novel locality-informed Grey Wolf clustering approach by [10]. In which, the proposed approach used the entropy for evaluating the clustering quality which showed merits over other well-known clustering algorithms. Nonetheless, the entropy validation measure was also used in [5] for automatic clustering based on an imperialist competitive algorithm. [16] suggested a GA approach for clustering, which tested three different functions as objective functions. Where the utilized objective functions are the Euclidean distance, the Mahalanobis distance, and the entropy.

2.1.10 Hubert $\hat{\Gamma}$ Statistic

Hubert Γ index is a statistical measurement that finds the correlation between two matrices of size $(N \times N)$, which is proposed in [79]. A higher value of Hubert $\hat{\Gamma}$ index indicates a higher correlation similarity of matrices. Hubert $\hat{\Gamma}$ index is defined by Eq. 21, where X and Y are two matrices and $M\ is\ (N(N-1)/2)$.

$$\hat{\Gamma} = \frac{1}{M} \sum_{i=1}^{N-1} \sum_{j=i+1}^{N} X(i,j) \cdot Y(i,j) \tag{21}$$

A modified version of it has been proposed in [78] for tackling the validation of clustering problems. Where it mainly replaced the two matrices by a proximity matrix of the dataset P and the matrix Q. Each element in the proximity matrix represents the distance between the respective two data points of (i,j) indices. While each element in the Q matrix presents the distance of the two centroids that the corresponding two data points belong to. The modified version is shown in Eq. 22. Accordingly, a higher value of the modified Hubert $\hat{\Gamma}$ index denotes a higher clustering compactness. In other words, given two data points in different clusters and the distance between them is relatively close to the distance of their centroids that means higher compact clusters.

$$\hat{\Gamma} = \frac{1}{M} \sum_{i=1}^{N-1} \sum_{j=i+1}^{N} P(i,j) \cdot Q(i,j) \tag{22}$$

Several studies have used the Hubert $\hat{\Gamma}$ index or its modified version for clustering validation. In [193], authors suggested a GA-based clustering approach for feature selection, in which, authors utilized Hubert $\hat{\Gamma}$ statistics as fitness function. Moreover, [137] proposed a clustering approach based on Genetic Programming for the determination of level-of-service in urban street, in which, Hubert statistic with other metrics were used for the validation.

2.2 *Internal Measures*

Internal validation measures are used when the ground truth labels are not known. These indices depend only on the information in the data. The following subsections present the internal indices used with evolutionary clustering.

2.2.1 Calinski-Harabasz Index

Calinski-Harabasz (CH) index [26] is also known as the Variance Ratio Criterion, is an internal evaluation measure that is based on the Between Cluster Scatter Matrix (BCSM) and the Within Cluster Scatter Matrix (WCSM). A larger BCSM value results in a better clustering, while a larger WCSM means a worse one. BCSM and WCSM are defined by Eqs. (23 and 24), respectively [163]. Where n is the total number of data points, k is the number of clusters, z_i is the centroid of the current cluster, z_{tot} is the centroid of all the data points, x is a data point belongs to cluster ci, and $d(x, z_i)$ is the Euclidean distance between x and z_i. Hence, depending on BCSM and WCSM, Eq. 25 presents the mathematical formula of Calinski-Harabasz index.

$$BCSM = \sum_{i=1}^{k} n_i.d(z_i, z_{tot})^2 \tag{23}$$

$$WCSM = \sum_{i=1}^{k} \sum_{x \in c_i} d(x, z_i)^2 \tag{24}$$

$$CH_k = \frac{BCSM}{k-1} . \frac{n-k}{WCSM} \tag{25}$$

Calinski-Harabasz index was used in unsupervised image clustering for feature selection using differential evolution. It evaluated clustering in two experiments [68]. Results indicated that the differential evolution feature selection had better results than the other used methods. This index is easy to compute, and its value is larger when clusters are compact and well-separated. Additionally, this index was used in clustering using Flower Pollination Algorithm (FPA) as an optimization criterion. The results indicated that FPA-based clustering method had a better performance than the other used methods [106]. Calinski-Harabasz index was also used in achieving natural clustering by validating results of iterative evolutionary clustering Approach [131], where a multi-objective genetic algorithm was used to decide on the number of clusters.

2.2.2 PBM Index (I Index)

PBM Index, which is also known by I Index, is an internal validity measure that evaluates the combination of compactness and separation. It has three key factors: $\frac{1}{k}$, $\frac{E_1}{E_k}$, and D_K [133]. The larger value of index PBM means better clustering. Equation 26 illustrate the formula of PBM Index [122]. Whereas, Eq. 27 defines the formula of intra-cluster distance E_K, and Eq. 28 defines the formula of the inter-cluster distance D_K.

$$PBM = \left(\frac{1}{K} \times \frac{E_1}{E_K} \times D_K \right)^{\gamma} \tag{26}$$

$$E_K = \sum_{k=1}^{K} \sum_{j=1}^{n} u_{kj} D(z_k, x_j) \tag{27}$$

$$D_K = \max_{i,j=1}^{K} \{D(z_i, z_j)\} \tag{28}$$

where $E1$ is a constant, u_{ki} is the membership degree of the ith data point to the kth cluster, and $D(z_k, x_i)$ is the distance of the ith data point x_i from the kth cluster center z_k.

I Index was used in modified differential evolution based fuzzy clustering for pixel classification in remote sensing imagery. It evaluated the performance of the proposed method which is a Modified Differential Evolution based Fuzzy Clustering (MoDEFC) and compared its performance with five other clustering algorithms. Those algorithms were Differential Evolution based Fuzzy Clustering (DEFC), genetic algorithm based fuzzy clustering (GAFC), simulated annealing based fuzzy clustering (SAFC), Fuzzy C-means (FCM), and Average Linkage (AL). The results indicated that the new method (MoDEFC) outperformed the other methods [115].

2.2.3 Dunn Index

Dunn Index is an internal evaluation measure that detects the clusters that are compact and well-separated. It depends on the minimum inter-cluster distance and the maximum cluster size [53]. A larger value of this index means a better clustering. Equation 29 defines the formula of Dunn Index [122]. In literature, different improvements have been conducted on Dunn index due to its sensitivity for noise. Kim et al. in [24, 89] proposed several generalized Dunn indexes such as gD31↑, gD41↑, gD51↑, and others.

$$DN = \min_{1 \leq i \leq K} \left\{ \min_{1 \leq j \leq K, j \neq i} \left\{ \frac{\delta(C_i, C_j)}{\max_{1 \leq k \leq K} \{\Delta(C_k)\}} \right\} \right\} \tag{29}$$

where $\delta(C_i, C_j)$ is the distance between clusters C_i and C_j. Equation 30 defines this distance.

$$\delta(C_i, C_j) = \min_{x_i \in C_i, x_j \in C_j} \{D(x_i, x_j)\} \tag{30}$$

And $\Delta(C_i)$ is the diameter of cluster C_i, where the diameter is shown by Eq. 31.

$$\Delta(C_i) = \max_{x_i, x_k \in C_i} \{D(x_i, x_k)\} \tag{31}$$

Dunn index was used in automatic scientific document clustering to evaluate a new multi-objective document clustering approach that is based on differential evolution (DE) and self-organizing map (SOM). The proposed method was compared to several clustering algorithms using this Index. Those algorithms were a multi-objective DE-based clustering approach without using SOM-based operators (MODoc-clust), and an archived multi-objective simulated annealing algorithm (AMOSA). The results indicated that the proposed approach had a better performance [161]. Dunn index is one of the most famous indices for clustering evaluation, but it requires high computational power when we have a high-dimensional data and when the number of clusters is large. This index was also used to select the best number of clusters in an iterative evolutionary clustering method to solve the scalability issue [131]. Dunn index was used in a comparison study of validity indices [197]. Swarm intelligence-based algorithms were evaluated when using the Dunn index and other indices as fitness functions. Furthermore, in [107], authors suggested an improved version of the Dunn index to handle the problem of clustering large data in a reasonable time, which is called the Big Data Dunn (BD-Dunn) index.

2.2.4 Silhouette Coefficient

Silhouette Coefficient [156] is an internal validation index that takes a data point and checks how much similar it is to its cluster in comparison to the other clusters. In other words, it shows how strongly the data points in the cluster are gathered. If most of the data points have a large value; this means a good clustering. Silhouette Coefficient is determined by Eq. 32. Where (a) is the mean distance between a point and the points in that cluster, and (b) is the mean distance between a point and the points in the next nearest cluster.

$$s(i) = \frac{b(i) - a(i)}{\max\{a(i), b(i)\}} \tag{32}$$

Silhouette Coefficient was used to evaluate a new evolutionary clustering algorithm named adaptive sequential k-means (ASK). The new algorithm was compared to c-means and k-means, where the results indicated that it accomplished better than both [139]. The value of this index is better when the clusters are compact and well-separated. This index was also used to evaluate the performance of a proposed evolutionary clustering algorithm [111]. In [66], the Silhouette index was utilized for fitness evaluation of Particle Swarm Optimization (PSO) clustering method. Remarkably, [107] suggested a modified version of Silhouette index for approaching the problem of validating the clustering of big data.

2.2.5 Davies-Bouldin Index

Davies-Bouldin index is an internal evaluation measure that was introduced in 1979 [47]. It uses the average similarity between a cluster and its most similar one to reflect how good the clustering is. The smaller DB index value means better clustering. In Eq. 33, the similarity has been formed [122].

$$R_i = \max_{j, j \neq i} \left\{ \frac{s_i + s_j}{d_{ij}} \right\} \tag{33}$$

where s_i is the average distance between each point of the cluster and the centroid of that cluster, and d_{ij} is the distance between-cluster centroids i and j. Equation 34 defines Davies-Bouldin coefficient.

$$DB = \frac{1}{N}\sum_{i=1}^{N} R_i \tag{34}$$

Davies-Bouldin Index was used in genetic clustering for automatic image classification as a measure of the validity of the clusters [20]. Further, it was used to evaluate a proposed multi-objective document clustering approach [161]. In [130], the authors implemented an evolutionary clustering approach based on GA for identifying traffic similarity. Where the k-means algorithm is used with Davies-Bouldin index for fitness evaluation. Moreover, [66] used Davies-Bouldin index for assessing the fitness of the PSO algorithm for clustering.

2.2.6 Xie-Beni Index

Xie-Beni index [194] is an internal validity measure that evaluates the compactness and separation of clusters. It is defined as the ratio between compactness σ and the minimum separation sep of the clusters. A smaller value of this index means better clustering. Equation 35 and 36 define the compactness and separation. Equation 37 represents Xie-Beni index [122].

$$\sigma = \sum_{k=1}^{K}\sum_{i=1}^{n} u_{ki}^2 D^2(z_k, x_i) \tag{35}$$

$$sep = \min_{k \neq l}\{D^2(z_k, x_i)\} \tag{36}$$

$$XB = \frac{\sigma}{n \times sep} \tag{37}$$

where u_{ki} is the membership degree of the ith data point to the kth cluster, and $D(z_k, x_i)$ is the distance of the ith data point x_i from the kth cluster center z_k.

Xie-Beni index was used to evaluate a modified differential evolution based fuzzy clustering algorithm [115]. Even that this index is simple to implement, but its weakness is that separation decreases when the number of clusters are close to the number of data points. Further, its used as a validity index in an automatic image pixel clustering with an improved differential evolution approach [45]. Another application of this index is that it was used to evaluate an improved differential evolution based clustering algorithm in [46].

2.2.7 Bezdek's Partition Coefficient

Partition Coefficient [23], it is an internal validity measure that describes the global cluster variance and how much overlap is among clusters. The range of values for this index is ∈ [1/c, 1], where c is the cluster number. The best result of this validity index is at its maximum value. Partition Coefficient is explained in Eq. 38 [169]. Where U_{ij} is the membership degree of the point x_j in cluster I, and c is the cluster number.

$$PC(c) = \frac{1}{N}\sum_{i=1}^{c}\sum_{j=1}^{N} U_{ij}^2 \tag{38}$$

Partition Coefficient was used to evaluate a novel evolutionary fuzzy clustering algorithm to solve the fuzzy C-means optimization problem. It was evaluated on three datasets, and the results indicated that it had a better performance than the classical fuzzy C-means algorithm [135]. Partition coefficient strength is that it identifies the right number of clusters in most of the datasets, but it decreases as the cluster number increases. Another drawback is that it is sensitive to the fuzzifier.

2.2.8 Average Between Group Sum of Squares (ABGSS)

ABGSS index is a measure that describes how well-separated are the clusters [90]. It represents the average distance between-clusters centroids and the centroid of all the data points. Equation 39 defines the formula of ABGSS [122].

$$ABGSS = \frac{\sum_{g=1}^{K} n_g \cdot D^2(z_g, \bar{z})}{K} \tag{39}$$

where n_g is the number of points in cluster g and $D(z_g, \bar{z})$ is the distance between the centroid of the cluster g and the center of the dataset $\bar{z}$. Moreover, another index that is related to ABGSS index is the Average Within Group Sum of Squares index (AWGSS) [90]. This index describes how compact the clusters are instead of focusing on clusters separation. Equation 40 defines AWGSS index.

$$AWGSS = \sum_{g=1}^{K} \frac{\sum_{i=1}^{n_g} D^2(z_g, x_i)}{n_g} \tag{40}$$

where $D(z_g, x_i)$ is the distance between the centroid of the cluster g and a data point x_i.

However, similar previously established validity indices are the $Trace\_W$ and $TraceCovW$ [118]. Mainly, $Trace\_W$ computes the within-cluster sum of squares, while $TraceCovW$ is the covariance of the within-cluster sum of squares. Both $Trace\_W$ and $TraceCovW$ have been used in [132], where a clustering approach depending on a multi-objective GA algorithm was proposed for skylines detection in databases.

ABGSS index was used in a newly proposed multi-objective genetic algorithm for clustering. Where used as a fitness function [90]. The new algorithm was compared to the k-means algorithm, and the results indicated that it had a better performance.

2.2.9 Intra-cluster Entropy

Intra-cluster Entropy [151] is an internal validation index that computes the average purity of clusters, while it does not rely on class labels. This index is different from Sum of Squared Error measure. The larger value of this index is better, Eq. 41 presents the Intra-cluster Entropy [122].

$$H = \sum_{i=1}^{K} [(1 - H(C_i).g(z_i))]^{\frac{1}{k}} \tag{41}$$

where

$$H(C_i) = -[g(z_i)log_2 g(z_i) + (1 - g(z_i))log_2(1 - g(z_i))] \tag{42}$$

And $g(z_i)$ is the average similarity between a cluster center z_i and the data points belonging to cluster i. Equation 43 is the formula of this average similarity.

$$g(z_i) = \frac{1}{n} \sum_{j=1}^{n} \left[0.5 + \frac{CO(z_i, x_i)}{2} \right] \tag{43}$$

where CO is the cosine distance, defined by Eq. 44. Where d is the number of features.

$$CO(x_i, x_j) = \frac{\sum_{k=1}^{d} x_{ik}.x_{jk}}{\sqrt{\sum_{k=1}^{d} x_{ik}^2}.\sqrt{\sum_{k=1}^{d} x_{jk}^2}} \tag{44}$$

Intra-cluster Entropy was used to evaluate the performance of a multi-objective data clustering algorithm [151]. It was also used as a fitness function in an evolutionary multi-objective clustering for overlapping clusters detection [150]. The results indicated that the proposed method successfully identified the overlapping clusters. This index has a low computational cost and it is insensitive to outliers; which makes it an attractive measure to utilize.

2.2.10 Overall Cluster Deviation (Cluster Connectedness)

Overall Cluster Deviation, which is also known as Cluster Connectedness, is an internal validation index that measures the compactness of clusters [74]. It is calculated as the summed distances between the data points and their clusters center. The smaller value of this index means a better clustering. Equation 45 defines the foundation of

Overall Cluster Deviation [122].

$$Dev(C) = \sum_{C_k \in C} \sum_{x_i \in C_k} D(z_k, x_i) \tag{45}$$

where $D(z_k, x_i)$ is the distance of the ith data point x_i from the kth cluster center z_k.

Overall Cluster Deviation was used to evaluate the performance of an evolutionary multi-objective data clustering algorithm [151]. The experimental results indicated that the proposed algorithm optimizes both Intra-cluster entropy and Inter-cluster distance.

2.2.11 Edge Index

Edge Index is an index that was introduced by Shirakawa and Nagao in 2009 [170]. This index calculates the summed distances between clusters. A smaller value of this index means a better clustering and higher separation. It presented by Eq. 46 [122]. Where $\xi i, j = D(xi, xj)$ if $\nexists\, C_k : i \in C_k^j \in C_k$, and $\xi i, j = 0$ otherwise, $\mathbb{F}$ is the set of N nearest neighbors of ith point, and N is a user-defined integer number.

$$Edge(C) = -\sum_{i=1}^{n} \sum_{j \in \mathbb{F}} \xi i, j \tag{46}$$

Edge Index was used in a proposed multi-objective evolutionary image segmentation algorithm; as an objective to be optimized [170]. Edge index is a very important factor in image segmentation. The results indicated that this algorithm obtains various good solutions; hence, could be useful if applied to other kinds of images like medical images.

2.2.12 Cluster Separation

Cluster Separation [151], which is also known as Inter-cluster Distance, is an index that evaluates the average distance between-cluster centers. The larger value of this index means a better clustering, as shown by Eq. 47 [122]. Where $D(z_i, z_j)$ is the distance between the centroid of cluster i and the centroid of cluster j.

$$Sep(C) = \frac{2}{K(K-1)} \sum_{i=1}^{K} \sum_{j=1, j \neq i}^{K} D^2(z_i, z_j) \tag{47}$$

Cluster Separation was used to evaluate a novel multi-objective evolutionary clustering algorithm [151]. This index has a low computational cost and it is a reliable index to implement since it is insensitive to outliers.

2.2.13 Bayesian Information Criterion

Bayesian Information Criterion (BIC) [167], which is also known as Schwarz Information Criterion. It is an internal validity index that was first introduced by Schwarz, in 1978, to solve the problem of model selection. Therefore, this index is based on the likelihood function and a larger value of it means a better clustering. Equation 48 is the BIC index [32].

$$BIC = \sum_{i=1}^{K}\{-\frac{1}{2}n_i \log|\Sigma_i|\} - Nk(d + \frac{1}{2}d(d+1)) \tag{48}$$

where K is the number of clusters, n_i is the number of data points in cluster i, Σ_i is the sample covariance matrix, N is the number of points in the dataset and d is the number of dimensions. So $d + \frac{1}{2}d(d+1)$ represents the number of parameters for each cluster.

In [39], a new evolutionary clustering method was proposed. Where the new method relied on cuckoo search and k-means for optimizing the search on the web. The developed algorithm utilized for document clustering, while integrated the BIC measure for evaluating and optimizing the quality of results. Moreover, [38] conducted web documents clustering based on genetic programming and a modified BIC technique for fitness optimization. In [164], authors designed a method used in the context of structural flaws detection of bridges. Hence, the methodology is a GA-based method used to search for the optimal hyper-parameters for having the optimal number of clusters. Where BIC is used for assessing the quality of GA solutions.

2.2.14 Sym-Index

Sym-Index [159], which is also known as Symmetry Distance-based index, is an internal validation index that was introduced in 2007. It is based on the idea of finding different-size symmetric clusters, as illustrated by Eq. 49. Where K is the number of clusters and D_K is the maximum Euclidean distance between two centroids. D_k is defined by Eq. 50.

$$Sym(K) = \frac{1}{K} \times \frac{1}{\varepsilon_k} \times D_K \tag{49}$$

$$D_K = \max_{i,j=1}^{K} ||\bar{c_i} - \bar{c_j}|| \tag{50}$$

where $\bar{c_i}$ is the centroid of cluster i, and ε_k is defined in Eq. 51.

$$\varepsilon_k = \sum_{i=1}^{K} \sum_{\bar{x}_j \in u_i} \frac{\sum_{ii=1}^{knear} d_{ii}}{knear} d_e(\bar{x}_j, \bar{c}_i). \tag{51}$$

where *knear* is the k nearest neighbor and $d_e(\bar{x}_j, \bar{c}_i)$ is the Euclidean distance between the point $\bar{x}$ and the centroid $\bar{c}_i$.

Sym-Index was used in an evolutionary multi-objective automatic clustering approach as an objective function [206]. Two algorithms were proposed and compared with other well-known algorithms. The results indicated that the new algorithms outperformed the other algorithms in terms of computational cost, time, and the ability to divide the data into meaningful clusters without knowing the cluster numbers. This index was also used as an objective function in a multi-objective fuzzy clustering approach based on tissue-like membrane systems. The approach was compared with other multi-objective clustering techniques, and the results indicated that it had a better performance than the others [138]. In [160], authors proposed an evolutionary clustering approach for symmetry-based automatic clustering. In which, the proposed approach utilized PSO and DE algorithms, in which, the Sym-Index was the objective function for the evolutionary algorithms.

2.2.15 CS Index

CS Index [35], which is also known as Chou, Su, and Lai Index. It is an internal validity index that is based on the ratio of clusters diameters to the between-clusters separation. The smaller value of this index means better clustering, as represented in Eq. 52 [28]. Where d is a distance function, K is the number of clusters, and $|C_k|$ is the number of data points in cluster C_k.

$$CS = \frac{\frac{1}{K}\sum_{k=1}^{K}\frac{1}{|C_k|}\sum_{x_i\ nC_k}\max_{x_j\in C_k} d(x_i, x_j)}{\frac{1}{K}\sum_{k=1}^{K}\min_{k,k\neq k'} d(c_k, c_{k'})} \tag{52}$$

CS Index was used to evaluate a proposed differential evolution clustering algorithm for determining the number of clusters relying on k-Means [173]. The new method was compared with other k-means related studies using six UCI datasets. Accordingly, the results indicated that the proposed method had the smallest value of CS; which means better clustering. This index was also used in automatic clustering using metaheuristic algorithms for content-based image retrieval [17]. The results indicated that clustering using optimization algorithms had better results than clustering using only main image retrieval algorithms. Moreover, CS Index was used as an objective function in [83]. In which, authors used a Differential Memetic algorithm for clustering that adopted CS index as a fitness function. Where the results were better than other algorithms in terms of CS, Davies-Bouldin, and accuracy.

2.2.16 C-Index

C-Index [43] is an internal validity index that was introduced in 1970. This index is based on the ratio between the minimum and the maximum distances between data points (Eq. 53) [48].

$$C - Index = \frac{S_W - S_{min}}{S_{max} - S_{min}} \tag{53}$$

where S_W is the sum of the within-cluster distances. S_{min} is the sum of the N_W smallest distances between all the pairs of data points, and S_{max} is the sum of the N_W largest distances between all the pairs of data points. Equation 54 shows the formulation of N_W.

$$N_W = \sum_{k=1}^{K} \frac{n_k(n_k - 1)}{2} \tag{54}$$

In literature, different research studies have used C-Index within the evolutionary clustering. For instance, [40] developed two genetic-based clustering algorithms, in which, the Calinski and Harabasz, C-index, and trace-W indices were compared for having the optimal number of clusters. Moreover, [105] applied a multi-objective genetic algorithm for clustering, in which various clustering validation indices were utilized including the C-Index. While the objectives of the proposed strategy were to minimize the number of clusters and the within-cluster variation. While in [158], a canonical GA algorithm was suggested for clustering. Where the canonical GA algorithm was used for optimizing the (C) validity index as an objective function for the sake of obtaining the optimal clusters.

2.2.17 Ball Hall Index

Ball Hall (BH) index [19] is an internal validation index that measures the average distance between each data point and the centroid of the cluster it belongs to. The smaller value of this index means better clustering. Equation 55 demonstrates the formula of Ball Hall index, where m_i is the centroid of the cluster c_i [145].

$$BH = \frac{1}{N} \sum_{i=1}^{K} \sum_{x \in C_i} ||x - m_i||. \tag{55}$$

Ball Hall index was used in particle swarm clustering for fitness evaluation with computational centroids [145]. This index was compared with the other 13 indices as fitness functions. Where it achieved better performance when the clustering task became more complicated. This index was also used to choose the number of clusters [131]. Furthermore, authors in [157] proposed a clustering method based on the GA algorithm for grouping the graphical data. Where the ball hall index was used as a fitness function for GA algorithm.

2.2.18 BetaCV Measure

BetaCV measure is an indicator for the coefficient of variation. It finds the proportion of the average intra-cluster distance to the average inter-cluster distance. The smaller the BetaCV value the better the clustering quality. Given K clusters, $W(x, y)$ is the sum of all weights of the edges located in between x and y, N_{out} is the number of unique edges between clusters and N_{in} is the number of unique edges within the cluster. BetaCV is defined by Eq. 56.

$$BetaCV = \frac{N_{out} \sum_{i=1}^{k} W(C_i, C_i)}{N_{in} \sum_{i=1}^{k} W(C_i, \overline{C_i})} \tag{56}$$

In [127], an interpretation of the PSO algorithm for evolutionary clustering was maintained. Where the BetaCV and Dunn indices were adopted for evaluating the fitness of the proposed algorithm. The results exhibited merits against the other clustering algorithms. Further, [126] performed a fuzzy rule interpolation based on the GA clustering approach. In their proposed methodology several clustering validation measures were deployed for evaluating the quality of GA individuals. The experimented fitness functions were Dunn, Davies-Bouldin, BetaCV, and Ball Hall indexes. The obtained outcomes showed that Dunn and Davies-Bouldin achieved the most accurate results.

2.2.19 Other Evolutionary Clustering Quality Indices

Among all mentioned indices, yet there are more used in the context of evolutionary clustering even that they are not widely used. For instance, Fisher's discriminant ratio is a measure of separability. A higher value of it indicates better separability. The mathematical model of Fisher's discriminant ratio is shown in Eq. 57, where k is the number of clusters, c is the centroid and σ is the cluster's variance.

$$Fisher's\ Ratio = \sum_{i=1}^{k} \sum_{j=1}^{k} \frac{||\overline{c_i} - \overline{c_j}||}{\overline{\sigma}_i^2 - \overline{\sigma}_j^2} \tag{57}$$

Sinha and Jana [174] investigated the use of Fisher's discriminant ratio for validating the results of a GA-based clustering approach, which showed efficient results in comparison with other clustering algorithms.

Another measure is the Hartigan index. Hartigan index is a well-regarded measure for searching the optimal number of clusters that are proposed in [75]. It stands on the within-cluster squared-sum of the distance between the points and the centroid. In Eq. 58, the $(n - k - 1)$ is the correction index, k is the clusters, n is the number of items in the data, and $W(k)$ is a matrix of the sum of squared distances.

$$Hartigan = \frac{W(k)}{W(k+1)} - 1 \cdot (n-k-1) \tag{58}$$

In [30], a fuzzy clustering approach was proposed relying on PSO algorithm. In which, several clustering validity measures were employed to find the optimal number of clusters including the Hartigan index. According to their reported results, Hartigan index achieved promising performance results.

The S_Dbw validity index not just quantify the separation and compactness of clusters, but also the density. It was proposed in [69]. In which, both the separation and compactness were represented by the inter and intra clusters variances, respectively. While the density is determined by the number of neighbors within a hyper-sphere. A lower value of S_Dbw index is preferable for better clustering quality. Equation 59 shows that the S_Dbw index is a summation of the scattering and density. Where the scattering and density are formulated in (Eq. 60, Eq. 61), respectively. Where $\sigma(v_i)$ is the variance of a cluster, $\sigma(x)$ is the variance of the dataset, u_{ij} is the central point of the link that connects the centroids v_i and v_j.

$$S\_Dbw = scattering + density \tag{59}$$

$$scattering = \frac{1}{k}\sum_{i=1}^{k} \frac{||\sigma(v_i)||}{||\sigma(x)||} \tag{60}$$

$$density = \frac{1}{k(k-1)}\sum_{i=1}^{k}\left(\sum_{j=1}^{k} \frac{density(u_{ij})}{Max\ (density(v_i), density(v_j))}\right) \tag{61}$$

Authors in [206] implemented an automatic clustering approach based on multi-objective evolutionary algorithm. Where different clustering evaluation measures were utilized including the S_Dbw index.

The SD validity index [70] depends on the mean scattering and the separation of clusters, as represented in Eq. 62. The scattering is as represented in Eq. 60, while the distance is the distance between the centroids of clusters as given by Eq. 63. A lower value of the index indicates better quality.

$$SD\ validity = scattering + distance \tag{62}$$

$$distance = \frac{Max(||v_j - v_i||)}{Min(||v_j - v_i||)}\sum_{i=1}^{k}\left(\sum_{z=1,z\neq t}^{k} ||v_t - v_z||\right)^{-1} \tag{63}$$

Even that SD validity index is not that widely used, but [105] showed an implementation of the multi-objective GA algorithm for clustering. In which, SD validity index was utilized for examining the results alongside other measures.

Furthermore, the R-squared index is often used when looking for determining the number of clusters within the hierarchical clustering approaches. R-squared is computed by finding the proportion of the sum of squares between clusters (SSB) to the sum of squares of the whole dataset (SST). Where the SST is the total sum of the SSB and the sum of squares within clusters (SSW) (Eq. 64). In which, the sum of squares is given by $\sum_{i=1}^{n}(x_i - \overline{x})^2$. The value of R-squared is within the range of [0, 1]; a value of 0 denotes that there is no difference between the clusters [188]. R-squared has been utilized in [166] for measuring the quality of a clustering strategy. In which, [166] presented a hybrid combination of Whale optimization and Moth Flame optimization for feature selection-based on clustering. The proposed approach is maintained in the situation of a breast cancer diagnosis.

$$R - squared = \frac{SSB}{SSB + SSW} \tag{64}$$

The COP-index is a relatively newly proposed index [67] that is designed to find the best cluster of hierarchical clustering. COP-index depends on the intra-cluster variance and inter-cluster variance (Eq. 65). P^Y is a partition Y, X is the dataset, d(x,y) is the Euclidean distance, intra(C)=$1/|C| \sum_{x \in C} d(x, \overline{C})$, and inter(C)= $Min_{x_i \notin C}\ Max_{x_j \in C}\ d(x_i, x_j)$.

$$COP(P^Y, X) = \frac{1}{|Y|} \sum_{C \in P^Y} |C| \frac{intra(C)}{inter(C)} \tag{65}$$

In [123], COP-index was used and compared with different clustering validation measures to assess the clustering quality of greedy-based evolutionary algorithm for clustering.

Nonetheless, Wint-index is a weighted inter and intra measurement developed in [179]. It involves the calculation of within-cluster distance (intra(C_i)) and the between-cluster distance (inter(C_i, C_j)) (Eq. 66). The intra(C_i) =$2/n_i(n_i - 1) \sum_{x,y \in C_i} d(x, y)$, inter($C_i, C_j$)=$1/n_i n_j \sum_{x \in C_i, y \in C_j} d(x, y)$. Given C is a cluster, k the number of clusters, and n is the data items.

$$Wint(k) = (1 - \frac{2k}{n}) \left(1 - \frac{\sum_{i=1}^{k} \frac{n_i}{n - n_i} \sum_{j=1, j \neq i}^{k} n_j \cdot inter(C_i, C_j)}{\sum_{i=1}^{k} n_i \cdot intra(C_i)}\right) \tag{66}$$

Fukuyama-Sugeno (FS) index quantifies the compactness and separation of clusters, it was created in 1989, in [62]. In Eq. 67, the first expression denotes compactness, while the second is an indicator for separation. Given that x is the data, c is the centroid, u is the degree of membership, and m is a controlling parameter of the fuzziness of the membership.

$$FS = \sum_{i=1}^{n} \sum_{j=1}^{n} u_{ij}^{m} ||x_i - c_j|| - \sum_{i=1}^{n} \sum_{j=1}^{n} u_{ij}^{m} ||c_j - \overline{c}|| \tag{67}$$

Chen and Ludwig [31] implemented Wint index and Fukuyama-Sugeno index for the evaluation of a PSO-based clustering approach. As well as, [51] utilized Wint and Fukuyama-Sugeno indices for the validation of clustering algorithm based on Artificial Bee Colony (ABC) and PSO.

Krzanowski-Lai (KL) index [92] stands on the squared distance of the points from their corresponding centroid. At the optimal value of this index the optimal number of clusters is reached. Equations (68 and 69) illustrates the formulation Krzanowski-Lai index, in which m is the number of features.

$$KL - index(k) = |\frac{diff(k)}{diff(k+1)}| \tag{68}$$

$$diff(k) = (k-1)^{2/m} \sum_{i=1}^{k-1} \sum_{x \in C_i} d^2(x, \overline{x_i}) - k^{2/m} \sum_{i=1}^{k} \sum_{x \in C_i} d^2(x, \overline{x_i}) \tag{69}$$

In [121], a fuzzy-GA clustering algorithm was developed for identifying the level-of-service of urban streets, where Krzanowski-lai index, c-index, Hargitan, weighted inter-intra measure, and R-squared were all maintained.

Ray-Turi index [148] was suggested to find the optimal number of clusters. Ray-Turi index is defined by Eq. 70, where the numerator presents the intra distance.

$$Ray - Turi = \frac{\frac{1}{n} \sum_{j=1}^{k-1} \sum_{x \in C_k} d(x, c_k)}{Min_{i \neq j}\ d(c_i, c_j)} \tag{70}$$

In [100], a k-means based PSO algorithm was developed for the detection of discontinuities within the rock mass avoiding such problems in the engineering sector. In the proposed approach the sum of all distances over all clusters was the fitness function, while the Ray-Turi index was the clustering validation index. Moreover, [63] used the inter-cluster distance, intra-cluster distance, and Ray-Turi index for validating the clustering results of the dynamic DE clustering algorithm.

Fuzzy Hypervolume [185] (FHV) is a validity index for fuzzy clustering. A smaller FHV value denotes a better quality of non-overlapping clusters. Interestingly, FHV does not depend on the sizes or shapes of the clusters. Primarily, the FHV index relies on the sum of the squared-root of the determinant of the covariance matrices (Eq. 71). In Eq. 71, det is the determinant and F_i is the fuzzy covariance matrix.

Few research studies have used fuzzy hypervolume for evolutionary clustering validation. However, in [36], an improved fuzzy c-mean algorithm was developed based on PSO algorithm and fuzzy c-means++. In which various validity measures were utilized for assessing the quality of the proposed method including the FHV.

$$FHV = \sum_{i=1}^{k} \sqrt{det(F_i)} \tag{71}$$

Furthermore, the Partition Coefficient and Exponential Separation (PCAES) index was proposed by Wu and Yang [192]. PCAES index measures the clustering quality by calculating a normalized fuzzy partition coefficient and an exponential separation measurement. PCAES is explained by Eq. 72. In which, u_M is $min_{1\le i \le c} \sum_{j=1}^{n} u_{ij}^2$, $\beta_T = (\sum_{i=1}^{c} ||z_i - \overline{z}||^2)/c$, and $\overline{z} = \sum_{j=1}^{n}(y_j/n)$.

$$PCAES = \sum_{i=1}^{c}\sum_{j=1}^{n}\frac{u_{ij}^2}{u_M} - \sum_{k=1}^{c} exp(-min_{k\neq i}||z_i - z_k||^2/\beta_T) \tag{72}$$

PCAES was used with evolutionary algorithms. In [31], a fuzzy-based PSO clustering algorithm was created for finding the optimal number of clusters, which involved PCAES for evaluation. Additionally, it was employed in a fuzzy-based chaotic PSO algorithm for clustering in [99].

Ratkowsky-Lance (RL) index [147] is a clustering quality measure that seeks for the best number of clusters. In literature, RL is identified as shown in Eq. 73, given that B is the sum of squares between the clusters for each data point, T is the total sum of squares of each data point. Although that RL was not widely used with evolutionary clustering algorithms, but [132] deployed it for the evaluation of a multi-objective GA clustering approach.

$$RL = \frac{average(\frac{B}{T})}{\sqrt{k}} \tag{73}$$

Moreover, Kwon's index [97] is a fuzzy clustering validation index. It was essentially proposed to alleviate the monotonically decreasing tendency with the increasing number of clusters. Kwon et al. defined the index as in Eq. 74, where $\overline{v} = \frac{1}{n}\sum_{j=1}^{n} x_j$. Kwon's index has been utilized for assessing two-level multi-objective GA clustering approach in [4].

$$Kwon's\ index = \frac{\sum_{j=1}^{n}\sum_{i=1}^{c} u_{ij}^2||x_j - v_i||^2 + \frac{1}{c}\sum_{i=1}^{c}||v_i - \overline{v}||^2}{min_{i\neq k}||v_i - v_k||^2} \tag{74}$$

Roughly speaking, clustering validation indices are vast, some are broadly-utilized and experimented in the context of evolutionary clustering, while other are less popular. Some of the rarely used are Marriot [109], Rubin [61], Cubic Clustering Criteria [165], Duda [52], Scott [168], and Friedman index [61]. Equations (75–80) present the conventional mathematical formulation for those indices, given that W is the within-cluster matrix, B is the between-clusters matrix, T is the total sum of squares matrix, $R^2 = 1 - (Trace(W)/Trace(T))$, p is the number of attributes, and $SSEWC_2$ is the sum of squared errors within-cluster when the data is grouped in two clusters, while SSE_1 is the sum of squared errors when the data represented by one cluster.

In [132], Marriot, Rubin, Duda, Scott, and Friedman were used for the validation of multi-objective GA algorithm for clustering. While [110] used the Cubic Clustering Criteria (CCC) for evaluating a clustering method that depends on a constructive GA algorithm for feature selection.

$$Marriot = k^2 W| \tag{75}$$

$$Rubin = \frac{|T|}{|W|} \tag{76}$$

$$CCC = ln\left(\frac{1 - E(R^2)}{1 - R^2}\right) \cdot \frac{\sqrt{\frac{np}{2}}}{(0.001 + E(R^2))^{1.2}} \tag{77}$$

$$Duda = \frac{SSEWC_2}{SSE_1} \tag{78}$$

$$Scott = nlog(\frac{|T|}{|W|}) \tag{79}$$

$$Friedman = Trace(W^{-1}B) \tag{80}$$

Table 1 presents a summary of clustering validation indices used within the era of evolutionary computation. It shows the indices names, their respective applications, and their references.

2.3 Other Non-evolutionary CVIs

Over the past decades, massive research studies have been conducted particularly for improving the evaluation of clustering. In previous sections and according to Fig. 2 in P. 27, an explanation of clustering quality measures has been conducted showing the extent of the integration with evolutionary algorithms.

Nowadays, plenty of clustering indices have been designed, which return to the early of the 60s. Yet, the evaluation of clustering is still an active research area. Figure 2 shows several indices that are not used with evolutionary clustering. Some of them were designed, especially for fuzzy clustering, whereas some others focus on optimizing the quality while maintaining the optimal number of clusters. To the best of our knowledge, Table 2 represents the clustering validation indices that are not used in the field of evolutionary clustering.

Table 1 A synopsis of evolutionary clustering quality measures; where they utilized and their related articles

Index	Type	Application	Related papers
CH index	Int	Data clustering, feature selection, determining number of clusters	[68, 106, 131]
PBM index	Int	Fuzzy image clustering	[115]
Dunn index	Int	Document clustering, determining the number of clusters, feature selection	[93, 131, 161, 197]
Silhouette	Int	Acoustic emission clustering, multi-objective optimization, feature selection, medical data clustering	[66, 93, 111, 139, 187]
Davies-Bouldin index	Int	Document clustering, traffic monitoring, feature selection, image clustering	[20, 37, 66, 93, 161]
Xie-Beni index	Int	Remote sensing imagery and fuzzy clustering, image clustering	[45, 46, 115]
Bezdek's P.Coefficient	Int	Fuzzy clustering	[135]
ABGSS	Int	Multi-objective optimization	[90]
Intra-cluster entropy	Int	Multi-objective clustering	[150, 151]
Cluster deviation	Int	Multi-objective clustering	[112, 151]
Edge index	Int	Image segmentation	[170]
Cluster separation	Int	Multi-objective clustering	[102, 151]
BIC	Int	Web search optimization, bridges flaws detection	[38, 39, 164]
Sym-index	Int	Evolutionary multi-objective automatic clustering and fuzzy clustering	[138, 160, 206]
CS-index	Int	Determining the number of clusters, image retrieval	[17, 83, 173]
C-index	Int	Data clustering, determining the number of clusters	[40, 105, 158, 187]
Ball Hall index	Int	Data clustering, determining the number of clusters, clustering graphical data	[131, 145, 157]
Improved Hubert Γ	Int	Feature selection, deciding the level-of-service	[137, 193]
BetaCV	Int	Data clustering, fuzzy rule interpolation	[126, 127]
Fisher's ratio	Int	Data clustering	[174]
Hartigan	Int	Fuzzy data clustering	[30]
S_Dbw index	Int	Automatic data clustering	[206]
SD validity	Int	Data clustering	[105]
R-squared	Int	Medical feature selection, image clustering	[37, 166]
COP-index	Int	Data clustering	[123]
Wint index	Int	Data clustering	[31, 51]
FS index	Int	Data clustering	[31, 51]
KL index	Int	Level of service in urban cities	[121]
Ray-Turi index	Int	Detecting discontinuities in rock, data clustering	[63, 100]
FHV	Int	Fuzzy data clustering	[36]
PCAES	Int	Fuzzy data clustering	[31, 99]
RL index	Int	Data clustering	[132]
Kwon's index	Int	Data clustering	[4]
Rand index	Ext	Automatic clustering, adaptive clustering, data clustering	[5, 81, 95, 196]
Adjusted Rand index	Ext	Clustering stability, evolutionary spectral clustering, automatic clustering	[5, 77, 195]
Mutual information	Ext	Multi-objective clustering of dynamic networks, clustering stability	[56, 77]
V-measure	Ext	Medical data clustering	[155]
Fowlkes M. Scores	Ext	Data clustering, feature selection, cancer diagnosis	[73, 166, 197]
Jaccard index	Ext	Overlapping clustering, community detection	[14, 33]
C. Dice index	Ext	Image segmentation, multi-objective clustering	[111, 181]
Purity	Ext	Data clustering	[8, 10, 57]
Entropy	Ext	Data clustering, automatic clustering	[5, 10, 16]

Table 2 Internal and external clustering validation indices not used within evolutionary clustering

Type	Measure	References	Formula
Internal	WB index	[203]	WB(M) = $M \cdot SSW/SSB$
	Gamma index	[18]	GI = $(s(+) - s(-))/(s(+) + s(-))$
	Negative entropy increment	[98]	NEI = $\frac{1}{2}\sum_{c_k \in C} p(c_k)log\lvert\sum_{c_k}\rvert - \frac{1}{2}log\lvert\sum_X\rvert - \sum_{c_k \in C} p(c_k)log\ p_{c_k}$
	Graded distance index	[85]	GDI = $\frac{\sum_{i=1}^{N}(u_{i,1stMax} - u_{i,2ndMax})}{N} - \frac{c}{N}$
	Score index	[162]	S(C) = $1 - (1/(e^{e^{bcd(C)+wcd(C)}})$
	SV index	[200]	SV(k,V,X) = $\frac{S_k}{V_k} = \frac{\sum_{i=1}^{k} Min_{j \in k, i \neq j}\ d(v_i, v_j)}{\sum_{i=1}^{k} Max_{x_j \in C_i}\ d(x_j, v_i)}$
	OS index	[200]	OS(k,X) = $\frac{O_k}{S_k} = \frac{\sum_{i=1}^{n}\sum_{x_j \in C_i} o_{x_j}}{\sum_{i=1}^{k} Min_{j \in k, i \neq j}\ d(v_i, v_j)}$
	McClain and Rao index	[116]	MC = $\frac{avg\left(\sum_{q}^{k=1}\sum_{i=1}^{nk}\sum_{j=i+1}^{nk} d_{ij}\right)}{avg\left(\sum_{k=1}^{q}\sum_{i \in C_k}\sum_{l=k+1}^{q}\sum_{j \in C_l} d_{ij}\right)}$
	Root-Mean-Squared STD	[152]	RMSSTD = $\sqrt{\frac{\sum_{i=1}^{p} SS_i}{\sum_{i=1}^{p} df_i}}$
	Banfeld-Raftery index	[48]	BR = $\sum_{k=1}^{k} n_k log\left(\frac{Tr(WG^k)}{n_k}\right)$
	Tau index	[153]	TI = $\frac{s(+)-s(-)}{[(n_d(n_d-1)/2-t)((n_d(n_d-1)/2)]^{1/2}}$
	G_plus index	[153]	G+ = $2s(-)/(n_d(n_d - 1))$
	Ksq_DetW index	[48]	KDI = $K^2(\lvert WG\rvert)$
	Log_Det_Ratio index	[48]	LDR = $Nlog\left(\frac{\lvert T\rvert}{\lvert WG\rvert}\right)$
	Log_SS_Ratio index	[48]	LSR = $log\left(\frac{BGSS}{\lvert WGSS\rvert}\right)$
	Point-Biserial index	[117]	PBI = $(\overline{d}_b - \overline{d}_w)(f_w . f_b/n_d^2)^{1/2}/S_d$
	VSC index	[149]	V_SC (c,U) = $Sep^N(c, U) + Comp^N(c, U)$
	Wemmert-Gancarski index	[48]	WI = $\frac{1}{N}\sum_{k=1}^{k} Max\{0, n_k - \sum_{i \in I_k} R(M_i)\}$
	Gap statistics	[183]	Gap = $\frac{1}{B}\sum_{b=1}^{B} log\ W_{qb} - log\ W_q$
	VCN measure	[205]	VCN = $\frac{bd(j)-wd(j)}{Max(bd(j), wd(j))}$
	KCE index	[82]	KCE = $k \times J$
	STR index	[177]	STR = $[E(K) - E(K-1)] \cdot [D(K+1) - D(K)]$
	NK criterion	[184]	NK = $\sum_{i=1}^{N} f_i(x, m_i)$
	Zhang index	[202]	Z_c(U,V) = $\frac{Var_{N,c}(U,V)}{Sep_N(c,U)}$
	Frey index	[60]	F = $\frac{\overline{d}_{v_{j+1}} - \overline{d}_{v_j}}{\overline{d}_{s_{j+1}} - \overline{d}_{s_j}}$
	Beale index	[21]	BI = $\left(\frac{W_m - (W_k + W_l)}{(W_k + W_l)}\right) / \left(\left(\frac{n_m - 1}{n_m - 2}\right) * 2^{2/p} - 1\right)$
	Pseudot2 index	[52]	Pdot = $B_{kl} / \frac{(W_k + W_l)}{n_k + n_l - 2}$
	In-Group Proportion (IGP)	[86]	IGP = $(\#k \mid class(k){=}class(k^N){=}j)/(\#k \mid class(k) = j)$
External	Mirkin metric	[120]	M = $\sum_k n_k^2 + \sum_{k'} n_{k'}^{'2} - 2\sum_k\sum_{k'} n_{kk'}^2$
	Wallace index	[191]	WLI = $W_{A \to B} = \frac{a}{a+b},\ W_{B \to A} = \frac{a}{a+c}$
	Kulczynski index	[48]	KuI = $\frac{1}{2}(Precision + Recall)$
	McNemar index	[48]	McI = $\frac{yn-ny}{\sqrt{yn+ny}}$
	Phi index	[48]	Phi = $\frac{yy \cdot nn - yn \cdot ny}{(yy+yn)(yy+ny)(yn+nn)(ny+nn)}$
	Rogers-Tanimoto index	[48]	RTI = $\frac{yy+nn}{yy+nn+2(ny+yn)}$
	Russel-Rao index	[48]	RRI = $\frac{1}{N_T}\sum_{i<j} X_1(i,j)X_2(i,j)$
	Sokal-Sneath indices	[48]	SSI = $\frac{yy+nn}{yy+nn+\frac{1}{2}(ny+yn)}$

3 Fitness Evaluation Indices

The objective of evolutionary clustering is to optimize the clustering process by optimizing a fitness function. The aim of the fitness (objective) functions is to evaluate the quality of the clustering. Broadly speaking, evolutionary objective functions for clustering can be classified into distance measures and similarity measures. The following sub-sections review thoroughly the used fitness functions with evolutionary clustering approaches by presenting a description of each corresponding measure and in which applications it used.

3.1 Root Mean Squared Error

Distance metrics calculate the space between two points in m-dimensional space. While in clustering, distance measures find the space between two data instances of m features. Distance functions in cluster analysis are a kind of similarity measurement. They used to quantify how much the objects are alike during the clustering process. Where the smaller the distance implies much similar data points. The Sum of Squared Errors (SSE) is a measurement used to optimize the obtained clusters throughout the search process. Equation 81 describes the formula of SSE. It represents the summation of the squared distance of all data points and each respective centroid. The clustering technique that results in the smallest SSE value is considered as the best clustering approach.

$$SSE = \sum_{j=1}^{k} \sum_{i=1}^{|C_j|} dist(n_i, c_j)^2 \tag{81}$$

In literature, several distance measures were implemented for the calculation of the SSE metric. Hence, this sub-section surveys the utilized distance functions for evaluating the validity of evolutionary clustering.

3.1.1 Euclidean Distance

The Euclidean distance between two points (a, b) represents the length of the line that links them together. The distance between a and b with m dimensions is the square root of the sum of the squared difference between (a) and (b). Given that (a) and (b) are located in the Euclidean space, then the Euclidean distance is illustrated by Eq. 82 in regard to the Cartesian coordinate system.

$$dist(a, b) = \sqrt{(a_1 - b_1)^2 + (a_2 - b_2)^2 + \cdots + (a_m - b_m)^2} \tag{82}$$

The Euclidean distance is a metric that obeys certain mathematical properties. First, the distance is a non-negative value. Second, the distance from the point to itself is zero. Third, the distance satisfies the symmetry property, where dist(a, b) = dist(b, a). Lastly, the distance follows the triangle inequality property. Thus, $dist(a, b) \leq dist(a, c) + dist(c, b)$ [72].

Several research studies have implemented the SSE metric for validating the goodness of the clustering process. For instance, [113, 124] proposed a GA algorithm with k-means for clustering. In which, the sum of squared Euclidean distances was the fitness function. Also, [94] implemented a two-stage clustering approach for order clustering. In which, the proposed approach is a combination of neural networks, GA and PSO. Where the sum of Euclidean distances was the fitness function. Furthermore, [96] designed a clustering approach that is a hybrid of differential evolution and k-means. Where the sum of squared errors was the fitness criterion. [50] developed an improved adaptive GA into fuzzy c-means. In order to optimize the initial centroids by optimizing the SSE based on the Euclidean distance. While [186] introduced the parallel k-bat algorithm for clustering big high-dimensional datasets. In which, the sum of mean squared Euclidean distance was the objective function. In addition, in [134], the authors utilized a spiral cuckoo search algorithm for clustering. The proposed approach was used for spam detection, where SSE metric was a fitness function. [174] used the SSE measure for the validation of k-means based on the GA clustering method. Nonetheless, [31] used a weighted within group sum of squared error as a fitness function. Where the proposed method was a hybrid of fuzzy c-means and PSO algorithm for finding the optimal number of clusters.

3.1.2 Mahalanobis Distance

Mahalanobis distance is a measurement for calculating the distance between two vectors in a multivariate space. One of the advantages of Mahalanobis distance is the adoption of a covariance matrix which leads for quantifying the similarity between the vectors. However, it has been known to have a high computational cost [59]. Given a data point ($\vec{a}$) and a centroid ($\vec{b}$), then Mahalanobis distance is defined by Eq. 83. In which, C^{-1} is the inverse covariance matrix [64].

$$d_{ab} = \sqrt{(\vec{a} - \vec{b})C^{-1}(\vec{a} - \vec{b})^T} \tag{83}$$

Authors in [174] proposed an evolutionary clustering approach for distributed datasets. The proposed approach is a hybrid of GA algorithm and k-means, where Mahalanobis distance is integrated as a fitness function. While in [65], the authors proposed a data filling algorithm using a combination of k-means and information entropy. In which, Mahalanobis distance is used for clustering instead of Euclidean distance. Moreover, [29] developed a clustering approach for topic categorization. Where the squared Mahalanobis distance was the clustering criterion.

3.1.3 Minkowski Distance

Minkowski distance is a generalization form of Euclidean distance. Often known as L^p norm metric. It is formulated in Eq. 84. When the value of $p = 1$, then it is Manhattan distance (City-Block), while $p = 2$ is the Euclidean distance and $p = \infty$ is the Chebyshev distance [72].

$$dist(a, b) = \sqrt[p]{\sum_{i=1}^{n} |a_i - b_i|^p}, \quad where\ p \geq 1 \tag{84}$$

In [84], the authors proposed an improved clustering approach for multimedia databases. In which, Minkowski distance is used for clustering validation. In addition, [76] implemented a hierarchical clustering approach for the detection of drug toxicity using biomarker genes. The proposed algorithm examined the integration of Minkowski distance for assessing the goodness of clusters. Whereas, [88] proposed an incremental clustering approach for time-series datasets. The proposed algorithm is a combination of two clustering algorithms; k-means and fuzzy c-means. As well as, four distance measures including the Minkowski algorithm.

3.1.4 Cosine Distance

The cosine distance is the angular distance between two vectors of the same length. Roughly, it is defined by (1-cosine similarity) and given by Eq. 85. The cosine similarity for two vectors is calculated by finding the cosine of the enclosed angle between them. Smaller angles mean higher similarity score, while $\theta \geq 90$ means no similarity. The cosine distance is a value $\in [0, 1]$, where value (1) means highly distant vectors and (0) means much closer vectors. Figure 3 presents the cosine distance of two vectors.

$$cosine\ distance(a, b) = 1 - \frac{a \cdot b}{||a|| \cdot ||b||} \tag{85}$$

where $(a.b)$ corresponds to the dot product and equals to $\sum_{i=0}^{m}(a_i \times b_i)$, while $||a||$ is the Euclidean norm of vector (a) and equals to $\sqrt{\sum a_i^2}$.

Several studies have implemented the cosine distance for the calculation of SSE fitness function. [104] designed an evolutionary clustering approach using k-means and GA algorithm. In which, the cosine distance is used as a fitness function. In [175], authors proposed a PSO approach for clustering. The proposed approach implemented the average of cosine distance as a fitness score for optimizing the clustering. Further, [199] implemented a clustering approach for power curve modeling of wind turbines. The clustering approach utilized different clustering algorithms including k-means. As well as, they deployed several fitness functions such as the Euclidean, Cosine, and City-Block distances. Also, [15] authors formulated

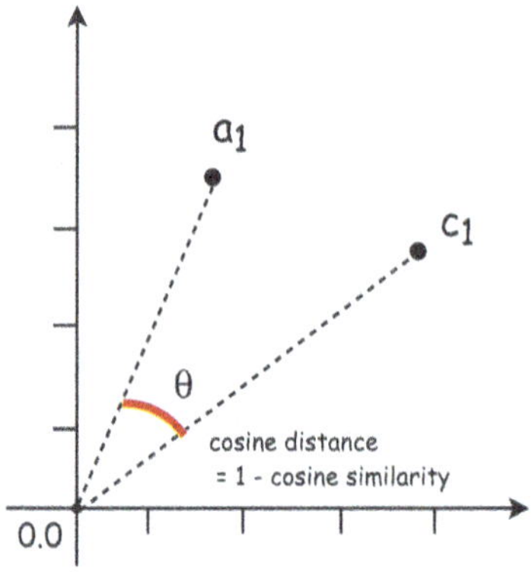

Fig. 3 A representation of cosine distance between point (a_1) and centroid (c_1)

a clustering approach for time analysis of air pollution using k-means. In which, various distances were used involving the Euclidean, City-Block, Cosine, and correlation distances. Nonetheless, [128] utilized k-means for color-based segmentation of images. In their approach, authors implemented different distance measures for SSE metric including the cosine distance.

3.2 (Intra and Inter) Cluster Distances

The Intra-Cluster distance is concerned with quantifying the similarity of points within a single cluster. As the objective of a clustering problem is maximizing the similarity among data points; the closer the points the more similar they are. Hence, the Intra-Cluster distance is considered as a minimization problem. Authors in [25] suggested three criteria for calculating the Intra-Cluster distance. Mainly, they are the complete diameter, the average diameter, and the centroid diameter. In the case of the complete diameter, the Intra-Cluster distance is characterized by the length of the link between the farthest two points in the cluster. For the average diameter, the distance is represented by the average of distances among all data points in the cluster. While the centroid diameter is characterized by the average distance between all data points and the centroid. Even that, there are other strategies different than the diameter such as the radius, the variance, and the variance multiplied by the SSE. Variance is a statistical measure describes the dispersion of data. A high variance value means the data points are very close to their mean value. The variance is presented by Eq. 86, where N is the number of data points, $\overline{X}$ is the mean value [72].

$$\sigma^2 = \frac{1}{N}\sum_{i=1}^{N}(x_i - \overline{x})^2 \tag{86}$$

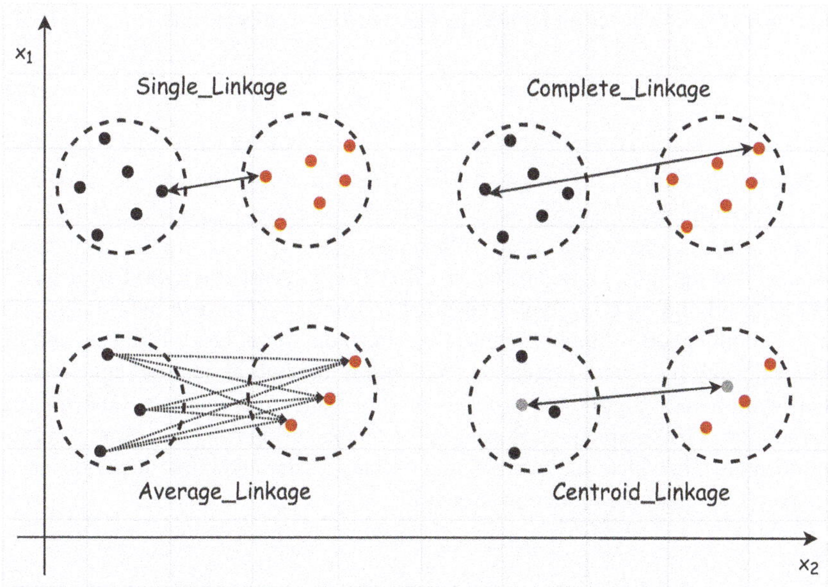

Fig. 4 A description of the linkage measures between two clusters represented by the red and black colors in 2-dimensional space. Depicting the single, the complete, the average, and centroid Inter-Cluster measures

Zang et al. [201] implemented a new clustering method based on the GA algorithm for improving spectral clustering. Authors used clustering variance ratio as a fitness function which finds the ratio of the within-cluster variance and the between-cluster variance. Further, [136] utilized GA, PSO, and DE for partitional clustering, in which, several fitness criteria have been adopted based heavily on the variance measure. They include the Marriotte criterion, the trace-within criterion, and the variance ratio criterion.

Inter-Cluster distance reflects how much the clusters are dissimilar. The more they are separated from each other the more dissimilar the clusters are. Finding the Inter-Cluster distances can be performed using the linkage criteria. There are several Linkage measures such as the single-linkage, complete-linkage, average-linkage, centroid or median linkage and Ward's method [72]. Figure 4 illustrates the difference between the single, complete, average, and centroid measures. The single-linkage measure finds the pair of data with the minimum distance as determined by $(Min \ \ (|x - x'|), \ \ x \in C_i, \ \ x' \in C_j)$ given two clusters (C_i and C_j) and two data points (x and x').

Regarding the complete-linkage criterion, the distance is a maximization metric represented by the length of the line segment between the two farthest neighbor points $(Max \ \ (|x - x'|))$. While the average-linkage distance is the average distance between all pairs of data represented by Eq. 87, where n_i and n_j are the number of points in clusters C_i and C_j, respectively. Whereas, the centroid-linkage computes the center of each cluster then finds the distance between centroids.

$$d(C_i, C_j) = \frac{1}{n_i \times n_j} \sum_{x \in C_i,\ x' \in C_j} |x - x'| \tag{87}$$

Nguyen and Kuo [129] proposed a fuzzy-GA clustering for categorical data. In the proposed approach, single-objective and multi-objective functions were deployed for assessing the clustering fitness. Where the used objectives are the inter-cluster distance and the intra-cluster distance. Also, [81] suggested a hybrid strategy for clustering relying on Grey Wolf and Whale optimizers. The new hybrid approach used a combination of the inter-cluster distance, the intra-cluster distance, and the cluster density for fitness evaluation. As a result, the hybrid method outperformed other evolutionary algorithms in terms of F-measure, Jaccard, and Rand indices. Furthermore, [108] implemented a clustering method using Ant colony and Ant Lion optimizations, in which, the combination of inter-cluster and intra-cluster distances were used for the fitness function structure.

3.3 Jaccard Distance

Jaccard Distance (JD) is a metric that satisfies the properties of the conventional metrics including the triangle inequality [72]. JD metric is used to measure the dissimilarity between two samples of data that constitutes subtracting the Jaccard Similarity (JS) from 1, as represented in Eq. 88 [72]. In which, JS is the Jaccard similarity between two given data sets (x and y).

$$JD = 1 - JS(x, y) \tag{88}$$

Evidently, as the name implies, the Jaccard similarity resembles how much the two sets (x,y) have common elements. It is defined by the ratio of the similar elements between the two sets to all similar and not similar elements. In other words, is the ratio of the intersection over the union of the two sets as in Eq. 89.

$$JS(x, y) = \frac{|x \cap y|}{|x \cup y|} \tag{89}$$

Both Jaccard distance and Jaccard similarity have been utilized for the fitness evaluation within evolutionary clustering. Andrade et al. [14] designed a biased random-key GA algorithm for overlapping clustering. Where the Jaccard similarity is used for assessing the fitness during the local search of clustering. While [104] proposed an evolutionary k-means based on the GA algorithm which adopted the Jaccard distance as a quality measure for the objective function. Whereas, in [103] authors suggested a new clustering and validity clustering technique for soft subspace clustering. The new algorithm depends on the adaptive evolutionary algorithm that implemented the Jaccard distance for clustering evaluation. Remarkably, incorporating the Jaccard distance exhibited a better performance in comparison to the results of Euclidean distance. Further, [1] designed a DE-based approach for clustering texts that used Jaccard similarity as a fitness function. Where it outperformed the Normalized Google Distance (NGD) similarity measure.

3.4 Compactness and Separation Measures of Clusters

Essentially cluster validation can be designed to maximize two objectives; the compactness and the separation. Compactness is a terminology that reflects how much the points are dense or close. In other words, it quantifies and minimizes the variance within a cluster. The lower the variance the higher the compactness and the better the clustering performance. On the other hand, the separation measures how much the created clusters are disconnected from each other. Separation can be computed based on different measures such as the distance or the variance. Evidently, having the clusters being well-separated from each other is an indicator of higher clustering quality.

Compactness and separation have been utilized for evaluating clustering fitness, but they also widely used within internal cluster validation. Most internal clustering validation measures depend on their implementation on the compactness, separation, or both.

Essentially, [187] implemented a clustering approach based on Principal Component Analysis (PCA) for identifying markers of a genotypic data. In which, the author utilized C-Index, Point Biserial, and Silhouette Indices as evaluation measures for the best number of clusters. Moreover, [37] used several evolutionary algorithms such as the ABC and Cuckoo Search for image clustering. In the proposed method, the Gap statistics has been utilized for evaluating the fitness of the used evolutionary algorithms, while Davies-Bouldin, Silhouette, Dunn, and R-Squared indices were used for evaluating the clustering results. Further, [112] created an adaptive k-determination clustering algorithm which adopted the connectivity and deviation as a fitness evaluation measures. Where the connectivity depends on a similarity measure to increase the compactness and, similarly, is the deviation that depends on a distance measure.

Interestingly, various research studies have implemented compactness and separation as a multi-objective fitness functions that targets maximizing the compactness and the separation. Authors in [27] have formulated the fitness function in a way that comprises both compactness and separation in the form of $(\alpha \times (separation/compactness))$, where it utilized for Fractional Lion Optimization for clustering. Generally speaking, clustering can be formulated as multi-objective optimization problem as given by Eq. 90. Where C is the cluster and (i, j) refer for clusters' numbers.

$$fitness = Maximize \begin{cases} f_1 = compactness(C_i) \\ f_2 = separation(C_i, C_j) \end{cases} \tag{90}$$

For example, [161] demonstrated the usage of multi-objective DE algorithm for clustering, where both PBM and Silhouette indices were handled as objective functions. While in [198] two objective functions utilized for clustering in the context of image segmentation. The clustering approach was based on multi-objective artificial immune learning that optimizes two measures; the cluster connectedness and the cluster scattering. Whereas [102] integrated three objectives for optimization including the intra-cluster dispersion, the inter-cluster separation, and the negative Shannon entropy alongside the Non-dominated Sorting GA algorithm (NSGA-III) for clustering.

3.5 Clusters Density

Generally, density-based methods are efficient for finding randomly-shaped clusters. Where the dense regions represent the potential clusters that are separated by the scattered regions. Primarily, the density of a data item i is characterized by the number of neighbors around it that satisfies a threshold value. Hence, density-based methods collect points with the high-density neighborhood to form the cluster. Often, the density of the neighborhood is quantified by a radius value of r. Figure 5 shows how the density measure affect the coherence of the candidate clusters.

On the contrary, density-based neighborhood methods are highly sensitive to the radius value. Therefore, one of the alternative approaches is the utilization of the kernel density estimation criterion. For instance, [13] proposed an evolutionary approach for clustering using the PSO algorithm and the density-based estimator for clusters validation. While [101] suggested a combination of density clustering algorithm and ACO algorithm for tackling the traveling salesman problem.

The previous sub-sections demonstrated different fitness evaluation functions for clustering approaches optimized by evolutionary search algorithms. However, there are other quality validation functions used to optimize the objective function of a clustering problem without the integration of evolutionary algorithms. Such objective functions are the Manhattan distance, Negated Average Association, Hamming similarity, Gaussian similarity, Spearman correlation, Pearson correlation, and Kendall-tau ranking. Table 3 shows the definition of those functions. In essence, [178] used Manhattan distance instead of Euclidean distance for optimizing Ward's clustering algorithm. Whereas, [34] implemented the negated average association measure and the association average as optimization objectives for spectral clustering. [182] implemented a clustering approach for detecting bias in machine learning algorithms. In which, k-means was deployed with the hamming distance function. Also, [144] utilized Gaussian similarity for clustering software modules and textual documents. While [171] developed a clustering approach based on the Spearman correlation coefficient for indoor-outdoor environmental detection. Further, [41] presented sev-

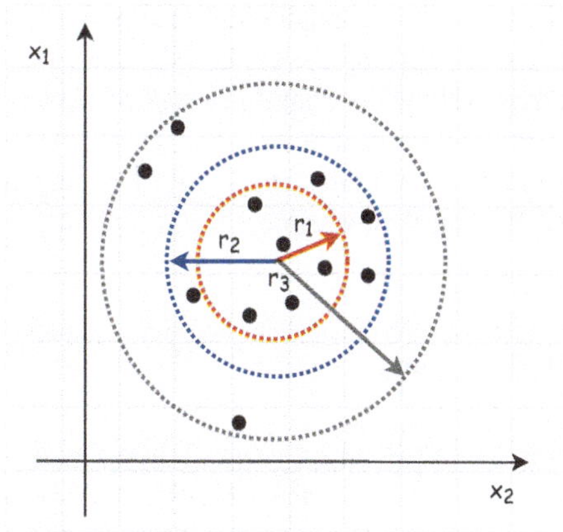

Fig. 5 An explanation of selecting a proper radius of a cluster might result in better cluster density. Setting r_2 as the radius of density maximizes the compactness of the cluster, while both r_1 and r_3 are improper choices

Table 3 A list of objective functions not used with evolutionary clustering

Function	References	Formula
Manhattan distance	[91]	dist(a, b) = $\sum_{i=1}^{n} \lvert a_i - b_i \rvert$
Negated average association	[34]	NAA = $Tr(W) - \sum_{i=1}^{k} \frac{assoc(V_i, V_i)}{\lvert V \rvert}$
Hamming distance	[71]	d(a, b) = $\sum_{k=0}^{n-1} (y_{a,k} \notin y_{b,k})$
Gaussian similarity	[190]	s(a ,b) = $exp(\frac{\lVert x_a - x_b \rVert}{2\sigma^2})$
Spearman correlation	[125]	$C_s = 1 - \frac{6 \sum d_i^2}{n(n^2-1)}$
Pearson correlation	[22]	$C_p(u, v) = \frac{\sum_i (r_{ui} - \bar{r}_u)(r_{vi} - \bar{r}_v)}{\sqrt{\sum_i (r_{ui} - \bar{r}_u)^2} \sqrt{\sum_i (r_v i - \bar{r}_v)^2}}$
Kendall-Tau ranking	[87]	$\tau = \frac{n_c - n_d}{n(n-1)/2}$

eral correlation metrics including Pearson for clustering electricity markets. Yet, in [44], Kendall-Tau has been adopted in a clustering criterion for preference rankings.

To this end, validating clustering is ubiquitous. Evidently, different problems can be tackled in different strategies depending on the nature of the problem and its context. Therefore, clustering validation under the realm of evolutionary algorithms or even generally is open-ended, where interested researchers can inspect and explore more.

4 Conclusion

In conclusion, this chapter introduced thoroughly the clustering validation indices generally and particularly for evolutionary clustering. It is included the internal and external clustering quality measures, as well as the objective (fitness) functions. Additionally, it presented the implementation areas of each respective index. As clustering is a fundamental unsupervised learning method with immense applications, quantifying the quality of clustering analysis is not trivial and has been studied extensively. This chapter covers in-depth clustering validation indices and clustering objective functions found in the literature. Hence, it is devoted as a reference point that facilitates the work of researchers who are interested in clustering and evolutionary clustering.

References

1. Abuobieda, Albaraa, Naomie Salim, Yogan Jaya Kumar, and Ahmed Hamza Osman. 2013. An improved evolutionary algorithm for extractive text summarization. In *Asian conference on intelligent information and database systems*, 78–89. Springer.
2. Al-Madi, Nailah, Ibrahim Aljarah, and Simone A. Ludwig. 2014. Parallel glowworm swarm optimization clustering algorithm based on mapreduce. In *2014 IEEE symposium on swarm intelligence (SIS)*, 1–8. IEEE.
3. Al Shorman, Amaal, Hossam Faris, and Ibrahim Aljarah. 2020. Unsupervised intelligent system based on one class support vector machine and grey wolf optimization for iot botnet detection. *Journal of Ambient Intelligence and Humanized Computing* 11 (7): 2809–2825.
4. Aldouri, Yamur K., Hassan Al-Chalabi, and Liangwei Zhang. 2018. Data clustering and imputing using a two-level multi-objective genetic algorithm (GA): A case study of maintenance cost data for tunnel fans. *Cogent Engineering* 5 (1): 1513304.
5. Aliniya, Zahra, and Seyed Abolghasem Mirroshandel. 2019. A novel combinatorial merge-split approach for automatic clustering using imperialist competitive algorithm. *Expert Systems with Applications* 117: 243–266.
6. Aljarah, Ibrahim, and Simone A. Ludwig. 2012. Parallel particle swarm optimization clustering algorithm based on mapreduce methodology. In *2012 fourth world congress on Nature and biologically inspired computing (NaBIC)*, 104–111. IEEE.
7. Aljarah, Ibrahim, and Simone A. Ludwig. 2013. Mapreduce intrusion detection system based on a particle swarm optimization clustering algorithm. In *2013 IEEE congress on evolutionary computation (CEC)*, 955–962. IEEE.
8. Aljarah, Ibrahim, and Simone A. Ludwig. 2013. A new clustering approach based on glowworm swarm optimization. In *2013 IEEE congress on evolutionary computation*, 2642–2649. IEEE.
9. Aljarah, Ibrahim, and Simone A. Ludwig. 2013. Towards a scalable intrusion detection system based on parallel pso clustering using mapreduce. In *Proceedings of the 15th annual conference companion on genetic and evolutionary computation*, 169–170. ACM.
10. Aljarah, Ibrahim, Majdi Mafarja, Ali Asghar Heidari, Hossam Faris, and Seyedali Mirjalili. 2019. Clustering analysis using a novel locality-informed grey wolf-inspired clustering approach. *Knowledge and Information Systems* 1–33.
11. Aljarah, Ibrahim, Majdi Mafarja, Ali Asghar Heidari, Hossam Faris, and Seyedali Mirjalili. 2020. Clustering analysis using a novel locality-informed grey wolf-inspired clustering approach. *Knowledge and Information Systems* 62 (2): 507–539.
12. Aljarah, Ibrahim, Majdi Mafarja, Ali Asghar Heidari, Hossam Faris, and Seyedali Mirjalili. 2020. Multi-verse optimizer: Theory, literature review, and application in data clustering. In *Nature-inspired optimizers*, 123–141. Springer.
13. Alswaitti, Mohammed, Mohanad Albughdadi, and Nor Ashidi Mat Isa. 2018. Density-based particle swarm optimization algorithm for data clustering. *Expert Systems with Applications* 91: 170–186.
14. Andrade, Carlos E., Mauricio G.C. Resende, Howard J. Karloff, and Flávio K. Miyazawa. 2014. Evolutionary algorithms for overlapping correlation clustering. In *Proceedings of the 2014 annual conference on genetic and evolutionary computation*, 405–412. ACM.

15. Arroyo, Ángel, Verónica Tricio, Álvaro Herrero, and Emilio Corchado. 2016. Time analysis of air pollution in a spanish region through k-means. In *International joint conference SOCO-16-CISIS-16-ICEUTE-16*, 63–72. Springer.
16. Auffarth, Benjamin. 2010. Clustering by a genetic algorithm with biased mutation operator. In *IEEE congress on evolutionary computation*, 1–8. IEEE.
17. Azarakhsh, Javad, and Zobeir Raisi. 2019. Automatic clustering using metaheuristic algorithms for content based image retrieval. In *Fundamental research in electrical engineering*, 83–99. Springer.
18. Baker, Frank B., and Lawrence J. Hubert. 1975. Measuring the power of hierarchical cluster analysis. *Journal of the American Statistical Association* 70 (349): 31–38.
19. Ball, Geoffrey H., and David J. Hall. 1965. Isodata, a novel method of data analysis and pattern classification. Technical report, Stanford Research Institute, Menlo Park, CA.
20. Bandyopadhyay, Sanghamitra, and Ujjwal Maulik. 2002. Genetic clustering for automatic evolution of clusters and application to image classification. *Pattern Recognition* 35 (6): 1197–1208.
21. Beale, E.M.L. 1969. Cluster analysis scientific control system.
22. Benesty, Jacob, Jingdong Chen, Yiteng Huang, and Israel Cohen. 2009. Pearson correlation coefficient. In *Noise reduction in speech processing*, 1–4. Springer.
23. Bezdek, James C. 2013. *Pattern recognition with fuzzy objective function algorithms*. Springer Science & Business Media.
24. Bezdek, James C., and Nikhil R. Pal. 1998. Some new indices of cluster validity.
25. Bolshakova, Nadia, and Francisco Azuaje. 2003. Cluster validation techniques for genome expression data. *Signal Processing* 83 (4): 825–833.
26. Caliński, Tadeusz, and Jerzy Harabasz. 1974. A dendrite method for cluster analysis. *Communications in Statistics-Theory and Methods* 3 (1): 1–27.
27. Chander, Satish, P. Vijaya, and Praveen Dhyani. 2018. Multi kernel and dynamic fractional lion optimization algorithm for data clustering. *Alexandria Engineering Journal* 57 (1): 267–276.
28. Charrad, Malika, Mohamed Ben Ahmed, Yves Lechevallier, and Gilbert Saporta. 2009. Determining the number of clusters in CROKI2 algorithm.
29. Chen, Junyang, Zhiguo Gong, and Weiwen Liu. 2019. A nonparametric model for online topic discovery with word embeddings. *Information Sciences* 504: 32–47.
30. Chen, Min, and Simone A. Ludwig. 2014. Fuzzy clustering using automatic particle swarm optimization. In *2014 IEEE international conference on fuzzy systems (FUZZ-IEEE)*, 1545–1552. IEEE.
31. Chen, Min, and Simone A. Ludwig. 2014. Particle swarm optimization based fuzzy clustering approach to identify optimal number of clusters. *Journal of Artificial Intelligence and Soft Computing Research* 4 (1): 43–56.
32. Chen, Scott Shaobing, and Ponani S. Gopalakrishnan. 1998. Clustering via the Bayesian information criterion with applications in speech recognition. In *Proceedings of the 1998 IEEE international conference on acoustics, speech and signal processing, ICASSP'98 (Cat. No. 98CH36181)*, vol. 2, 645–648. IEEE.
33. Cheng, Fan, Su Tingting Cui, Yunyun Niu Yansen, and Xingyi Zhang. 2018. A local information based multi-objective evolutionary algorithm for community detection in complex networks. *Applied Soft Computing* 69: 357–367.
34. Chi, Yun, Xiaodan Song, Dengyong Zhou, Koji Hino, and Belle L. Tseng. 2007. Evolutionary spectral clustering by incorporating temporal smoothness. In *Proceedings of the 13th ACM SIGKDD international conference on Knowledge discovery and data mining*, 153–162. ACM.
35. Chou, Chien-Hsing, Su Mu-Chun, and Eugene Lai. 2003. A new cluster validity measure for clusters with different densities. In *IASTED international conference on intelligent systems and control*, 276–281.
36. Chovancova, Olga, Lucia Piatrikova, and Adam Dudas. 2019. Improving fuzzy c-means algorithm using particle swarm optimization and fuzzy c-means++. In *2019 international conference on information and digital technologies (IDT)*, 173–179. IEEE.

37. Civicioglu, P., U.H. Atasever, C. Ozkan, E. Besdok, A.E. Karkinli, and A. Kesikoglu. 2014. Performance comparison of evolutionary algorithms for image clustering. *The International Archives of Photogrammetry, Remote Sensing and Spatial Information Sciences* 40 (7): 71.
38. Cobos, Carlos, Leydy Muñoz, Martha Mendoza, Elizabeth León, and Enrique Herrera-Viedma. 2012. Fitness function obtained from a genetic programming approach for web document clustering using evolutionary algorithms. In *Ibero-American conference on artificial intelligence*, 179–188. Springer.
39. Cobos, Carlos, Henry Muñoz-Collazos, Richar Urbano-Muñoz, Martha Mendoza, Elizabeth León, and Enrique Herrera-Viedma. 2014. Clustering of web search results based on the cuckoo search algorithm and balanced Bayesian information criterion. *Information Sciences* 281: 248–264.
40. Cowgill, Marcus Charles. 1993. Monte Carlo validation of two genetic clustering algorithms.
41. Cui, Tianyu, Francesco Caravelli, and Cozmin Ududec. 2018. Correlations and clustering in wholesale electricity markets. *Physica A: Statistical Mechanics and its Applications* 492: 1507–1522.
42. Czekanowski, Jan. 1932. Coefficient of racial likeness" und durchschnittliche differenz. *Anthropologischer Anzeiger* (H. 3/4): 227–249.
43. Dalrymple-Alford, E.C. 1970. Measurement of clustering in free recall. *Psychological Bulletin* 74 (1): 32.
44. D'Ambrosio, Antonio, and Willem J. Heiser. 2019. A distribution-free soft-clustering method for preference rankings. *Behaviormetrika* 46 (2): 333–351.
45. Das, Swagatam, and Amit Konar. 2009. Automatic image pixel clustering with an improved differential evolution. *Applied Soft Computing* 9 (1): 226–236.
46. Das, Swagatam, and Sudeshna Sil. 2010. Kernel-induced fuzzy clustering of image pixels with an improved differential evolution algorithm. *Information Sciences* 180 (8): 1237–1256.
47. Davies, David L., and Donald W. Bouldin. 1979. A cluster separation measure. *IEEE Transactions on Pattern Analysis and Machine Intelligence* 2: 224–227.
48. Desgraupes, Bernard. 2013. Clustering indices. *University of Paris Ouest-Lab Modal-X* 1: 34.
49. Dice, Lee R. 1945. Measures of the amount of ecologic association between species. *Ecology* 26 (3): 297–302.
50. Ding, Yi, and Fu Xian. 2016. Kernel-based fuzzy c-means clustering algorithm based on genetic algorithm. *Neurocomputing* 188: 233–238.
51. Duan, Ganglong, Hu Wenxiu, and Zhiguang Zhang. 2016. A novel data clustering algorithm based on modified adaptive particle swarm optimization. *International Journal of Signal Processing, Image Processing and Pattern Recognition* 9 (3): 179–188.
52. Duda, Richard O., and Peter E Hart. 1973. *Pattern classification and scene analysis*. Wiley.
53. Dunn, Joseph C. 1973. A fuzzy relative of the isodata process and its use in detecting compact well-separated clusters.
54. Faris, Hossam, Ibrahim Aljarah, and Ja'far Alqatawna. 2015. Optimizing feedforward neural networks using krill herd algorithm for e-mail spam detection. In *2015 IEEE Jordan conference on applied electrical engineering and computing technologies (AEECT)*, 1–5. IEEE.
55. Faris, Hossam, Ibrahim Aljarah, Seyedali Mirjalili, Pedro A. Castillo, and Juan J. Merelo. 2016. EvoloPy: An open-source nature-inspired optimization framework in Python. In *IJCCI (ECTA)*, 171–177.
56. Folino, Francesco, and Clara Pizzuti. 2010. A multiobjective and evolutionary clustering method for dynamic networks. In *2010 international conference on advances in social networks analysis and mining*, 256–263. IEEE.
57. Forestier, Germain, Cédric Wemmert, and Pierre Gançarski. 2010. Background knowledge integration in clustering using purity indexes. In *International conference on knowledge science, engineering and management*, 28–38. Springer.
58. Fowlkes, Edward B., and Colin L. Mallows. 1983. A method for comparing two hierarchical clusterings. *Journal of the American Statistical Association* 78 (383): 553–569.

59. Franco, Manuel, and Juana-María Vivo. 2019. Cluster analysis of microarray data. In *Microarray bioinformatics*, 153–183. Springer.
60. Frey, T., and H. Van Groenewoud. 1972. A cluster analysis of the D2 matrix of white spruce stands in Saskatchewan based on the maximum-minimum principle. *The Journal of Ecology* 873–886.
61. Friedman, Herman P., and Jerrold Rubin. 1967. On some invariant criteria for grouping data. *Journal of the American Statistical Association* 62 (320): 1159–1178.
62. Fukuyama, Yoshiki. 1989. A new method of choosing the number of clusters for the fuzzy c-mean method. In *Proceedings of the 5th fuzzy systems symposium, 1989*, 247–250.
63. Georgieva, Kristina S., and Andries P. Engelbrecht. 2013. Dynamic differential evolution algorithm for clustering temporal data. In *International conference on large-scale scientific computing*, 240–247. Springer.
64. Gnanadesikan, R., J.W. Harvey, and J.R. Kettenring. 1993. Mahalanobis metrics for cluster analysis. *Sankhyā: The Indian Journal of Statistics, Series A* 494–505.
65. Gong, Xiaofei, Jie Zhang, and Yijie Shi. 2018. Research on data filling algorithm based on improved k-means and information entropy. In *2018 IEEE 4th international conference on computer and communications (ICCC)*, 1774–1778. IEEE.
66. Guan, Chun, Kevin Kam Fung Yuen, and Frans Coenen. 2019. Particle swarm optimized density-based clustering and classification: Supervised and unsupervised learning approaches. *Swarm and Evolutionary Computation* 44: 876–896.
67. Gurrutxaga, Ibai, Iñaki Albisua, Olatz Arbelaitz, José I. Martín, Javier Muguerza, Jesús M. Pérez, and Iñigo Perona. 2010. SEP/COP: An efficient method to find the best partition in hierarchical clustering based on a new cluster validity index. *Pattern Recognition* 43 (10): 3364–3373.
68. Gutoski, Matheus, Manassés Ribeiro, Nelson Marcelo Romero Aquino, Leandro Takeshi Hattori, André Eugênio Lazzaretti, and Heitor Silvério Lopes. 2018. Feature selection using differential evolution for unsupervised image clustering. In *International conference on artificial intelligence and soft computing*, 376–385. Springer.
69. Halkidi, Maria, and Michalis Vazirgiannis. 2001. Clustering validity assessment: Finding the optimal partitioning of a data set. In *Proceedings 2001 IEEE international conference on data mining*, 187–194. IEEE.
70. Halkidi, Maria, Michalis Vazirgiannis, and Yannis Batistakis. 2000. Quality scheme assessment in the clustering process. In *European conference on principles of data mining and knowledge discovery*, 265–276. Springer.
71. Hamming, Richard W. 1950. Error detecting and error correcting codes. *The Bell System Technical Journal* 29 (2): 147–160.
72. Han, Jiawei, Micheline Kamber, and Jian Pei. 2011. Getting to know your data. *Data Mining: Concepts and Techniques* 3 (744): 39–81.
73. Hancer, Emrah. 2020. A new multi-objective differential evolution approach for simultaneous clustering and feature selection. *Engineering Applications of Artificial Intelligence* 87: 103307.
74. Handl, Julia, and Joshua Knowles. 2007. An evolutionary approach to multiobjective clustering. *IEEE Transactions on Evolutionary Computation* 11 (1): 56–76.
75. Hartigan, John A. 1975. *Clustering algorithms*. New York, NY: Wiley Inc.
76. Hasan, Mohammad Nazmol, Masuma Binte Malek, Anjuman Ara Begum,Moizur Rahman, Md. Mollah, and Nurul Haque. 2019. Assessment of drugs toxicity and associated biomarker genes using hierarchical clustering. *Medicina* 55 (8): 451.
77. He, Zhenfeng, and Chunyan Yu. 2019. Clustering stability-based evolutionary k-means. *Soft Computing* 23 (1): 305–321.
78. Hubert, Lawrence, and Phipps Arabie. 1985. Comparing partitions. *Journal of Classification* 2 (1): 193–218.
79. Hubert, Lawrence, and James Schultz. 1976. Quadratic assignment as a general data analysis strategy. *British Journal of Mathematical and Statistical Psychology* 29 (2): 190–241.

80. Jaccard, Paul. 1901. Étude comparative de la distribution florale dans une portion des alpes et des jura. *Bulletin de la Societe Vaudoise des Sciences Naturelles* 37: 547–579.
81. Jadhav, Amolkumar Narayan, and N. Gomathi. 2018. WGC: Hybridization of exponential grey wolf optimizer with whale optimization for data clustering. *Alexandria Engineering Journal* 57 (3): 1569–1584.
82. Jauhiainen, Susanne, and Tommi Kärkkäinen. 2017. A simple cluster validation index with maximal coverage. In *European symposium on artificial neural networks, computational intelligence and machine learning*. ESANN.
83. Jiang, Lei, and Datong Xie. 2018. An efficient differential memetic algorithm for clustering problem. *IAENG International Journal of Computer Science* 45 (1).
84. Jiang, Xiaoping, Chenghua Li, and Jing Sun. 2018. A modified k-means clustering for mining of multimedia databases based on dimensionality reduction and similarity measures. *Cluster Computing* 21 (1): 797–804.
85. Joopudi, Sreeram, Suraj S. Rathi, Shankar Narasimhan, and Raghunathan Rengaswamy. 2013. A new cluster validity index for fuzzy clustering. *IFAC Proceedings Volumes* 46 (32): 325–330.
86. Kapp, Amy V., and Robert Tibshirani. 2006. Are clusters found in one dataset present in another dataset? *Biostatistics* 8 (1): 9–31.
87. Kendall, Maurice G. 1938. A new measure of rank correlation. *Biometrika* 30 (1/2): 81–93.
88. Khobragade, Sneha, and Preeti Mulay. 2018. Enhance incremental clustering for time series datasets using distance measures. In *International conference on intelligent computing and applications*, 543–556. Springer.
89. Kim, Minho, and R.S. Ramakrishna. 2005. Some new indexes of cluster validity. *Pattern Recognition Letters* 26 (15): 2353–2363.
90. Kirkland, Oliver, Victor J. Rayward-Smith, and Beatriz de la Iglesia. 2011. A novel multi-objective genetic algorithm for clustering. In *International conference on intelligent data engineering and automated learning*, 317–326. Springer.
91. Klamroth, Kathrin. 2002. Measuring distances. In *Single-facility location problems with barriers*, 3–14. Springer.
92. Krzanowski, Wojtek J., and Y.T. Lai. 1988. A criterion for determining the number of groups in a data set using sum-of-squares clustering. *Biometrics* 23–34.
93. Kumar, Lalit, and Kusum Kumari Bharti. 2019. An improved BPSO algorithm for feature selection. In *Recent trends in communication, computing, and electronics*, 505–513. Springer.
94. Kuo, R.J., and L.M. Lin. 2010. Application of a hybrid of genetic algorithm and particle swarm optimization algorithm for order clustering. *Decision Support Systems* 49: 451–462.
95. Kurada, Ramachandra Rao, and Karteeka Pavan Kanadam. 2019. A novel evolutionary automatic clustering technique by unifying initial seed selection algorithms into teaching–learning-based optimization. In *Soft computing and medical bioinformatics*, 1–9. Springer.
96. Kwedlo, Wojciech. 2011. A clustering method combining differential evolution with the k-means algorithm. *Pattern Recognition Letters* 32 (12): 1613–1621.
97. Kwon, Soon H. 1998. Cluster validity index for fuzzy clustering. *Electronics Letters* 34 (22): 2176–2177.
98. Lago-Fernández, Luis F., and Fernando Corbacho. 2010. Normality-based validation for crisp clustering. *Pattern Recognition* 43 (3): 782–795.
99. Li, Chaoshun, Jianzhong Zhou, Pangao Kou, and Jian Xiao. 2012. A novel chaotic particle swarm optimization based fuzzy clustering algorithm. *Neurocomputing* 83: 98–109.
100. Li, Yanyan, Qing Wang, Jianping Chen, Liming Xu, and Shengyuan Song. 2015. K-means algorithm based on particle swarm optimization for the identification of rock discontinuity sets. *Rock Mechanics and Rock Engineering* 48 (1): 375–385.
101. Liao, Erchong, and Changan Liu. 2018. A hierarchical algorithm based on density peaks clustering and ant colony optimization for traveling salesman problem. *IEEE Access* 6: 38921–38933.
102. Liu, Chao, Yuanrui Li, Qi Zhao, and Chenqi Liu. 2019. Reference vector-based multi-objective clustering for high-dimensional data. *Applied Soft Computing* 78: 614–629.

103. Liu, Chao, Jing Xie, Qi Zhao, Qiwei Xie, and Chenqi Liu. 2019. Novel evolutionary multi-objective soft subspace clustering algorithm for credit risk assessment. *Expert Systems with Applications* 138: 112827.
104. Liu, Chuanren, Tianming Hu, Yong Ge, and Hui Xiong. 2012. Which distance metric is right: An evolutionary k-means view. In *Proceedings of the 2012 SIAM international conference on data mining*, 907–918. SIAM.
105. Liu, Yimin, Tansel Özyer, Reda Alhajj, and Ken Barker. 2005. Integrating multi-objective genetic algorithm and validity analysis for locating and ranking alternative clustering. *Informatica* 29 (1).
106. Łukasik, Szymon, Piotr A. Kowalski, Małgorzata Charytanowicz, and Piotr Kulczycki. 2016. Clustering using flower pollination algorithm and Calinski-Harabasz index. In *2016 IEEE congress on evolutionary computation (CEC)*, 2724–2728. IEEE.
107. Luna-Romera, José María, Jorge García-Gutiérrez, María Martínez-Ballesteros, and José C. Riquelme Santos. 2018. An approach to validity indices for clustering techniques in big data. *Progress in Artificial Intelligence* 7 (2): 81–94.
108. Mageshkumar, C., S. Karthik, and V.P. Arunachalam. 2019. Hybrid metaheuristic algorithm for improving the efficiency of data clustering. *Cluster Computing* 22 (1): 435–442.
109. Marriott, F.H.C. 1971. Practical problems in a method of cluster analysis. *Biometrics* 501–514.
110. Martarelli, Nádia Junqueira, and Marcelo Seido Nagano. 2018. A constructive evolutionary approach for feature selection in unsupervised learning. *Swarm and Evolutionary Computation* 42: 125–137.
111. Martínez-Peñaloza, María-Guadalupe, Efrén Mezura-Montes, Nicandro Cruz-Ramírez, Héctor-Gabriel Acosta-Mesa, and Homero-Vladimir Ríos-Figueroa. 2017. Improved multi-objective clustering with automatic determination of the number of clusters. *Neural Computing and Applications* 28 (8): 2255–2275.
112. Matake, Nobukazu, Tomoyuki Hiroyasu, Mitsunori Miki, and Tomoharu Senda. 2007. Multiobjective clustering with automatic k-determination for large-scale data. In *Proceedings of the 9th annual conference on genetic and evolutionary computation*, 861–868. ACM.
113. Maulik, Ujjwal, and Sanghamitra Bandyopadhyay. 2000. Genetic algorithm-based clustering technique. *Pattern Recognition* 33 (9): 1455–1465.
114. Maulik, Ujjwal, and Sanghamitra Bandyopadhyay. 2002. Performance evaluation of some clustering algorithms and validity indices. *IEEE Transactions on Pattern Analysis and Machine Intelligence* 24 (12): 1650–1654.
115. Maulik, Ujjwal, and Indrajit Saha. 2009. Modified differential evolution based fuzzy clustering for pixel classification in remote sensing imagery. *Pattern Recognition* 42 (9): 2135–2149.
116. McClain, John O., and Vithala R. Rao. 1975. CLUSTISZ: A program to test for the quality of clustering of a set of objects. *JMR, Journal of Marketing Research (pre-1986)* 12 (000004): 456.
117. Milligan, Glenn W. 1980. An examination of the effect of six types of error perturbation on fifteen clustering algorithms. *Psychometrika* 45 (3): 325–342.
118. Milligan, Glenn W., and Martha C. Cooper. 1985. An examination of procedures for determining the number of clusters in a data set. *Psychometrika* 50 (2): 159–179.
119. Milligan, Glenn W., and Martha C. Cooper. 1986. A study of the comparability of external criteria for hierarchical cluster analysis. *Multivariate Behavioral Research* 21 (4): 441–458.
120. Mirkin, Boris. 1998. Mathematical classification and clustering: From how to what and why. In *Classification, data analysis, and data highways*, 172–181. Springer.
121. Mohapatra, Smruti Sourava, Prasanta Kumar Bhuyan, and K.V. Rao. 2012. Genetic algorithm fuzzy clustering using GPS data for defining level of service criteria of urban streets.
122. Mukhopadhyay, Anirban, Ujjwal Maulik, and Sanghamitra Bandyopadhyay. 2015. A survey of multiobjective evolutionary clustering. *ACM Computing Surveys (CSUR)* 47 (4): 61.
123. Muravyov, Sergey, Denis Antipov, Arina Buzdalova, and Andrey Filchenkov. 2019. Efficient computation of fitness function for evolutionary clustering. *MENDEL* 25: 87–94.
124. Murthy, Chivukula A., and Nirmalya Chowdhury. 1996. In search of optimal clusters using genetic algorithms. *Pattern Recognition Letters* 17: 825–832.

125. Myers, Leann, and Maria J. Sirois. 2004. Spearman correlation coefficients, differences between. In *Encyclopedia of statistical sciences*, 12.
126. Naik, Nitin, Ren Diao, and Qiang Shen. 2015. Choice of effective fitness functions for genetic algorithm-aided dynamic fuzzy rule interpolation. In *2015 IEEE international conference on fuzzy systems (FUZZ-IEEE)*, 1–8. IEEE.
127. Nerurkar, Pranav, Aruna Pavate, Mansi Shah, and Samuel Jacob. 2019. Performance of internal cluster validations measures for evolutionary clustering. In *Computing, communication and signal processing*, 305–312. Springer.
128. Neumann, Aneta, and Frank Neumann. 2018. On the use of colour-based segmentation in evolutionary image composition. In *2018 IEEE congress on evolutionary computation (CEC)*, 1–8. IEEE.
129. Nguyen, Thi Phuong Quyen, and R.J. Kuo. 2019. Partition-and-merge based fuzzy genetic clustering algorithm for categorical data. *Applied Soft Computing* 75: 254–264.
130. Oujezsky, Vaclav, and Tomas Horvath. 2018. Traffic similarity observation using a genetic algorithm and clustering. *Technologies* 6 (4): 103.
131. Ozyer, Tansel, and Reda Alhajj. 2006. Achieving natural clustering by validating results of iterative evolutionary clustering approach. In *2006 3rd international IEEE conference intelligent systems*, 488–493. IEEE.
132. Özyer, Tansel, Ming Zhang, and Reda Alhajj. 2011. Integrating multi-objective genetic algorithm based clustering and data partitioning for skyline computation. *Applied Intelligence* 35 (1): 110–122.
133. Pakhira, Malay K., Sanghamitra Bandyopadhyay, and Ujjwal Maulik. 2004. Validity index for crisp and fuzzy clusters. *Pattern Recognition* 37 (3): 487–501.
134. Pandey, Avinash Chandra, and Dharmveer Singh Rajpoot. 2019. Spam review detection using spiral cuckoo search clustering method. *Evolutionary Intelligence* 12 (2): 147–164.
135. Pantula, Priyanka D., Srinivas S. Miriyala, and Kishalay Mitra. 2019. A novel ANN-fuzzy formulation towards evolution of efficient clustering algorithm. In *2019 Fifth Indian control conference (ICC)*, 254–259. IEEE.
136. Paterlini, Sandra, and Thiemo Krink. 2006. Differential evolution and particle swarm optimisation in partitional clustering. *Computational Statistics & Data Analysis* 50 (5): 1220–1247.
137. Patnaik, Ashish Kumar, and Prasanta Kumar Bhuyan. 2016. Application of genetic programming clustering in defining LOS criteria of urban street in Indian context. *Travel Behaviour and Society* 3: 38–50.
138. Peng, Hong, Peng Shi, Jun Wang, Agustín Riscos-Núñez, and Mario J. Pérez-Jiménez. 2017. Multiobjective fuzzy clustering approach based on tissue-like membrane systems. *Knowledge-Based Systems* 125: 74–82.
139. Pomponi, Eraldo, and Alexei Vinogradov. 2013. A real-time approach to acoustic emission clustering. *Mechanical Systems and Signal Processing* 40 (2): 791–804.
140. Qaddoura, Raneem, Hossam Faris, and Ibrahim Aljarah. 2020. An efficient clustering algorithm based on the k-nearest neighbors with an indexing ratio. *International Journal of Machine Learning and Cybernetics* 11 (3): 675–714.
141. Qaddoura, Raneem, Hossam Faris, Ibrahim Aljarah, and Pedro A. Castillo. 2020. EvoCluster: An open-source nature-inspired optimization clustering framework in Python. In *International conference on the applications of evolutionary computation (part of EvoStar)*, 20–36. Springer.
142. Qaddoura, R., H. Faris, and I. Aljarah. 2020. An efficient evolutionary algorithm with a nearest neighbor search technique for clustering analysis. *Journal of Ambient Intelligence and Humanized Computing* 1–26.
143. Qaddoura, R., H. Faris, I. Aljarah, J. Merelo, and P. Castillo. 2020. Empirical evaluation of distance measures for nearest point with indexing ratio clustering algorithm. In *Proceedings of the 12th international joint conference on computational intelligence*, vol. 1, 430–438. NCTA. ISBN 978-989-758-475-6, https://doi.org/10.5220/0010121504300438.
144. Radhakrishna, Vangipuram, Chintakindi Srinivas, and C.V. GuruRao. 2014. A modified gaussian similarity measure for clustering software components and documents. In *Proceedings of*

the international conference on information systems and design of communication, 99–104. ACM.

145. Raitoharju, Jenni, Kaveh Samiee, Serkan Kiranyaz, and Moncef Gabbouj. 2017. Particle swarm clustering fitness evaluation with computational centroids. *Swarm and Evolutionary Computation* 34: 103–118.
146. Rand, William M. 1971. Objective criteria for the evaluation of clustering methods. *Journal of the American Statistical Association* 66 (336): 846–850.
147. Ratkowsky, D.A., and G.N. Lance. 1978. Criterion for determining the number of groups in a classification.
148. Ray, Siddheswar, and Rose H. Turi. 1999. Determination of number of clusters in k-means clustering and application in colour image segmentation. In *Proceedings of the 4th international conference on advances in pattern recognition and digital techniques*, 137–143. Calcutta, India.
149. Rezaee, Babak. 2010. A cluster validity index for fuzzy clustering. *Fuzzy Sets and Systems* 161 (23): 3014–3025.
150. Ripon, Kazi Shah Nawaz, and Mia Nazmul Haque Siddique. 2009. Evolutionary multi-objective clustering for overlapping clusters detection. In *2009 IEEE congress on evolutionary computation*, 976–982. IEEE.
151. Ripon, K.S. Nawaz, Chi-Ho Tsang, Sam Kwong, and Man-Ki Ip. 2006. Multi-objective evolutionary clustering using variable-length real jumping genes genetic algorithm. In *18th international conference on pattern recognition (ICPR'06)*, vol. 1, 1200–1203. IEEE.
152. Ritz, Christian, and Ib Skovgaard. 2005. Module 2: Cluster analysis. *Master of Applied Statistics*, 1–20. http://www2.imm.dtu.dk/~pbb/MAS/ST116/module02/module.pdf.
153. Rohlf, F.James. 1974. Methods of comparing classifications. *Annual Review of Ecology and Systematics* 5 (1): 101–113.
154. Rosenberg, Andrew, and Julia Hirschberg. 2007. V-measure: A conditional entropy-based external cluster evaluation measure. In *Proceedings of the 2007 joint conference on empirical methods in natural language processing and computational natural language learning (EMNLP-CoNLL)*, 410–420.
155. Ross, Edith M., and Florian Markowetz. 2016. OncoNEM: Inferring tumor evolution from single-cell sequencing data. *Genome Biology* 17 (1): 69.
156. Rousseeuw, Peter J. 1987. Silhouettes: A graphical aid to the interpretation and validation of cluster analysis. *Journal of Computational and Applied Mathematics* 20: 53–65.
157. Roy, Parthajit, and J.K. Mandal. 2016. An SVD based real coded genetic algorithm for graph clustering.
158. Runkler, Thomas A., and James C. Bezdek. 2019. Optimizing the c index using a canonical genetic algorithm. In *International conference on the applications of evolutionary computation (part of EvoStar)*, 287–298. Springer.
159. Saha, Sriparna, and Sanghamitra Bandyopadhyay. 2008. Application of a new symmetry-based cluster validity index for satellite image segmentation. *IEEE Geoscience and Remote Sensing Letters* 5 (2): 166–170.
160. Saha, Sriparna, and Ranjita Das. 2018. Exploring differential evolution and particle swarm optimization to develop some symmetry-based automatic clustering techniques: Application to gene clustering. *Neural Computing and Applications* 30 (3): 735–757.
161. Saini, Naveen, Sriparna Saha, and Pushpak Bhattacharyya. 2019. Automatic scientific document clustering using self-organized multi-objective differential evolution. *Cognitive Computation* 11 (2): 271–293.
162. Saitta, Sandro, Benny Raphael, and Ian F.C. Smith. 2007. A bounded index for cluster validity. In *International workshop on machine learning and data mining in pattern recognition*, 174–187. Springer.
163. Saitta, Sandro, Benny Raphael, and Ian F.C. Smith. 2008. A comprehensive validity index for clustering. *Intelligent Data Analysis* 12 (6): 529–548.
164. Santos, Adam, Eloi Figueiredo, Moisés Silva, Reginaldo Santos, Claudomiro Sales, and João C.W.A. Costa. 2017. Genetic-based EM algorithm to improve the robustness of gaussian

mixture models for damage detection in bridges. *Structural Control and Health Monitoring* 24 (3): e1886.

165. Sarle, W.S. 1983. *SAS technical report A-108, cubic clustering criterion, 56*. Cary, NC: SAS Institute Inc.
166. Sayed, Gehad Ismail, Ashraf Darwish, and Aboul Ella Hassanien. 2019. Binary whale optimization algorithm and binary moth flame optimization with clustering algorithms for clinical breast cancer diagnoses. *Journal of Classification* 1–31.
167. Schwarz, Gideon, et al. 1978. Estimating the dimension of a model. *The Annals of Statistics* 6 (2): 461–464.
168. Scott, Allen J., and Symons, Michael J. 1971. Clustering methods based on likelihood ratio criteria. *Biometrics* 387–397.
169. sheng Li, Chun. 2011. The improved partition coefficient. *Procedia Engineering* 24: 534–538.
170. Shirakawa, Shinichi, and Tomoharu Nagao. 2009. Evolutionary image segmentation based on multiobjective clustering. In *2009 IEEE congress on evolutionary computation*, 2466–2473. IEEE.
171. Shtar, Guy, Bracha Shapira, and Lior Rokach. 2019. Clustering Wi-Fi fingerprints for indoor-outdoor detection. *Wireless Networks* 25 (3): 1341–1359.
172. Shukri, Sarah, Hossam Faris, Ibrahim Aljarah, Seyedali Mirjalili, and Ajith Abraham. 2018. Evolutionary static and dynamic clustering algorithms based on multi-verse optimizer. *Engineering Applications of Artificial Intelligence* 72: 54–66.
173. Silva, Jesús, Omar Bonerge Pineda Lezama, Noel Varela, Jesús García Guiliany, Ernesto Steffens Sanabria, Madelin Sánchez Otero, and Vladimir Álvarez Rojas. 2019. U-control chart based differential evolution clustering for determining the number of cluster in k-means. In *International conference on green, pervasive, and cloud computing*, 31–41. Springer.
174. Sinha, Ankita, and Prasanta K. Jana. 2018. A hybrid mapreduce-based k-means clustering using genetic algorithm for distributed datasets. *The Journal of Supercomputing* 74 (4): 1562–1579.
175. Song, Wei, Wei Ma, and Yingying Qiao. 2017. Particle swarm optimization algorithm with environmental factors for clustering analysis. *Soft Computing* 21 (2): 283–293.
176. Sørensen, Thorvald Julius. 1948. *A method of establishing groups of equal amplitude in plant sociology based on similarity of species content and its application to analyses of the vegetation on Danish commons*. I kommission hos E. Munksgaard.
177. Starczewski, Artur. 2017. A new validity index for crisp clusters. *Pattern Analysis and Applications* 20 (3): 687–700.
178. Strauss, Trudie, and Michael Johan von Maltitz. 2017. Generalising ward's method for use with Manhattan distances. *PloS One* 12 (1): e0168288.
179. Strehl, Alexander. 2002. *Relationship-based clustering and cluster ensembles for high-dimensional data mining*. PhD thesis.
180. Strehl, Alexander, and Joydeep Ghosh. 2002. Cluster ensembles—a knowledge reuse framework for combining multiple partitions. *Journal of Machine Learning Research* 3: 583–617.
181. Tan, Teck Yan, Li Zhang, Chee Peng Lim, Ben Fielding, Yu. Yonghong, and Emma Anderson. 2019. Evolving ensemble models for image segmentation using enhanced particle swarm optimization. *IEEE Access* 7: 34004–34019.
182. Thomson, Robert, Elie Alhajjar, Joshua Irwin, and Travis Russell. 2018. Predicting bias in machine learned classifiers using clustering. In *Annual social computing, behavior prediction, and modeling-behavioral representation in modeling simulation conference*.
183. Tibshirani, Robert, Guenther Walther, and Trevor Hastie. 2001. Estimating the number of clusters in a data set via the gap statistic. *Journal of the Royal Statistical Society: Series B (Statistical Methodology)* 63 (2): 411–423.
184. Tinós, Renato, Zhao Liang, Francisco Chicano, and Darrell Whitley. 2016. A new evaluation function for clustering: The NK internal validation criterion. In *Proceedings of the genetic and evolutionary computation conference 2016*, 509–516. ACM.
185. Trauwaert, E., P. Rousseeuw, and L. Kaufman. 1993. Fuzzy clustering by minimizing the total hypervolume. In *Information and classification*, 61–71. Springer.

186. Tripathi, Ashish Kumar, Kapil Sharma, and Manju Bala. 2018. Dynamic frequency based parallel k-bat algorithm for massive data clustering (DFBPKBA). *International Journal of System Assurance Engineering and Management* 9 (4): 866–874.
187. van Heerwaarden, Joost, T.L. Odong, and F.A. van Eeuwijk. 2013. Maximizing genetic differentiation in core collections by PCA-based clustering of molecular marker data. *Theoretical and Applied Genetics* 126 (3): 763–772.
188. Vazirgiannis, Michalis. 2009. *Clustering validity, 388–393*. Boston, MA: Springer, US.
189. Vinh, Nguyen Xuan, Julien Epps, and James Bailey. 2010. Information theoretic measures for clusterings comparison: Variants, properties, normalization and correction for chance. *Journal of Machine Learning Research* 11: 2837–2854.
190. Von Luxburg, Ulrike. 2007. A tutorial on spectral clustering. *Statistics and Computing* 17 (4): 395–416.
191. Wallace, David L. 1983. A method for comparing two hierarchical clusterings: Comment. *Journal of the American Statistical Association* 78 (383): 569–576.
192. Wu, Kuo-Lung, and Miin-Shen Yang. 2005. A cluster validity index for fuzzy clustering. *Pattern Recognition Letters* 26 (9): 1275–1291.
193. Wu, Yi-Leh, Cheng-Yuan Tang, Maw-Kae Hor, and Wu Pei-Fen. 2011. Feature selection using genetic algorithm and cluster validation. *Expert Systems with Applications* 38 (3): 2727–2732.
194. Xie, Xuanli Lisa, and Gerardo Beni. 1991. A validity measure for fuzzy clustering. *IEEE Transactions on Pattern Analysis & Machine Intelligence* 8: 841–847.
195. Xu, Kevin S., Mark Kliger, and Alfred O. Hero. 2010. Evolutionary spectral clustering with adaptive forgetting factor. In *2010 IEEE international conference on acoustics, speech and signal processing*, 2174–2177. IEEE.
196. Xu, Kevin S., Mark Kliger, and Alfred O. Hero III. 2014. Adaptive evolutionary clustering. *Data Mining and Knowledge Discovery* 28 (2): 304–336.
197. Xu, Rui, Jie Xu, and Donald C. Wunsch. 2012. A comparison study of validity indices on swarm-intelligence-based clustering. *IEEE Transactions on Systems, Man, and Cybernetics, Part B (Cybernetics)* 42 (4): 1243–1256.
198. Yang, Dongdong, Xiaowei Zhang, Lintao Lv, and Wenzhun Huang. 2018. An automatic SAR image segmentation framework by multi-objective clustering and artificial immune learning. In *2018 international conference on mathematics, modelling, simulation and algorithms (MMSA 2018)*. Atlantis Press.
199. Yesilbudak, Mehmet. 2018. Implementation of novel hybrid approaches for power curve modeling of wind turbines. *Energy Conversion and Management* 171: 156–169.
200. Žalik, Krista Rizman, and Borut Žalik. 2011. Validity index for clusters of different sizes and densities. *Pattern Recognition Letters* 32 (2): 221–234.
201. Zang, Wenke, Zhenni Jiang, and Liyan Ren. 2017. Improved spectral clustering based on density combining DNA genetic algorithm. *International Journal of Pattern Recognition and Artificial Intelligence* 31 (04): 1750010.
202. Zhang, Yunjie, Weina Wang, Xiaona Zhang, and Yi Li. 2008. A cluster validity index for fuzzy clustering. *Information Sciences* 178 (4): 1205–1218.
203. Zhao, Qinpei, and Pasi Fränti. 2014. WB-index: A sum-of-squares based index for cluster validity. *Data & Knowledge Engineering* 92: 77–89.
204. Zhao, Ying, and George Karypis. 2002. Evaluation of hierarchical clustering algorithms for document datasets. In *Proceedings of the eleventh international conference on Information and knowledge management*, 515–524. ACM.
205. Zhou, Shibing, and Xu Zhenyuan. 2018. A novel internal validity index based on the cluster centre and the nearest neighbour cluster. *Applied Soft Computing* 71: 78–88.
206. Zhu, Shuwei, Lihong Xu, and Erik D. Goodman. 2019.Evolutionary multi-objective automatic clustering enhanced with quality metrics and ensemble strategy. *Knowledge-Based Systems* 105018.

A Grey Wolf-Based Clustering Algorithm for Medical Diagnosis Problems

Raneem Qaddoura, Ibrahim Aljarah, Hossam Faris, and Seyedali Mirjalili

Abstract Evolutionary and swarm intelligence algorithms are used as optimization algorithms for solving the clustering problem. One of the most popular optimization algorithms is the Grey Wolf Optimizer (GWO). In this chapter, we use GWO on seven medical data sets to optimize the initial clustering centroids represented by the individuals of each population at each iteration. The aim is to minimize the distances between instances of the same cluster to predict certain diseases and medical problems. The results show that solving the clustering task using GWO outperforms the other well-regarded evolutionary and swarm intelligence clustering algorithms, by converging toward enhanced solutions having low dispersion from the average values, for all the selected data sets.

Keywords Data clustering · Nature-inspired algorithms · Metaheuristics · Medical applications · Grey wolf optimizer · Evolutionary computation · EvoCluster · EvoloPy

---

R. Qaddoura
Information Technology, Philadlphia University, Amman, Jordan
e-mail: rqaddoura@philadelphia.edu.jo

I. Aljarah (✉) · H. Faris
King Abdullah II School for Information Technology, The University of Jordan, Amman, Jordan
e-mail: i.aljarah@ju.edu.jo

H. Faris
e-mail: hossam.faris@ju.edu.jo

S. Mirjalili
Center for Artificial Intelligence Research and Optimization, Torrens University Australia, Fortitude Valley, Brisbane, Australia
e-mail: ali.mirjalili@laureate.edu.au

I. Aljarah et al. (eds.), *Evolutionary Data Clustering: Algorithms and Applications*, Algorithms for Intelligent Systems,
https://doi.org/10.1007/978-981-33-4191-3_3

1 Introduction

Clustering is an unsupervised learning problem, which gathers similar instances to the same cluster by analyzing the patterns of the features of these instances [10–12, 26]. The clustering task can be solved using partitional, density-based, hierarchical clustering [27, 56, 60], and nature-inspired clustering algorithms [4, 59, 63].

Grey Wolf Optimizer (GWO) [47] as one of the nature-inspired clustering algorithms, is a popular algorithm that is used for optimization problems and can be used for solving the clustering task [37, 67, 70].

Clustering applications include financial risk analysis [34], document categorization [14, 43], dental radiography segmentation [55], information retrieval [18, 58], image processing [33, 36, 66], search engines [40], academics [52], and pattern recognition [39, 62]. Some other applications related to biological and medical applications include bioinformatics [24], cancerous data [30, 65], and drug activity prediction [5], and many others [3, 9, 22, 23].

In this chapter, we use GWO, which is customized for clustering as part of the EvoCluster framework [57], to optimize the centroids of the clustering solution for medical data sets, by minimizing the sum of squared error (SSE) of the instances of each cluster, to predict certain diseases and medical problems. The clustering solutions are represented by the individuals of the population at each iteration. The experiments are conducted on seven popular data sets, and the results are compared with other well-known evolutionary and swarm intelligence clustering algorithms.

The remainder of the chapter is organized as follows: Section 2 summarizes the related work in the field of evolutionary and swarm intelligence clustering. Section 3 describes GWO in detail and how it is customized to solve the clustering task. Section 4 presents the results of conducting the experiments on well-known medical data sets. Finally, Sect. 5 concludes the work.

2 Related Work

Nature-inspired algorithms try to solve the optimization problems by applying natural operations on candidate solutions [53, 54]. Fuad [25] categorized the nature-inspired algorithms into evolutionary algorithms (EA) and swarm intelligence (SI). The most popular EAs are the genetic algorithm (GA) [16], genetic programming (GP) [35], multi-verse optimizer (MVO) [46], evolutionary strategies (ES) [28], and differential evolution (DE) [63]. In contrast, many SIs exist including particle swarm optimization (PSO) [20], firefly algorithm (FFA) [68], ant colony optimization (ACO) [19], cuckoo search (CS) [69], artificial bee colony (ABC) [32], moth-flame optimization (MFO) [44], whale optimization algorithm (WOA) [45], and grey wolf optimizer (GWO) [47].

These algorithms are used to solve different optimization problems including the clustering problem. Many studies can be found in the literature concerning clustering with evolutionary and swarm intelligence algorithms which are reviewed in several surveys. The authors of [27] reviewed the studies of determining the number of clusters using evolutionary algorithms, while the authors of [51] reviewed nature-inspired algorithms of partitional clustering. In addition, authors of [49, 50] reviewed the evolutionary clustering algorithms in a two-parts survey. Faris et al. [21] reviewed the recent variants and applications of GWO and presented the clustering applications of using GWO as part of their review.

In 2014, Mirjalili et al. [47] implemented GWO which becomes a very popular swarm intelligence algorithm for solving the optimization problems. Many studies were observed since then which are classified into updating mechanisms, new operators, encoding scheme of the individuals, and population structure and hierarchy as per the review study by Faris et al. [21].

Some studies can be found on GWO for performing the clustering task by optimizing the centroids of the clustering solution. Grey Wolf Algorithm Clustering (GWAC) [37], GWO with Powell local optimization (PGWO) [70], Grey Wolf Optimizer and K-means (GWO-KM) [67], and K-means with GWO (K-GWO) are clustering algorithms which work on enhancing the centroids decision for the individuals in the population of the GWO at each iteration. In addition, the authors in [48] have developed mGWO which balances between the exploratory and exploitative behavior of the algorithm to generate enhanced results. An enhanced GWO (EGWO) was proposed by [64], which is a hybrid combination of the hunting process of GWO and the binomial crossover and lévy flight steps. Another hybrid algorithm named WGC was proposed by [29] as a combination of the GWO and the WOA algorithms with a newly formulated fitness function.

GWO are used in some clustering applications including satellite image segmentation [31] and wireless sensor networks [1, 48]. However, many other clustering applications can be experimented using the GWO as a clustering algorithm which is the main goal of this chapter. Thus, we experiment GWO as a clustering algorithm on medical data sets.

3 Grey Wolf Optimizer for Clustering

This section presents the GWO algorithm details which are used to optimize general problems. This section also includes the details of optimizing the centroids selection of the clustering solution using GWO.

3.1 GWO Algorithm

GWO is inspired by the social behavior of grey wolf packs which includes the following:

- Social hierarchy of leadership: The leader of the pack is referred to as the alpha wolf. Beta wolves support the alpha wolf in decision-making and substitute them in case of death or illness [21, 47]. The remaining hierarchy of leadership includes delta and omega wolves.
- Group hunting: It includes tracking, chasing, pursuing, encircling, harassing, and attacking the prey [47].

The social behavior of wolves is translated into an optimization algorithm by mathematically modeling the behavior.

The hierarchy of leadership is obtained by considering the best solution in a population as the alpha wolf (α), the following best solutions are considered as beta wolf (β) and delta wolf (δ), respectively. Consequently, the remaining solutions in the population are considered as omega wolves (ω).

The hunting behavior of the wolves is obtained by creating a set of solutions of grey wolves for the first iteration of the algorithm. For the following iterations, the position of the prey is predicted by the alpha, beta, and delta wolves. Candidate solutions converge toward the prey, which results in exploiting the search space, and diverge from the prey, which results in exploring fitter solutions in the search space when the values of A are considered between -1 and 1, or the values of A are considered larger than 1 or smaller than -1, respectively. This can be further explained by applying the following processes:

- Encircling: Equations 1, 2, 3, 4 simulate this process. Equation 1 represents a vector calculated according to the position vector of the prey and the current position vector of the wolf. Equation 2 represents the next position vector of the wolf calculated according to the current position vector of the wolf and the vector calculated in Eq. 1. Equations 3 and 4 represents the coefficient vectors A and C, respectively.

$$D = |C.X_p(t) - X(t)| \tag{1}$$

$$X(t+1) = X(t) - A.D \tag{2}$$

$$C = 2.r_2 \tag{3}$$

$$A = 2a.r_1 - a \tag{4}$$

Where t is the current iteration, X and X_p are the positions of the grey wolf and the prey, respectively, r_1 and r_2 are random vectors of values between 0 and 1, and a is a vector which linearly decreases from 2 to 0.

- Hunting: The next position vector of the omega wolves are calculated by Eq. 5 according to the position vectors of the alpha, beta, and delta wolves. The values of the X_1, X_2, X_3 are calculated by Eqs. 6, 7, and 8, respectively.

$$X(t+1) = (X_1 + X_2 + X_3)/3 \tag{5}$$

$$X_1 = X_\alpha - A_1.(|C_1.X_\alpha - X|) \tag{6}$$

$$X_2 = X_\beta - A_2.(|C_2.X_\beta - X|) \tag{7}$$

$$X_3 = X_\delta - A_3.(|C_3.X_\delta - X|) \tag{8}$$

  Where X_α, X_β, and X_δ are the position vectors of the alpha, beta, and delta wolves, X and $X(t+1)$ are the current and next position vectors of the wolves, respectively. A_1 & C_1, A_2 & C_2, and A_3 & C_3 are coefficient vectors for alpha, beta, and delta wolves, respectively.
- Attacking (exploitation): Decreasing the value of a from 2 to 0 causes a consequent decrease of the value of A from 1 to -1 which results in moving toward the prey and finally attacking the prey when the prey stops moving. This technique results in the exploitative behavior of searching for the fitter preys.
- Searching (exploration): Grey wolves diverge from the prey by considering values of A less than -1 or larger than 1. In addition, the random values of weights of C, which are between 0 and 2, provide different effects of the position of the prey. Both techniques show the exploratory behavior of the search space and local optima avoidance of the solutions.

3.2 *Clustering with GWO*

The main idea of clustering with GWO is to optimize the centroids selection of the clustering solutions. This can be achieved by selecting an initial population of solutions, which represents the individuals of the initial centroids. Then, the instances are assigned to the closest centroids and the evaluation of each individual is achieved by calculating the sum of squared error (SSE) of the distances between the centroids and the corresponding instances at each cluster. For the following iterations, GWO optimizes the candidate solutions, according to the aforementioned processes, which are discussed in the previous section. The fittest solution of the last iteration is returned to reflect the solution achieved by applying the algorithm.

The pseudocode of performing the clustering task using GWO can be observed by Algorithm 1. The algorithm accepts the data set instances, the number of clusters (k), the number of iterations (#iterations), and the number of individuals (#individuals) as parameter values. The initial population is created in line 2 which represents the initial centroids of the candidate individuals. Then, the algorithm repeats until the value of iterations is reached which is presented in lines 3–10. For each iteration, instances of each solution are assigned to the closest cluster and the SSE value of each individual is calculated. These operations are presented in lines 4–7. The algorithm recognizes the alpha, beta, delta, and omega individuals in line 8. Then, the Encircling, Hunting, Attacking, and Searching processes are performed in line 9. The algorithm terminates in line 11 by returning the alpha individual of the last iteration.

Algorithm 1 Clustering with GWO

```
1: procedure CGWO(#ITERATIONS, #INDIVIDUALS, K, INSTANCES)
2:    Create initial population of individuals
3:    repeat
4:       for all individuals ∈ population do
5:          Assign instances to the closest centroids
6:          Calculate SSE
7:       end for
8:       Point out alpha, beta, delta, and omega individuals
9:       Perform Encircling, Hunting, Attacking, and Searching processes
10:   until #iterations is reached
11:   Return alpha individual
12: end procedure
```

4 Experimental Results

This section discusses the data sets used in the experiments, the parameter selection for each algorithm, the fitness function used in the experiments, and a detailed analysis of the obtained results.

4.1 Data Sets

Seven medical data sets are used for the conducted experiments, which are gathered from the UCI machine learning repository[1] [17] and KEEL.[2] Table 1 shows the number of classes, instances, and features for each data set.

The description of these data sets are given as follows:

- **Appendicitis[2]**: Represents the presence or absence of appendicitis for 106 patients depending on 7 measures.
- **Blood[1]**: This data set is gathered from the Blood Transfusion Service Center in Hsin-Chu City in Taiwan. It represents the status of denoting blood in March 2007,

[1] https://archive.ics.uci.edu/ml/.

[2] https://sci2s.ugr.es/keel/.

Table 1 Characteristics of Data sets

ID	Name	#Classes	#Instances	#Features	Type
1	Appendicitis	2	106	7	Real
2	Blood	2	748	5	Real
3	Diagnosis II	2	120	6	Real
4	Heart	2	270	13	Real
5	Liver	2	345	7	Real
6	Vertebral II	2	310	6	Real
7	Vertebral III	3	310	6	Real

based on 4 attributes of 748 donors. The attributes include the months since last donation, total number of donations, total blood donated in c.c., and months since first donation.

- **Diagnosis II**[1]: Represents the presence or absence of nephritis of renal pelvis origin for 120 patients depending on 6 symptoms.
- **Heart**[1]: Represents the presence or absence of a heart disease for 270 patients depending on 13 attributes.
- **Liver**[1]: Represents the presence or absence of liver disorders that might be caused by alcohol consumption for 345 patients depending on 7 attributes.
- **Vertebral II**[1]: The data set is gathered by Dr. Henrique da Mota inCentre MAmedico-Chirurgical de RAadaptation des Massues, Lyon, France. It represents the presence or absence of vertebral problems which are classified into normal or abnormal vertebral for 310 patients depending on 6 attributes
- **Vertebral III**[1]: The same as the Vertebral II data set but abnormal vertebral are split into Disk Hernia and Spondylolisthesis forming 3 classes in addition to the normal vertebral.

4.2 Algorithms and Parameters Selection

Recent and well-known algorithms are used to compare the results achieved by running the CGWO algorithm. These algorithms include CSSA, CGA, CPSO, and CWOA which represent the clustering variation of the original algorithms which are SSA, GA, PSO, and WOA, respectively. They use a population of individuals which represent the clustering solutions of centroids that are optimized across iterations.

The number of classes for each data set is passed as a parameter to each algorithm. The population size of 50 and the iteration value of 100 are also considered for all algorithms. In addition, Roulette Wheel selection mechanism and the values of 0.8 and 0.001 for the crossover and mutation probabilities, respectively, are considered for CGA which are extensively found in the literature [13, 15, 41, 42, 61]. Table 2

Table 2 Parameters setting

Algorithm	Parameter	Value
CGA	Crossover probability	0.8
	Mutation probability	0.001
CPSO	Vmax	6
	wMax	0.9
	wMin	0.2
	c1	2
	c2	2
CGWO	a	From 2 to 0
CSSA	No. iterations	100
	No. population size	50
	No. of leaders	0.5
CWOA	a	From 2 to 0

shows the selection of parameters for these algorithms. The population/swarm size is unified for all algorithms and set to 50, while the number of iterations is set to 100.

4.3 *Fitness Function*

We use the Sum of Squared Error (SSE) [38] to evaluate the convergence of the algorithms toward the optimal solution, which indicates compact clusters. The SSE is the sum of the squares of the euclidean distance between an instance and the corresponding centroid of its cluster. The objective is to minimize the value of SSE to obtain optimized clustering results. SSE is calculated by the following equation [38]:

$$SSE = \sum_{i=1}^{k} \sum_{j=1}^{J} ||p_j - c_i||^2 \tag{9}$$

where k is the number of clusters J is the number of points for a cluster i, p_j is the jth point in cluster i, c_i is the centroid of cluster i.

4.4 *Results and Discussion*

The experiments are conducted on the selected data sets for 30 independent runs, then the averages of the results are reported. Table 3 shows the performance of CGWO against other recent and well-known algorithms which are CSSA, CGA, CPSO, and

Table 3 Average values of SSE of different algorithms for 30 independent runs

Dataset	CGWO	CSSA	CGA	CPSO	CWOA
Appendicitis	17.88	19.63	21.11	21.13	22.18
Blood	36.8	46.67	43.27	41.55	39.39
Diagnosis II	106.11	106.43	109.21	106.76	108.7
Heart	259.18	259.39	284.64	295.33	301.15
Liver	24.12	27.82	30.02	29.83	31.71
Vertebral2	21.33	25.18	26.8	26.4	29.59
Vertebral3	18.85	20.95	26.58	21.07	27.29

CWOA. It is observed from the table that CGWO outperforms the other algorithms, having the lowest SSE value compared to the other algorithms for all the selected data sets. It has a recognizable lower value than CWOA and CGA for most of the data sets. It also has close values for Diagnosis II and heart data sets compared to CSSA.

The convergence curves of the algorithms for the selected data sets are also shown in Fig. 1. The convergence curve represents the average SSE value for 30 runs for each of the algorithms. It shows the tendency of each algorithm for finding an enhanced solution by minimizing the SSE value across 100 iterations. It is observed from the figure that CGWO has recognizable optimized solutions during the course of iterations for Liver, Blood, Vertebral 2, and Appendicitis data sets. It slightly improves the solution for Diagnosis II data set compared to CSSA and CPSO algorithms. It competes CSSA in optimizing the solution during the course of iterations for Vertebral II, Vertebral III, Diagnosis II, and Heart data sets, but finally reaches a better solution than CSSA for these data sets.

Furthermore, the box plots of the SSE values for each data set are shown in Fig. 2 to access the stability of CGWO. The box plot shows the interquartile range, average SSE value, best SSE values, and worst SSE value [6] for the 30 runs of every algorithm in comparison.

It is observed from the figure that CGWO has the most compacted box compared to the other algorithms for all the data sets, which indicates low standard deviation and stability of the algorithm. Specifically, it has almost the same values for SSE for different runs of CGWO for Liver and Blood data sets having very low values of standard deviation. It also has a recognizable minimal SSE value compared to the other algorithms for most of the data sets.

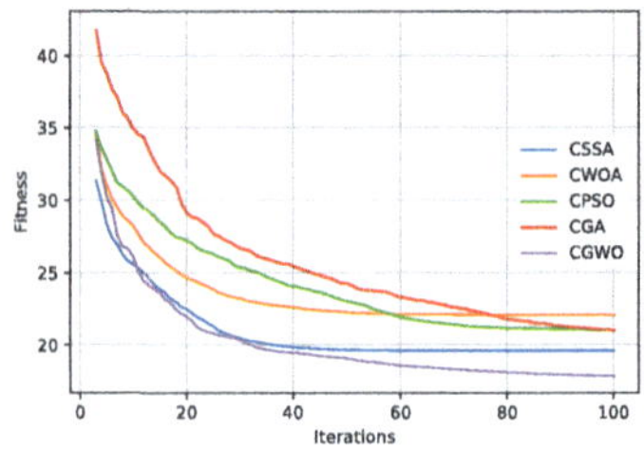

(a) Appendicitis

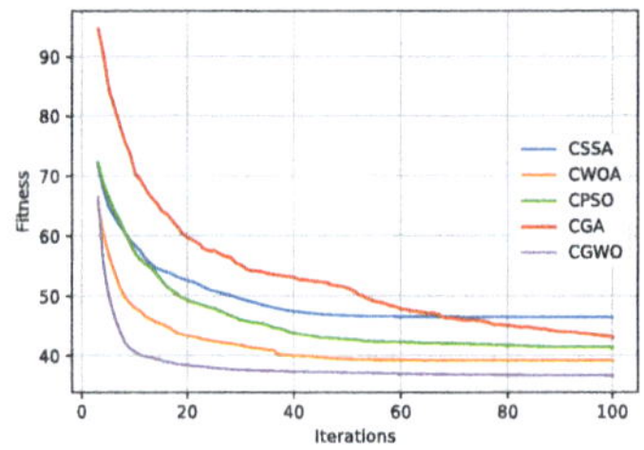

(b) Blood

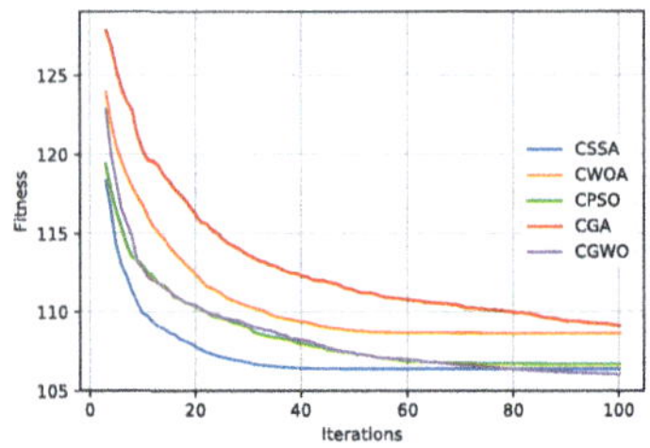

(c) Diagnosis II

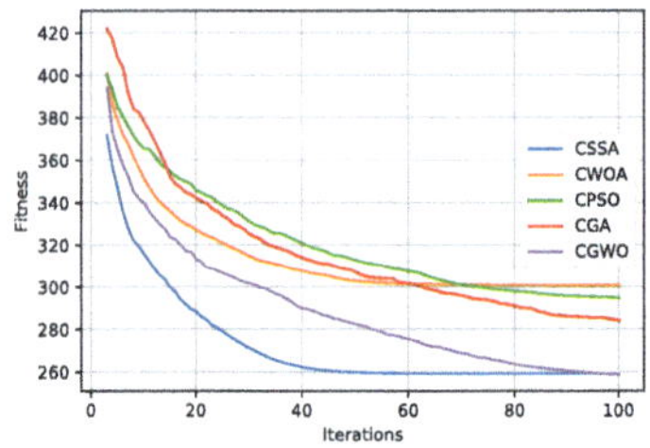

(d) Heart

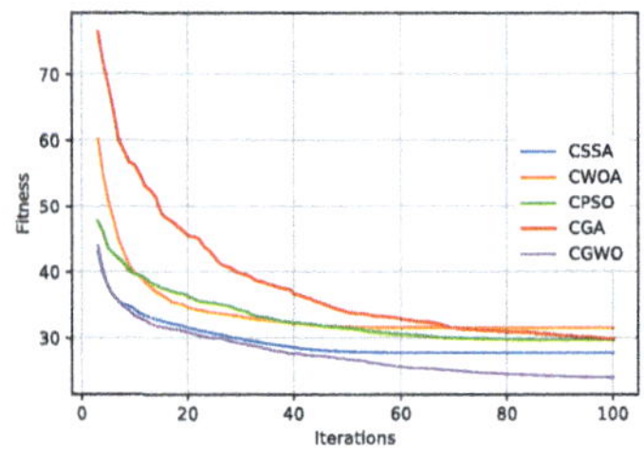

(e) Liver

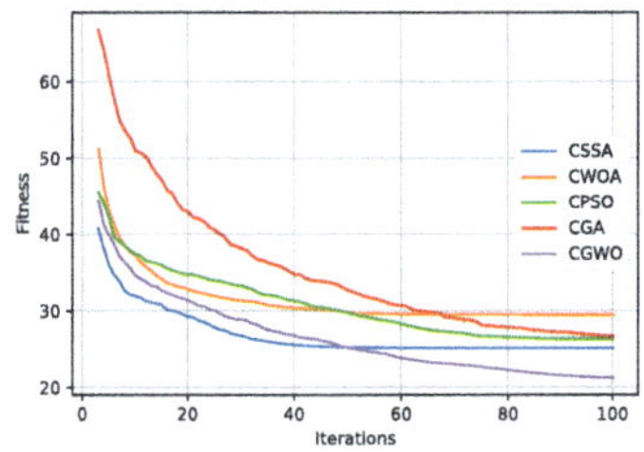

(f) Vertebral II

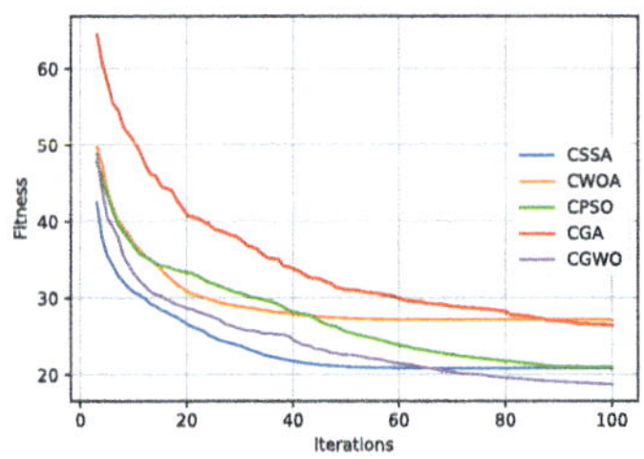

(g) Vertebral III

Fig. 1 Convergence curve for CSSA, CWOA, CPSO, CGA, and CGWO using SSE objective function for **a** Appendicitis; **b** Blood; **c** Diagnosis II; **d** Heart; **e** Liver; **f** Vertebral II; and **g** Vertebral III

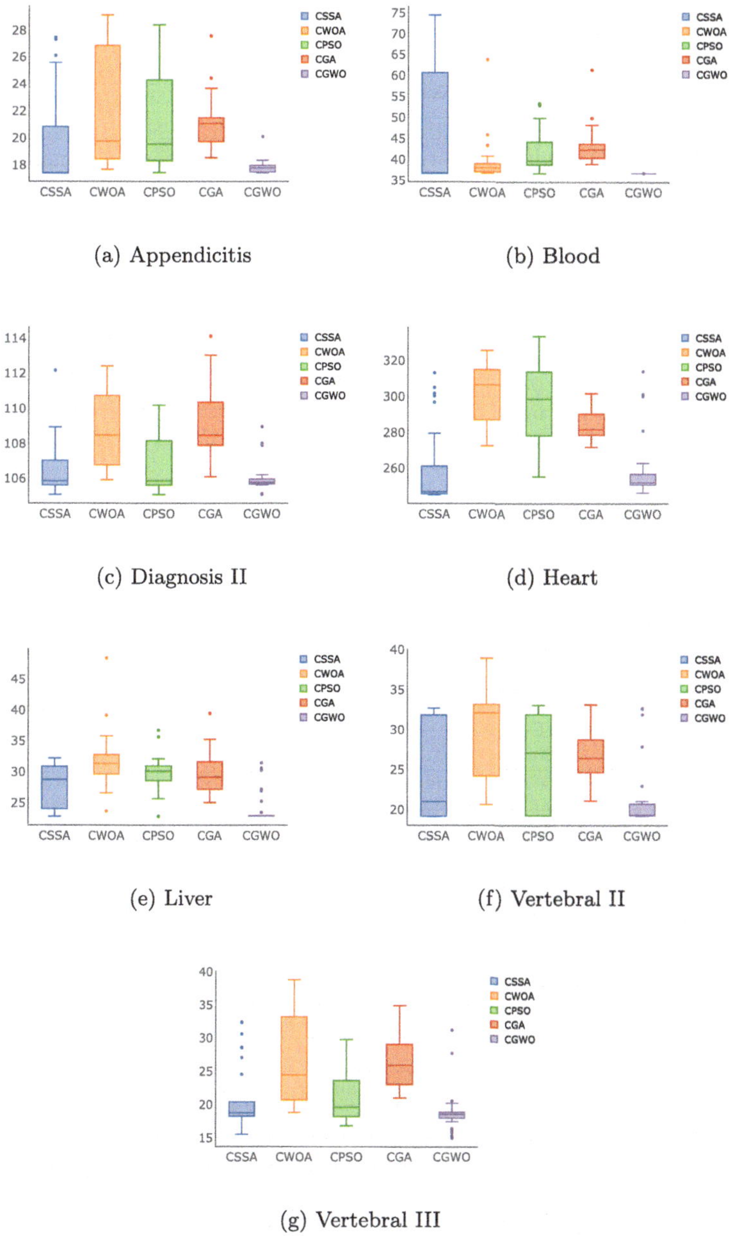

Fig. 2 Box plot for CSSA, CWOA, CPSO, CGA, and CGWO using SSE objective function for **a** Appendicitis; **b** Blood; **c** Diagnosis II; **d** Heart; **e** Liver; **f** Vertebral II; and **g** Vertebral III

5 Conclusion

In this chapter, we have applied GWO, which is customized for clustering as part of the EvoCluster framework, as an optimizer for solving the clustering task for seven medical data sets. The individuals at each iteration represent the initial centroids that are used to cluster the data set instances into certain diseases and medical problems.

The results show that applying GWO on the selected medical data sets causes a recognizable convergence of the algorithm toward enhanced solutions compared to other well-regarding evolutionary and swarm intelligence algorithms. In addition, the box plots of thirty independent runs for each algorithm show that lower dispersion from the average values is achieved using GWO compared to the other algorithms.

For future work, we plan to investigate other clustering applications using the same customized clustering algorithm of GWO.

References

1. Al-Aboody, N.A., and H.S. Al-Raweshidy. 2016. Grey wolf optimization-based energy-efficient routing protocol for heterogeneous wireless sensor networks. In *2016 4th International Symposium on Computational and Business Intelligence (ISCBI)*, pp. 101–107. IEEE.
2. Al-Madi, Nailah, Ibrahim, Aljarah, and Simone A. Ludwig. 2014. Parallel glowworm swarm optimization clustering algorithm based on mapreduce. In *2014 IEEE Symposium on Swarm Intelligence*, pp. 1–8. IEEE.
3. Al Shorman, Amaal, Hossam, Faris, and Ibrahim, Aljarah. 2020. Unsupervised intelligent system based on one class support vector machine and grey wolf optimization for iot botnet detection. *Journal of Ambient Intelligence and Humanized Computing* 11(7):2809–2825.
4. Alam, Shafiq, Gillian, Dobbie, Yun Sing, Koh, Patricia, Riddle, and Saeed Ur, Rehman. 2014. Research on particle swarm optimization based clustering: a systematic review of literature and techniques. *Swarm and Evolutionary Computation* 17:1–13.
5. Alhalaweh, Amjad, Ahmad Alzghoul, and Waseem Kaialy. 2014. Data mining of solubility parameters for computational prediction of drug-excipient miscibility. *Drug Development and Industrial Pharmacy* 40 (7): 904–909.
6. Aljarah, Ibrahim, Al-Zoubi, AlaM, Hossam, Faris, Mohammad A. Hassonah, Seyedali, Mirjalili, and Heba, Saadeh. 2018. Simultaneous feature selection and support vector machine optimization using the grasshopper optimization algorithm. *Cognitive Computation*, pp. 1–18.
7. Aljarah, Ibrahim, and Simone A. Ludwig. 2012. Parallel particle swarm optimization clustering algorithm based on mapreduce methodology. In *2012 Fourth World Congress on Nature and Biologically Inspired Computing (NaBIC)*, pp. 104–111. IEEE.
8. Aljarah, Ibrahim, and Simone A. Ludwig. 2013. Mapreduce intrusion detection system based on a particle swarm optimization clustering algorithm. In *2013 IEEE congress on evolutionary computation*, pp. 955–962. IEEE.
9. Aljarah, Ibrahim, and Simone A. Ludwig. 2013. A new clustering approach based on glowworm swarm optimization. In *2013 IEEE congress on evolutionary computation*, pp. 2642–2649. IEEE.
10. Aljarah, Ibrahim, and Simone A. Ludwig. 2013. Towards a scalable intrusion detection system based on parallel pso clustering using mapreduce. In *Proceedings of the 15th annual conference companion on genetic and evolutionary computation*, pp. 169–170.
11. Aljarah, Ibrahim, Majdi, Mafarja, Ali Asghar, Heidari, Hossam, Faris, and Seyedali, Mirjalili. 2020. Clustering analysis using a novel locality-informed grey wolf-inspired clustering approach. *Knowledge and Information Systems*, 62(2):507–539.

12. Aljarah, Ibrahim, Majdi, Mafarja, Ali Asghar, Heidari, Hossam, Faris, and Seyedali, Mirjalili. 2020. Multi-verse optimizer: theory, literature review, and application in data clustering. In *Nature-Inspired Optimizers*, pp. 123–141. Berlin: Springer.
13. Beg, A.H. and M.d. Zahidul, Islam. 2015. Clustering by genetic algorithm-high quality chromosome selection for initial population. In *2015 IEEE 10th conference on industrial electronics and applications (ICIEA)*, pp. 129–134. IEEE.
14. Brodić, Darko, Alessia, Amelio, and Zoran N. Milivojević. 2017. Clustering documents in evolving languages by image texture analysis. *Applied Intelligence*, 46(4):916–933.
15. Chang, Dong-Xia, Xian-Da Zhang, and Chang-Wen Zheng. 2009. A genetic algorithm with gene rearrangement for k-means clustering. *Pattern Recognition* 42 (7): 1210–1222.
16. Davis, Lawrence. 1991. Handbook of genetic algorithms.
17. Dheeru, Dua, and Efi Karra, Taniskidou. 2017. UCI machine learning repository, 2017.
18. Djenouri, Youcef, Asma, Belhadi, Philippe, Fournier-Viger, and Jerry Chun-Wei, Lin. 2018. Fast and effective cluster-based information retrieval using frequent closed itemsets. *Information Sciences*, 453:154–167.
19. Marco Dorigo and Luca Maria Gambardella. 1997. Ant colony system: a cooperative learning approach to the traveling salesman problem. *IEEE Transactions on Evolutionary Computation* 1 (1): 53–66.
20. Eberhart, Russell, and Kennedy, James. 1995. A new optimizer using particle swarm theory. In *MHS'95. Proceedings of the sixth international symposium on micro machine and human science*, pp. 39–43. IEEE.
21. Faris, Hossam, Ibrahim, Aljarah, Mohammed Azmi, Al-Betar, and Seyedali, Mirjalili. 2018. Grey wolf optimizer: a review of recent variants and applications. *Neural Computing and Applications*, 30(2):413–435.
22. Faris, Hossam, Ibrahim, Aljarah, and Ja'far, Alqatawna. Optimizing feedforward neural networks using krill herd algorithm for e-mail spam detection. In *2015 IEEE Jordan conference on applied electrical engineering and computing technologies (AEECT)*, pp. 1–5. IEEE.
23. Faris, Hossam, Ibrahim, Aljarah, Seyedali, Mirjalili, Pedro A. Castillo, and Juan Julián Merelo, Guervós. 2016. Evolopy: An open-source nature-inspired optimization framework in python. In *IJCCI (ECTA)*, pp. 171–177.
24. Frank, Eibe, Mark, Hall, Len, Trigg, Geoffrey, Holmes, and Ian H. Witten. 2004. Data mining in bioinformatics using weka. *Bioinformatics*, 20(15):2479–2481.
25. Fuad, Muhammad Marwan Muhammad. 2019. Applying nature-inspired optimization algorithms for selecting important timestamps to reduce time series dimensionality. *Evolving Systems* 10 (1): 13–28.
26. Han, Jiawei, Jian Pei, and Micheline, Kamber. 2011. *Data mining: concepts and techniques*. Elsevier.
27. Hancer, Emrah, and Dervis Karaboga. 2017. A comprehensive survey of traditional, merge-split and evolutionary approaches proposed for determination of cluster number. *Swarm and Evolutionary Computation* 32: 49–67.
28. Hansen, Nikolaus, and Stefan, Kern. 2004. Evaluating the cma evolution strategy on multimodal test functions. In *International conference on parallel problem solving from nature*, pp. 282–291. Berlin: Springer.
29. Jadhav, Amolkumar Narayan, and N. Gomathi. 2018. Wgc: hybridization of exponential grey wolf optimizer with whale optimization for data clustering. *Alexandria engineering journal*, 57(3):1569–1584.
30. Jang, Ho, Youngmi, Hur, and Hyunju, Lee. 2016. Identification of cancer-driver genes in focal genomic alterations from whole genome sequencing data. *Scientific Reports* 6.
31. Kapoor, Shubham, Irshad, Zeya, Chirag, Singhal, and Satyasai Jagannath, Nanda. 2017. A grey wolf optimizer based automatic clustering algorithm for satellite image segmentation. *Procedia Computer Science*, 115:415–422.
32. Karaboga, Dervis. 2005. An idea based on honey bee swarm for numerical optimization. Technical report, Technical report-tr06, Erciyes University, Engineering Faculty, Computer, 2005.

33. Khan, Zubair, Jianjun Ni, Xinnan Fan, and Pengfei Shi. 2017. An improved k-means clustering algorithm based on an adaptive initial parameter estimation procedure for image segmentation. *International Journal Of Innovative Computing Information and Control* 13(5):1509–1525.
34. Kou, Gang, Yi Peng, and Guoxun Wang. 2014. Evaluation of clustering algorithms for financial risk analysis using mcdm methods. *Information Sciences* 275: 1–12.
35. Koza, John R., and John R. Koza. 1992. *Genetic programming: on the programming of computers by means of natural selection*, vol. 1. MIT Press.
36. Kumar, Sushil, Millie Pant, Manoj Kumar, and Aditya Dutt. 2018. Colour image segmentation with histogram and homogeneity histogram difference using evolutionary algorithms. *International Journal of Machine Learning and Cybernetics* 9 (1): 163–183.
37. Kumar, Vijay, Jitender Kumar, Chhabra, and Dinesh, Kumar. 2017. Grey wolf algorithm-based clustering technique. *Journal of Intelligent Systems*, 26(1):153–168.
38. Lee, C.-Y., and E.K. Antonsson. 2000. Dynamic partitional clustering using evolution strategies. In *Industrial Electronics Society, 2000. IECON 2000. 26th Annual Confjerence of the IEEE*, vol. 4, pp. 2716–2721. IEEE.
39. Liu, Anan, Yuting, Su, Weizhi, Nie, and Mohan S. Kankanhalli. 2017. Hierarchical clustering multi-task learning for joint human action grouping and recognition. *IEEE Trans. Pattern Anal. Mach. Intell.* 39(1):102–114.
40. Liu, Ting, Charles, Rosenberg, and Henry A. Rowley. 2007. Clustering billions of images with large scale nearest neighbor search. In *Applications of Computer Vision, 2007. WACV'07. IEEE Workshop on*, pp. 28–28. IEEE.
41. Liu, Yongguo, Wu Xindong, and Yidong Shen. 2011. Automatic clustering using genetic algorithms. *Applied Mathematics and Computation* 218 (4): 1267–1279.
42. Maulik, Ujjwal, and Sanghamitra Bandyopadhyay. 2000. Genetic algorithm-based clustering technique. *Pattern Recognition* 33 (9): 1455–1465.
43. Mei, Jian-Ping, Yangtao Wang, Lihui Chen, and Chunyan Miao. 2017. Large scale document categorization with fuzzy clustering. *IEEE Transactions on Fuzzy Systems* 25 (5): 1239–1251.
44. Mirjalili, Seyedali. 2015. Moth-flame optimization algorithm: A novel nature-inspired heuristic paradigm. *Knowledge-Based Systems* 89: 228–249.
45. Mirjalili, Seyedali, and Andrew Lewis. 2016. The whale optimization algorithm. *Advances in Engineering Software* 95: 51–67.
46. Mirjalili, Seyedali, Seyed Mohammad, Mirjalili, and Abdolreza, Hatamlou. 2016. Multi-verse optimizer: a nature-inspired algorithm for global optimization. *Neural Computing and Applications*, 27(2):495–513.
47. Mirjalili, Seyedali, Seyed Mohammad, Mirjalili, and Andrew, Lewis. 2014. Grey wolf optimizer. *Advances in Engineering Software*, 69:46–61.
48. Mittal, Nitin, Urvinder, Singh, and Balwinder Singh, Sohi. 2016. Modified grey wolf optimizer for global engineering optimization. *Applied Computational Intelligence and Soft Computing*, 2016:8.
49. Mukhopadhyay, Anirban, Ujjwal, Maulik, Sanghamitra, Bandyopadhyay, and Carlos A. Coello. 2014. Survey of multiobjective evolutionary algorithms for data mining: Part ii. *IEEE Transactions on Evolutionary Computation*, 18(1):20–35.
50. Mukhopadhyay, Anirban, Ujjwal Maulik, Sanghamitra Bandyopadhyay, and Carlos Artemio Coello Coello. 2013. A survey of multiobjective evolutionary algorithms for data mining: Part i. *IEEE Transactions on Evolutionary Computation* 18 (1): 4–19.
51. Satyasai Jagannath Nanda and Ganapati Panda. 2014. A survey on nature inspired metaheuristic algorithms for partitional clustering. *Swarm and Evolutionary computation* 16: 1–18.
52. Oyelade, O.J. O.O. Oladipupo, and I.C. Obagbuwa. 2010. Application of k means clustering algorithm for prediction of students academic performance. arXiv:1002.2425.
53. Qaddoura, R., H. Faris, and I. Aljarah. 2020. An efficient evolutionary algorithm with a nearest neighbor search technique for clustering analysis. *Journal of Ambient Intelligence and Humanized Computing* 1–26.
54. Qaddoura, R., H. Faris, I. Aljarah, J. Merelo, and P. Castillo. 2020. Empirical evaluation of distance measures for nearest point with indexing ratio clustering algorithm. In *Proceedings*

of the 12th International Joint Conference on Computational Intelligence - Volume 1: NCTA, ISBN 978-989-758-475-vol. 6, pp. 430–438. https://doi.org/10.5220/0010121504300438.
55. Qaddoura, Raneem, Waref, Al Manaseer, Mohammad A.M. Abushariah, and Mohammad Aref, Alshraideh. 2020. Dental radiography segmentation using expectation-maximization clustering and grasshopper optimizer. *multimedia Tools and Applications*.
56. Qaddoura, Raneem, Hossam Faris, and Ibrahim Aljarah. 2020. An efficient clustering algorithm based on the k-nearest neighbors with an indexing ratio. *International Journal of Machine Learning and Cybernetics* 11 (3): 675–714.
57. Qaddoura, Raneem, Hossam, Faris, Ibrahim, Aljarah, and Pedro A. Castillo. 2020. Evocluster: An open-source nature-inspired optimization clustering framework in python. In *International Conference on the Applications of Evolutionary Computation (Part of EvoStar)*, pp. 20–36. Berlin: Springer.
58. Sharma, Manorama, G.N. Purohit, and Saurabh, Mukherjee. 2018. Information retrieves from brain mri images for tumor detection using hybrid technique k-means and artificial neural network (kmann). In *Networking Communication and Data Knowledge Engineering*, pp. 145–157. Berlin: Springer.
59. Sheikh, Rahila H., Mukesh M. Raghuwanshi, and Anil N. Jaiswal. 2008. Genetic algorithm based clustering: a survey. In *First International Conference on Emerging Trends in Engineering and Technology*, pp. 314–319. IEEE.
60. Shukri, Sarah, Hossam Faris, Ibrahim Aljarah, Seyedali Mirjalili, and Ajith Abraham. 2018. Evolutionary static and dynamic clustering algorithms based on multi-verse optimizer. *Engineering Applications of Artificial Intelligence* 72: 54–66.
61. Siddiqi, Umair F., and Sadiq M. Sait. 2017. A new heuristic for the data clustering problem. *IEEE Access*, 5:6801–6812.
62. Silva, Samuel, Rengan, Suresh, Feng, Tao, Johnathan, Votion, and Yongcan, Cao. 2017. A multi-layer k-means approach for multi-sensor data pattern recognition in multi-target localization. arXiv:1705.10757.
63. Storn, Rainer, and Kenneth Price. 1997. Differential evolution-a simple and efficient heuristic for global optimization over continuous spaces. *Journal of Global Optimization* 11 (4): 341–359.
64. Tripathi, Ashish Kumar, Kapil, Sharma, and Manju, Bala. 2018. A novel clustering method using enhanced grey wolf optimizer and mapreduce. *Big Data Research*, 14:93–100.
65. Wang, X.Y., and Jon M. Garibaldi. 2005. A comparison of fuzzy and non-fuzzy clustering techniques in cancer diagnosis. In *Proceedings of the 2nd International Conference in Computational Intelligence in Medicine and Healthcare, BIOPATTERN Conference, Costa da Caparica, Lisbon, Portugal*, vol. 28.
66. Yadav, Ms Chandni, Ms Shrutika, Zele, Ms Tejashree, Patil, Ms Vishakha, Bombadi, and Mr Tushar, Chaudhari. 2018. Automatic blood cancer detection using image processing. *Cell*, 4(03).
67. Yang, Hongguang, and Jiansheng Liu. 2015. A hybrid clustering algorithm based on grey wolf optimizer and k-means algorithm. *J Jiangxi Univ Sci Technol* 5: 015.
68. Yang, Xin-She. 2009. Firefly algorithms for multimodal optimization. In *International symposium on stochastic algorithms*, pp. 169–178. Berlin: Springer.
69. Yang, Xin-She, and Suash, Deb. 2009. Cuckoo search via lévy flights. In *2009 World Congress on Nature & Biologically Inspired Computing (NaBIC)*, pp. 210–214. IEEE.
70. Zhang, Sen, and Yongquan, Zhou. 2015. Grey wolf optimizer based on powell local optimization method for clustering analysis. *Discrete Dynamics in Nature and Society* 2015.

EEG-Based Person Identification Using Multi-Verse Optimizer as Unsupervised Clustering Techniques

Zaid Abdi Alkareem Alyasseri, Ammar Kamal Abasi, Mohammed Azmi Al-Betar, Sharif Naser Makhadmeh, João P. Papa, Salwani Abdullah, and Ahamad Tajudin Khader

Abstract Recently, electroencephalogram (EEG) signal provides great potential for identification systems. Many studies have shown that the EEG introduces unique, universal features and natural robustness for spoofing attacks. The EEG represents a graphic recording of the electrical activity of the brain that can be measured by placing sensors (electrodes) at different locations on the scalp. This chapter proposes a new technique using unsupervised clustering and optimization techniques for user identification-based EEG signals. The proposed method employs four algorithms which are Genetic Algorithm (GA), Multi-verse Optimizer (MVO), Particle Swarm Optimization (PSO) and the *k*-means algorithm. A standard EEG motor imagery

Z. A. A. Alyasseri (✉)
ECE Department, Faculty of Engineering, University of Kufa, Najaf, Iraq
e-mail: zaid.alyasseri@uokufa.edu.iq

Z. A. A. Alyasseri · S. Abdullah
Faculty of Information Science and Technology, Center for Artificial Intelligence, Universiti Kebangsaan Malaysia, 43600 Bangi, Selangor, Malaysia
e-mail: salwani@ftsm.ukm.my

A. K. Abasi · S. N. Makhadmeh · A. T. Khader
School of Computer Sciences, Universiti Sains Malaysia, George Town, Pulau Pinang, Malaysia
e-mail: ammar_abasi@student.usm.my

S. N. Makhadmeh
e-mail: m_shareef_cs@yahoo.com

A. T. Khader
e-mail: tajudin@usm.my

M. Al-Betar
Department of Information Technology - MSAI, College of Engineering and Information Technology, Ajman University, Ajman, United Arab Emirates
e-mail: mohbetar@bau.edu.jo

IT Department, Al-Huson University College, Al-Balqa Applied University, Irbid, Jordan

J. P. Papa
Department of Computing, São Paulo State University - UNESP, Bauru, Brazil
e-mail: papa@fc.unesp.br

I. Aljarah et al. (eds.), *Evolutionary Data Clustering: Algorithms and Applications*, Algorithms for Intelligent Systems,
https://doi.org/10.1007/978-981-33-4191-3_4

dataset is used to evaluate the proposed method's performance, and its results are evaluated using four criteria: (i) Precision, (ii) Recall, (iii) F-Score and (v) Purity. It is worth mentioning that this work is one of the first to employ optimization methods with unsupervised clustering methods for person identification using EEG. As a conclusion, the MVO algorithm achieved the best results compared with GA, PSO and k-means. Finally, the proposed method can draw future directions to apply to different research areas.

Keywords EEG · Biometric · Multi-verse optimizer · Evolutionary computation · Meta-heuristics algorithms · Clustering · Auto-regressive

1 Introduction

Electroencephalogram (EEG) is a visualization of brain neural activity measured from scalp that reflects the variations of the voltage that originate from ionic currents in the brain's neurons [14, 18, 39, 53]. EEG signals could therefore produce most of the brain activity information needed and be collected through non-invasive or invasive methods [10, 19, 48]. The key distinction between these approaches is that the intrusive technique entails the deployment of electrode sets embedded in the brain, for example, an eastern group-brain oncology computer interface for coordination of arm movement [49]. Berger [24] introduced the first non-invasive method for recording human brain activities using brain signals (EEG).

Researchers have expanded the technique from Han's for flexible applications in the last few decades. For instance, brain signals are successfully applied in medical for many purposes such as to treat, rehabilitate and re-establish patients. The EEG was also utilized in non-medical purposes, for example, self-regulation and education, the neuromarketing and advertising sector as well as for the neuroergonomics and the smart environment [6, 36]. EEG signals have effectively been utilized as a new identification and authentication technique for security and authorization purposes [6, 17, 19, 38, 39]. The researchers concluded that EEG signals are distinctive and powerful that are also hard to fudge, and also, recognize as a latest biometric technique [48, 50].

Asanza et al. [21] introduced a new EEG pre-processing method for noise removal. The signals were collected from two electrodes which are Left Occipital (LO) and Right Occipital (RO) using the EMOTIV headset with 14 channels. The proposed method used the k-means algorithm for clustering purposes right after feature extraction using frequency and temporal characteristics. Lit later on, the same authors applied five different clustering algorithms, say that k-means, Hierarchical k-medoids, BDSCAN, and Spectral to detect motor and imaginary motor tasks [20]. They used a standard EEG dataset with 25 healthy persons using 64 channels. In the feature extraction phase, they extracted spectral density, being the best results for detecting motor activity obtained by k-means and k-medoids, while hierarchical clustering achieved the best results for imaginary tasks of hands.

Recently, there have been major development relative with traditional techniques for the EEG-based user identification with the supervised classification and optimization methods [11, 15]. Zaid Alyaseri has developed a novel technique for the detection of users based on the EEG signal [17]. This method used a novel technique called MOFPA-WT. The MOFPA-WT method used a multi-objective FPA and the Wavelet to derive EEG features. Several variations of EEG energy features from the EEG sub-bands have been extracted. Later, this work has been extended with more features extracted from the EEG using WT decomposition process to 10 levels [18]. A number of measures such as accuracy, true acceptance rate, false acceptance rate and F-score were used to test the proposed method. The MOFPA-WT method extracted several time-main features such as mean, entropy, standard deviation, energy the logarithm of the energy, absolute energy and Resting Energy Expenditure (REE) [16]. The performance results were evaluated using accuracy, sensitivity, specificity, false acceptance rate and F-Ore. The MOFPA-WT method was comparable to some edge-cutting techniques with prospective outcomes. Moreover, the optimization technique has shown successfully significant in the field of EEG-based user identification.

The literature refers that there are a few works which are applying unsupervised clustering techniques with for EEG-based biometric person identification. The main reason for that is because most researchers prefer to use the classification methods to obtain good results compared to the methods of clustering, which does not contain a training phase, but the new data is determined to its cluster by calculating the distance between the centre of clusters and then the new data will join the closest one. Recently, the metaheuristic-based algorithms are utilized for EEG, they are normally classify into three main categories: Trajectory-based Algorithms (TAs), Evolutionary-based Algorithms (EAs) and Swarm Intelligence (SI) [15]. In general, all these types of metaheuristic algorithms are nature-inspired with common characteristics such as stochastic behaviours, set more than one control parameter to deal with the problem [32, 40].

Recently, the metaheuristic-based algorithms are utilized for EEG, they are normally classified into three main categories: Trajectory-based Algorithms (TAs), Evolutionary-based Algorithms (EAs) and Swarm Intelligence (SI) [13, 42]. In general, all these types of metaheuristic algorithms are nature-inspired with common characteristics such as stochastic behaviours, set more than one control parameter to deal with the problem.

The first type of metaheuristic algorithms is TAs; in the initial phase, the TAs are started with a single solution (one solution), and based on the neighbouring-moves operators, this initial solution is then evolved, or moved over the generations until a local optimal solution, which is the best region in the search space is reached from these algorithms. An example of TAs is Self-Organizing Maps (SOMs) and β-hill climbing [7, 41]. In the same context, there are techniques which are related with the problem nature (i.e. clustering), these techniques utilized the same mechanism of evolving the solution during the search spaces like K-means, K-medoids. Although TAs can deeply search the search space region of the initial solution and reach local optima, they cannot navigate several search space regions simultaneously.

The second type of metaheuristic algorithms is EA, which is initiated with a group of provisional individuals called population. Generation after generation, the population is evolved on the basis of three main operators: *recombination* for mixing the individual features, *mutation* for diversifying the search and selection for utilising the survival-of-the-fittest principle [12]. The EA is stopped when no further evolution can be achieved. The main shortcoming of EAs is that although they can simultaneously navigate several areas in the search space, they cannot perform deep searching in each area to which they navigate. Consequently, EAs mostly suffer from premature convergence. EAs that have been successfully utilized for TDC include Genetic Algorithm (GA) [34], harmony search [30] and cuckoo search [55].

The last type of metaheuristic algorithms is SI; an SI algorithm is also initiated with a set of random solutions called a swarm. Iteration after iteration, the solutions in the swarm are reconstructed by means of attracting them by the best solutions that are so far found [44]. SI-based algorithms can easily converge prematurely. Several SI-based TDC are utilized, such as Particle Swarm Optimization (PSO) [25, 43], Grey Wolf Optimizer (GWO) [3] and artificial bee colony [35].

The Multi-Verse Optimizer (MVO) algorithm was recently proposed as a stochastic population-based algorithm [45] inspired by multi-verse theory [23]. The big bang theory [37] explains the origin of the universe to have been a massive explosion. According to this theory, the origin of everything in our universe requires one big bang. Multi-verse theory believes that more than one explosion (big bang) occurred, with each big bang creating a new and independent universe. This theory is modelled as an optimization algorithm with three concepts: white hole, black hole and wormhole, for performing exploration, exploitation and local search, respectively. The MVO algorithm presents many benefits over other algorithms. Just a few parameters are needed to be defined at the initial stage; no complex data mathematical derivation is possible. During the search, it can easily balance exploration with exploitation. It is also sound-and-complete, flexible, scalable, adaptable and simple. MVO has also been used for a number of optimization problems, such as optimising SVM parameters [8, 27], oil recovery [33], feature selection [4, 26] and text document clustering [1, 2, 5].

However, the literature refers that there are a few works that applied clustering techniques for EEG-based biometric person identification. Therefore, the main contributions of this paper are two-fold: (i) to introduce a hybrid approach composed of both unsupervised classification and optimization techniques for EEG-based biometric person identification; (ii) to evaluate the performance of some metaheuristic algorithms such as MVO, Genetic Algorithm (GA) and PSO together with k-means for EEG-based biometric person identification. The performance of the proposed is evaluated regarding four measurement factors: precision, recall, F-Score and purity. Our work includes 109 EEG signal respondents recorded from 64 channels on various cognitive tasks. In the extraction process, auto-regressive are derived from the original EEG signals with three separate coefficients (i.e. 5, 10 and 20 orders).

The organizer of this chapter is organized as follows. Section 2 provides a Preliminaries of the techniques considered in this work. Section 3 describes the proposed approach, and the results are discussed in Sect. 4. Finally, the conclusion and future works are stated in Sect. 5.

2 Preliminaries

In this section, we provided a brief background about the k-means, particle swarm optimization, genetic algorithm and multi-verse optimizer algorithms which will use in this work.

2.1 *k-Means*

The well-known k-means is a widely and simple utilized technique in clustering domain [29, 54]. The idea behind k-means clustering is to group the samples into fixed k clusters using their similarities. The Euclidean distance is one of the commonly used measures to determine such a degree of similarity. Also, the clusters are represented by their centroids. In each round of iteration, these centroids are recalculated and used as reference points to group samples around their closest k clusters [22]. The main idea of k-means is implemented in the following Algorithm 1

Algorithm 1 clustering method using k-means algorithm

1: **Inputs:** A reading t number of samples, and a k of clusters.
2: **Outputs:** Assign t to k.
3: determine an initial trials group with k clusters.
4: **while** achieved the final criterion or reach the stop condition **do**
5: for each K clusters compute new cluster centroid.
6: Generate a new groups by assigning each trial to its closest cluster centroid based on distance(trial; cluster centroid).
7: **end while**

2.2 *Principles for Multi-Verse Optimizer (MVO)*

The general information on MVO is provided in this subsection. Multi-version theory and mathematical modelling as well as inspiration.

2.2.1 Inspiration

MVO is a population-based algorithm [45] inspired by nature phenomena (multi-verse theory) [23]. According to multi-verse theory, universes connect and might even collide with each other. MVO engages and reformulates using three main concepts: white holes, black holes and wormholes. The probability is used for determining the inflation rate (corresponds to the objective function in the optimization context) for the universe, thereby allowing the universe to assign one of these holes. Given that the universe has a high inflation rate, the probability of a white hole existing increases. Meanwhile, a low inflation rate leads to increased probability of a black hole existing [45]. Regardless of the universe's inflation rate, wormholes move objects towards the best universe randomly [28, 33]. In the optimization of terms, the white and black hole concepts play a central role in managing the exploration phase. On another side, the wormhole concept ensures the phase of exploitation. In multi-verse theory, each universe corresponds to a solution in optimization theory. In that solution, each object in the universe corresponds to one of the decision variables. The multi-verse theory is the main inspiration of MVO, which states that more than one universe exists and the universes can interact with each other (i.e. exchange the objects) by three main concepts: (i) the white hole, (ii) the black hole and (iii) the wormhole. Each one of these concepts plays a different role in the MVO working mechanism [47].

Like any population-based optimization algorithms, MVO starts with a population of feasible solutions, and it has been designed to improve solutions by mapping them to different values based on their fitness function. In MVO, the decision variables may face some exchange between the solutions randomly regardless of the fitness function to keep population diversity and to skip the local optimum [28].

2.2.2 Mathematical Modelling of MVO

The black and white hole concepts in MVO are formulated for exploring search spaces, and the wormhole concept is formulated for exploiting search spaces. In other EAs, MVO is initiated by a population of individuals (universes). Thereafter, MVO improves these solutions until a stopping criterion. Figure 1 illustrates the conceptual model of the MVO and shows the movements of the objects between the universes via white/black hole tunnels. These hole tunnels are created between two universes on the basis of the inflation rate of each universe (i.e. one universe has a higher inflation rate than the other universes.). Objects move from universes with high inflation rates using white holes. These objects are received by universes with low inflation rates using black holes.

After a population of solutions is initiated, all solutions in MVO are sorted from high inflation rates to low ones. Thereafter, it visits the solutions one by one to attract these solutions to the best one. This is done under the assumption that the solution that has been visited has the black hole. As for the white holes, the roulette wheel mechanism is used for selecting one solution. The formulation of the population U is provided in Eq. (1).

$$U = \begin{bmatrix} Sol_1^1 & Sol_1^2 & \cdots & Sol_1^d \\ Sol_2^1 & Sol_2^2 & \cdots & Sol_2^d \\ \vdots & \vdots & \vdots & \vdots \\ Sol_n^1 & Sol_n^2 & \cdots & Sol_n^d \end{bmatrix}. \quad (1)$$

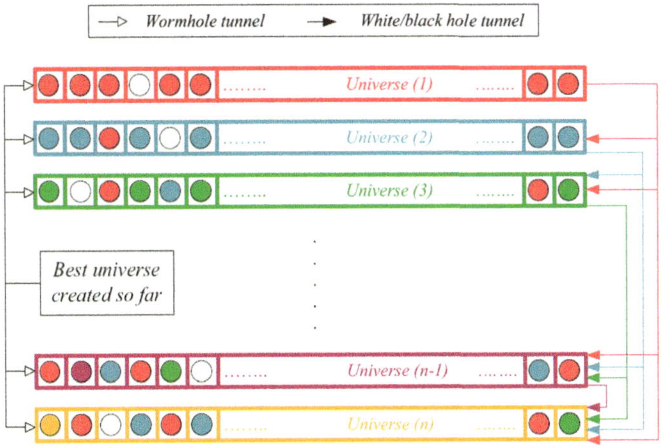

Fig. 1 Conceptual model of the MVO algorithm $I(U_1) > I(U_2 > \cdots > I(U_n)$

where n is the number of solutions (i.e. candidate universes), d is the number of decision variables (i.e. objects) and U is the population matrix of size $n \times d$ containing the set of universes. In solution i, the object j is generated randomly as follows:

$$Sol_i^j = lb_j + rand()\%((ub_j - lb_j) + 1) \quad \forall i \in (1, 2, \ldots, n) \wedge \forall j \in (1, 2, \ldots, d), \tag{2}$$

where $rand()$ is a function generating a discrete random number of the $(1, 2, \ldots, MAX\_INT)$ distribution, MAX_INT, where MAX_INT is the maximum integer number which the machine can produce, and $[lb_j, ub_j]$ signifies the j object's discrete lower and upper limits.

In each iteration, in solution j, which has the black hole, the decision variable i (i.e. $Soli^j$) can exchange the value from better solutions such that $Sol_i^j \in (Sol_i^1, Sol_i^2, \ldots, Sol_i^{j-1})$ or the value remains unchanged. This is formulated as shown in Eq. (3).

$$Sol_i^j = \begin{cases} Sol_k^j & z_1 < NoI(U_i), \\ Sol_i^j & z_1 \geq NoI(U_i), \end{cases} \tag{3}$$

where z_1 is function that generates a uniform random number between (0, 1), Sol_k^j represent the j_{th} decision variable of k_{th} solution selected by a roulette wheel selection, Sol_i^j represent the j_{th} decision variable of the i_{th} solution and $NoI(U_i)$ is normalized objective function of i_{th} solution. At the same time, regardless of the fitness function, the decision variables of the solution j perform random movements in respect of the optimal value for solutions diversity improvement in the MVO algorithm. This method is modelled using the following formula:

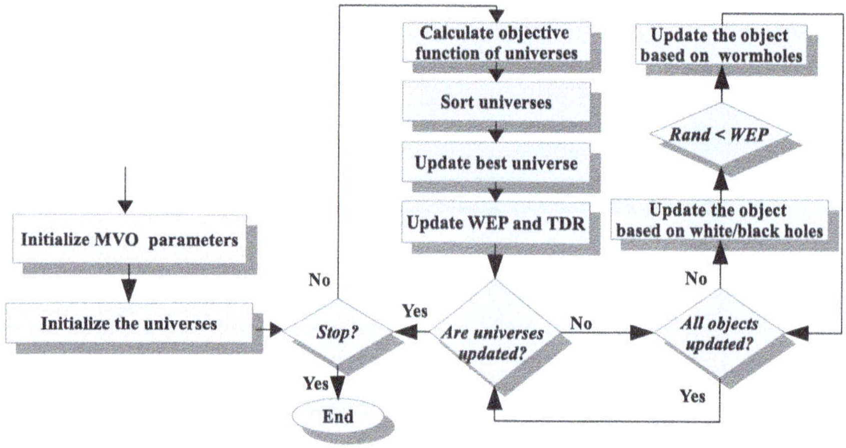

Fig. 2 Flowchart of MVO algorithm

$$Sol_i^j = \begin{cases} \begin{cases} Sol_j + TDR \times ((ub_j - lb_j) \times z_4 + lb_j) & z_3 < 0.5, \\ Sol_j - TDR \times ((ub_j - lb_j) \times z_4 + lb_j) & z_3 \geq 0.5, \end{cases} & z_2 < PWE \\ Sol_i^j & r_2 \geq PWE \end{cases} \quad (4)$$

where Sol_j represents the j_{th} location of the best universe in the population; (TDR) refers to the travelling distance rate and (PWE) refers to wormhole existence probability which are coefficient parameters; ubj and lbj are the upper and lower bounds, respectively; $z2$, $z3$ and $z4$ are random value (0, 1). The formulas for TDR and PWE are as follows:

$$PWE = min + l \times \left(\frac{max - min}{L} \right), \quad (5)$$

$$TDR = 1 - \frac{l^{1/p}}{L^{1/p}}, \quad (6)$$

where min and max are constant pre-defined values, l represents the current iteration, L is the maximum number of iterations and p is a constant that denotes the accuracy of exploitation over the iterations. The PWE coefficient values are smoothly increased during the iterations for increasing the attendance chance of wormholes in universes. Thus, the exploitation phase is stressed in every iteration. At the same time, TDR coefficient values decrease the distance of decision variables around the best solution. Therefore, the accuracy of local search is improved. A flowchart of the optimization procedure of MVO is shown in Fig. 2.

Algorithm 2 provides the pseudocodes of the MVO algorithm.

Algorithm 2 Pseudocodes and general steps of MVO algorithm

```
Initialize MVO parameters (Min, Max, Best solution = 0, LB, UB, Number of solutions, Number of dimensions
  (Dim), Number of iterations).
Create random solutions (Sol) based on LB, UB, Number of solutions, Dim.
while achieved the final criterion or reach the stop condition do
  Evaluate the objective function for every solutions.
  Sort_s=Ranking Solutions using fitness value.
  Normalize=Normalize the objective function for all solutions.
  Update the Best solution vector.
  for each solutions indexed by i except the best solution do
    Calculate PWE and TDR using Equations (5 and 6).
     Black_hole_index=i.
     for each object indexed by j do
       z1 = random([0, 1]).
       if z1 < Normalize(S_i) then
         White_hole_index = Roulette_Wheel_Selection .
         S(Black_hole_index,j)= Sort_s(White_hole_index,j).
       end if
       z2 = random([0, 1]).
       if z2 < Wormhole_existance_probability then
         z3 = random([0, 1]).
         z4 = random([0, 1]).
         if z3 < 0.5 then
           Update the Position of Solution using Equation 4 case one.
         else
           Update the Position of Solution using Equation 4 case two.
         end if
       end if

     end for
  end for
end while
Produce the optimal Solution.
```

3 Hybridizing EEG with MVO for Unsupervised Person Identification: Proposed Method

In this section, independent EEG-based user recognition method is proposed. The method suggested inclusive of five steps in which the output of each step is used to instruct the next step. The proposed method, which is described , is shown in Fig. 3.

3.1 Signal Acquisition

A standard EEG signal data set is used to acquire the EEG signal [31]. The EEG signals were obtained from 109 healthy people using a BCI2000 application program [52]. EEG signals from 64 electrodes (i.e. sensors) are then captured. Each subject runs 14 motor/imagery tasks, including neurological rehabilitation and brain-computer interface applications. Such activities are usually to visualize or imitate an action, such as eye opening and eye closing. The EEG signal input for each user, with

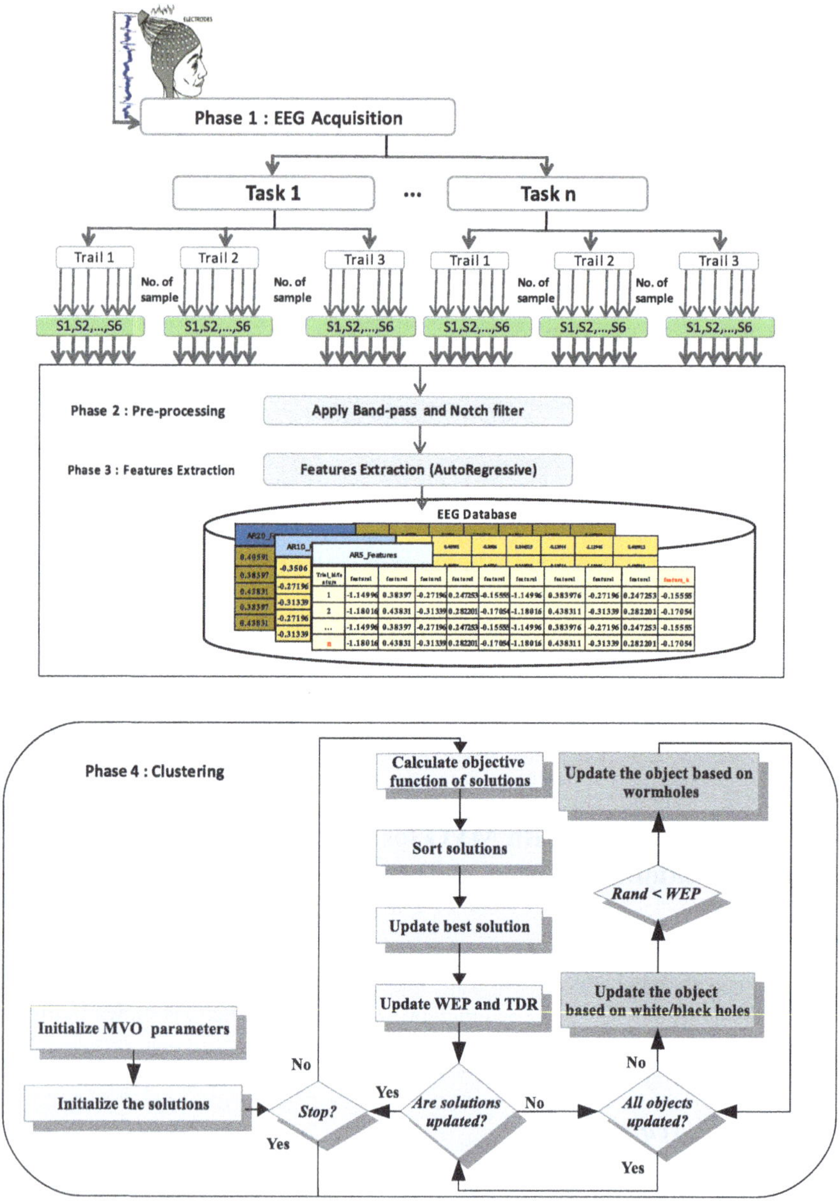

Fig. 3 EEG-based user identification system proposed in this work

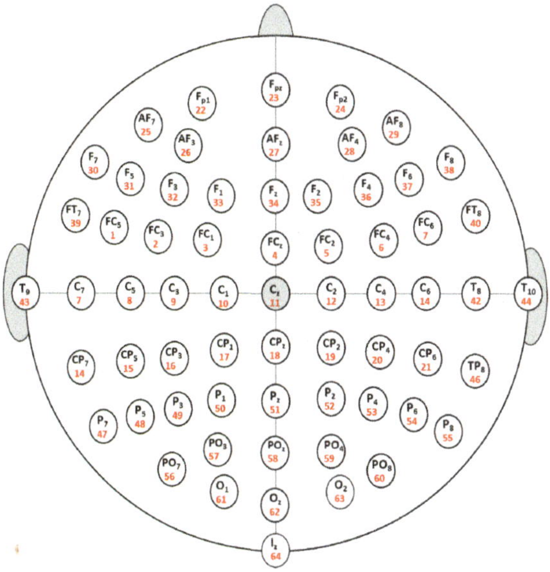

Fig. 4 Electrodes distribution for motor movement/imagery dataset

a length of one minute, hereby collected three times. Figure 4 shows the distribution of the EEG channels which are used in this work.

3.2 Signal Pre-processing

The input EEG signal is split into 6 segments of 10s each. We have used a bandpass and notch filter to de-noise, as the EEG signal can compromise throughout recording [53].

3.3 EEG Feature Extraction

Effective features are important to extract in every authentication system [51, 53]. This phase consequently focuses on extracting unique information from auto-regressive models with three different coefficients, i.e. 5, 10 and 20 orders that allow the clustering phase to achieve good results.[1] In this chapter, we applied the Yule-Walker approach to find the coefficients of the Auto-Regressive (AR) model by using the least square method criterion.

[1]Such numbers are suggested in the work of Rodrigues et al. [9, 50].

3.4 EEG Signal Clustering

The main role of any clustering method is gathering similar samples in the same cluster, and dissimilar ones in a different cluster. In this paper, different optimization techniques are considered to optimize k-means algorithm in the context of EEG-based person identification, as explained further.

3.4.1 Representation of Solutions

In the proposed approach, each possible solution is represented as vector $s \in \Re^n$, where n is number of dataset samples. The value of each element s_i falls within the range $[1, k]$, where k is any cluster number. Figure 5 shows an illustrative example of the proposed solution representation with the steps to obtain the optimal/near-optimal solution.

3.4.2 Fitness Function

As stated by Papa et al. [46], the k-means working mechanism is essentially an optimization problem, where the idea is to minimize the distance from each dataset sample to its closest centroid. Therefore, the fitness function used to evaluate each solution can be computed by the following steps:

1. calculate the cluster centroid for each cluster;
2. calculate the distance between each sample (task) and the cluster centroid associated with it; and
3. calculate average distance of samples to the cluster centroid (ADDC).

To calculate the clusters' centroids, we can use the following formula:

$$C_j = \frac{\sum_{i=1}^{n} s_i}{\sum_{i=1}^{n} N_{ji}}, \tag{7}$$

where C_j denotes the centroid of cluster j, $j \in \{1, 2, \ldots, k\}$, n is the number of dataset samples and $N_{n\times k}$ is a binary-valued matrix defined as follows:

$$N_{ki} = \begin{cases} 1 & \text{sample (task) } s_i \text{ belongs to the cluster } j \\ 0 & \text{otherwise.} \end{cases} \tag{8}$$

We used the average distance of samples to the cluster centroid measure, which is defined as follows:

$$ADDC = \frac{\sum_{i=1}^{k} \left(\frac{1}{n_i} \sum_{\forall s_j \in C_i} D(C_i, s_j) \right)}{k}, \tag{9}$$

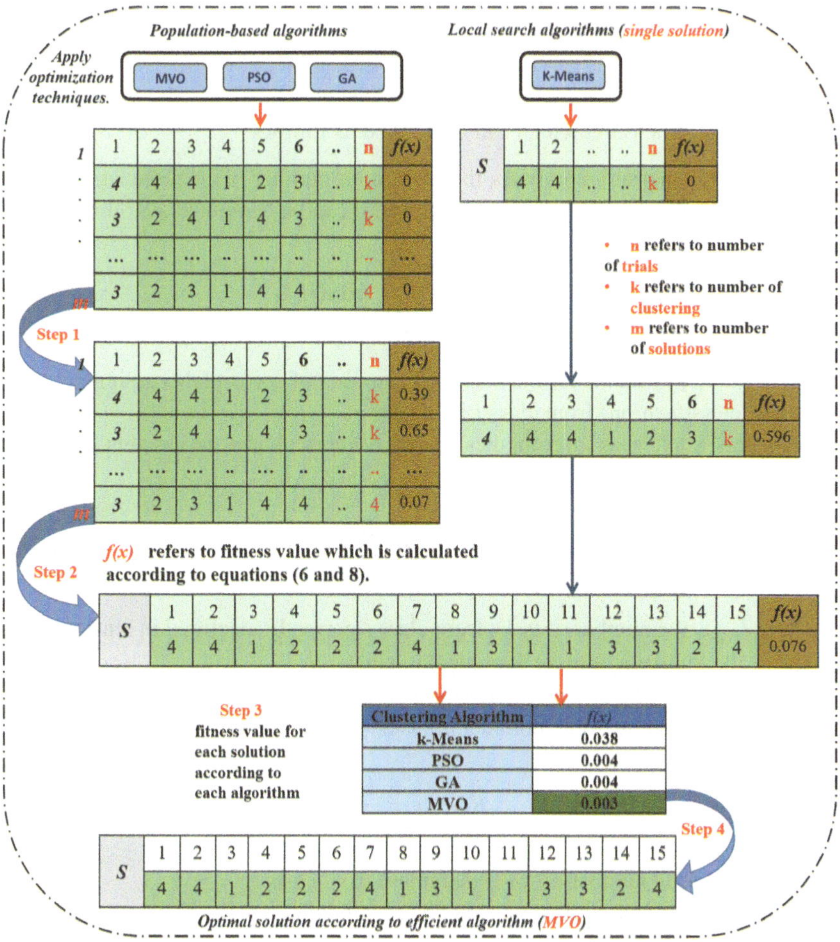

Fig. 5 Proposed approach for unsupervised EEG-based person identification using metaheuristic optimization

where $D(C_j, s_j)$ denotes the distance between the centroid of cluster j and sample s_j. Therefore, the main idea of this work is to minimize the value of ADDC for a given possible solution, i.e. to associate the "best cluster" to each dataset sample.

4 Results and Discussions

In this section, we present the evaluation measures, results and the discussion about the findings of our work.

4.1 Evaluation Measures

In order to evaluate the proposed approach, four measures are used Purity, Recall, Precision and F-measure, which can be defined as follows:

- **Purity**: This measure computes the maximum number of correct class for each cluster over all tasks in the cluster [54], and it can be computed as follows:

$$Purity = \frac{1}{n}\sum_{i=1}^{k} \max(i, j), \tag{10}$$

  where $\max(i, j)$ stands for the maximum number of correct label assignment (i.e. correct classification) from class i in cluster j.
- **Precision**: Such measure computes the ratio of correct label assignment from class i over the total number of tasks in cluster j [54]:

$$P(i, j) = \frac{L_{i,j}}{|C_j|}, \tag{11}$$

  where L_{i_j} is the number of tasks from class i correctly identified in cluster j, and $|C_j|$ is the numbers of tasks (samples) in cluster j.
- **Recall**: Such measure computes the ratio of tasks from cluster j correctly identified (i.e. $\hat{L}_{i,j}$) against the total numbers of tasks (samples) from the class i:

$$R(i, j) = \frac{\hat{L}_{i,j}}{T_i}, \tag{12}$$

  where T_i is total number of tasks (samples) from class i.
- **F-score**: It is combing the precision and recall defines as the F-score, which tells us the clustering quality. The larger its values, the better is the solution:

$$F(i, j) = \frac{2 \times P(i, j) \times R(i, j)}{P(i, j) + R(i, j)}. \tag{13}$$

  The overall F-score is computed by taking the weighted F-score for each class.

Table 1 shows the experiment results of 30 runs concerning standard k-means and its versions optimized by MVO, GA and PSO. The results are presented in terms of the best, average and worst fitness values. Besides, the MVO algorithm achieved the best results according to all measures, i.e. Purity, Recall, Precision and F-measure. In the best case, the MVO obtained 0.9643 with AR_5 (5-order auto-regressive model), 0.9740 with AR_{10} (10-order auto-regressive model) and 0.9804 with AR_{20} (20-order auto-regressive model). For the GA obtained 0.8098 with AR_5, 0.8145 with AR_{10} and 0.8179 with AR_{20}. For the PSO obtained 0.8128 with AR_5, 0.8158 with AR_{10} and 0.8158 with AR_{20} where the k-means obtained 0.8074 with AR_5, 0.8180 with AR_{10} and 0.6884 with AR_{20}.

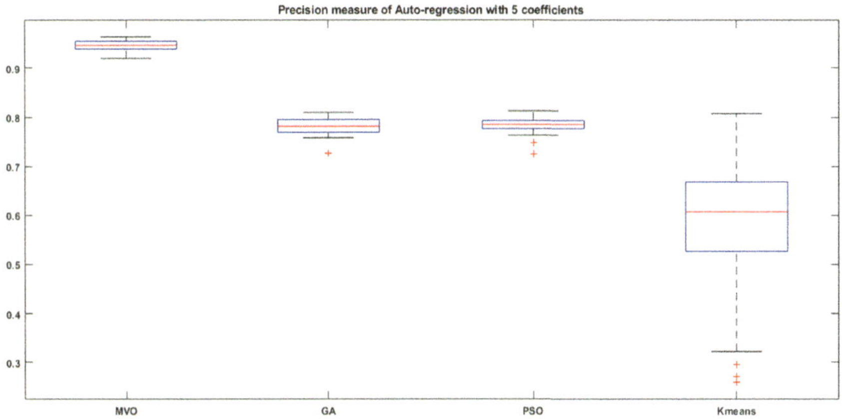

Fig. 6 Precision measure of the auto-regression model with 5 coefficients

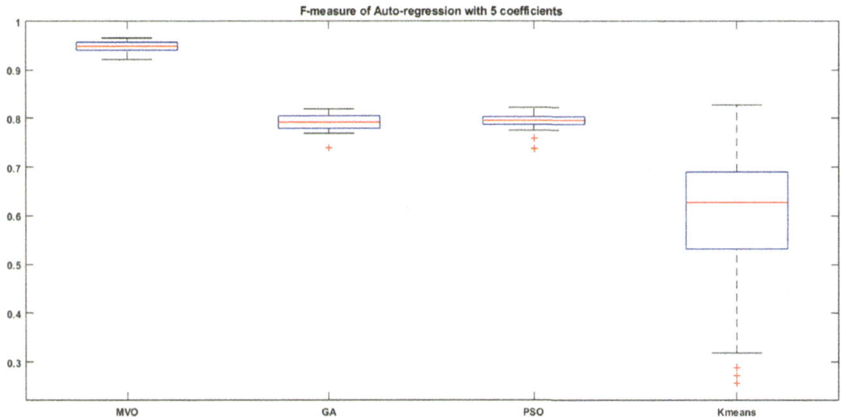

Fig. 7 F-measure of the auto-regression model with 5 coefficients

Such results are pretty very interesting, and it refers that the MVO algorithm is suitable for EEG biometric applications.

Figures 6, 7, 8 and 9 show the performance of the proposed methods. The results show the MOV obtained a great significant advantage using all criteria, while GA and PSO have closer results. Notice that *k*-means has achieved a fluctuation change in the results.

Table 1 Experimental results

Measures	Method/AR_5	MVO	k-means	PSO	GA
Precision	Best	**0.9643**	0.8074	0.8128	0.8098
Recall		**0.9665**	0.8487	0.8324	0.8283
F-Score		**0.9654**	0.8275	0.8225	0.8189
Purity		**0.9643**	0.8074	0.8128	0.8098
Measures	Method/AR_5	MVO	k-means	PSO	GA
Precision	Average	**0.9471**	0.5697	0.7836	0.7816
Recall		**0.9499**	0.6007	0.8047	0.8021
F-Score		**0.9485**	0.5845	0.7940	0.7917
Purity		**0.9471**	0.7088	0.7836	0.7816
Measures	Method/AR_5	MVO	k-means	PSO	GA
Precision	Worst	**0.9199**	0.2595	0.7255	0.7276
Recall		**0.9228**	0.2520	0.7504	0.7517
F-Score		**0.9214**	0.2557	0.7377	0.7394
Purity		**0.9199**	0.6334	0.7255	0.7276
Measures	Method/AR_{10}	MVO	k-means	PSO	GA
Precision	Best	**0.9740**	0.8180	0.8158	0.8145
Recall		**0.9751**	0.8661	0.8335	0.8328
F-Score		**0.9746**	0.8414	0.8246	0.8236
Purity		**0.9740**	0.8503	0.8158	0.8145
Measures	Method/AR_{10}	MVO	k-means	PSO	GA
Precision	Average	**0.9562**	0.5582	0.7850	0.7878
Recall		**0.9578**	0.5872	0.8050	0.8081
F-Score		**0.9570**	0.5719	0.79492	0.7978
Purity		**0.9562**	0.7186	0.7850	0.7878
Measures	Method/AR_{10}	MVO	k-means	PSO	GA
Precision	Worst	**0.9170**	0.2561	0.7570	0.7507
Recall		**0.9197**	0.2622	0.7791	0.7744
F-Score		**0.9184**	0.2591	0.7679	0.7624
Purity		**0.9170**	0.6123	0.7570	0.7507
Measures	Method/AR_{20}	MVO	k-means	PSO	GA
Precision	Best	**0.9804**	0.6884	0.8158	0.8179
Recall		**0.9811**	0.7096	0.8335	0.8345
F-Score		**0.9807**	0.6988	0.8246	0.8261
Purity		**0.9804**	0.8101	0.8158	0.8179
Measures	Method/AR_{20}	MVO	k-means	PSO	GA
Precision	Average	**0.9575**	0.4578	0.7836	0.7919
Recall		**0.9588**	0.4608	0.8037	0.8113
F-Score		**0.9581**	0.4587	0.7935	0.8015
Purity		**0.9575**	0.6964	0.7836	0.7919
Measures	Method/AR_{20}	MVO	k-means	PSO	GA
Precision	Worst	**0.9309**	0.1501	0.7570	0.7652
Recall		**0.9324**	0.1414	0.7791	0.7863
F-Score		**0.9317**	0.1534	0.7679	0.7756
Purity		**0.9309**	0.6126	0.7570	0.7652

Bold value indicates the best value

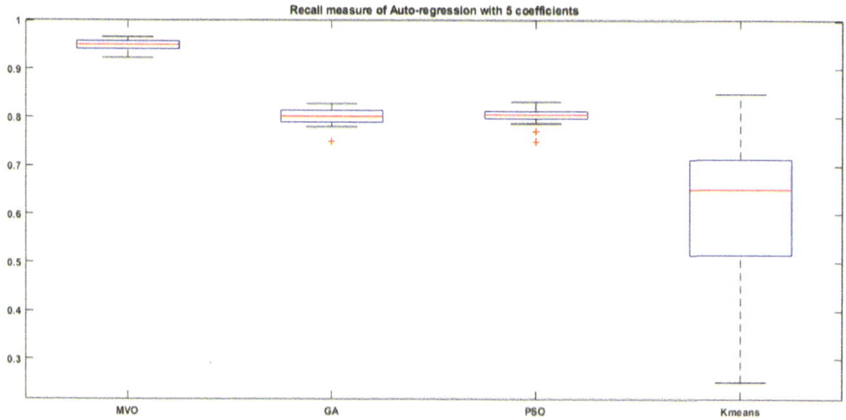

Fig. 8 Recall measure of the auto-regression model with 5 coefficients

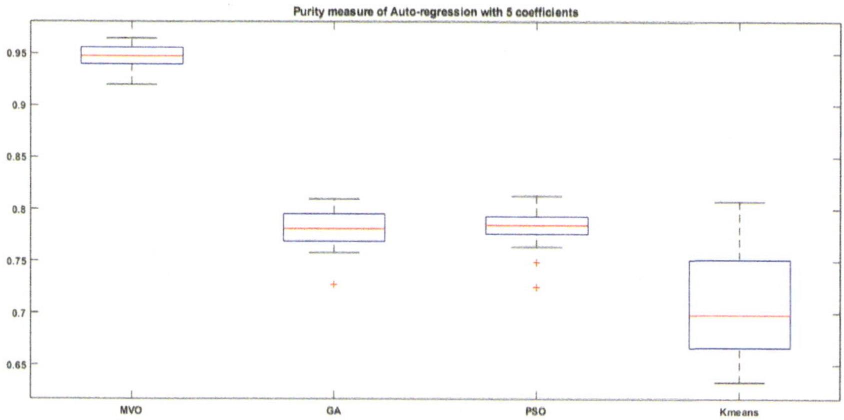

Fig. 9 Purity measure of the auto-regression model with 5 coefficients

Figure 10 summarizes the average performance over AR_5, AR_{10} and AR_{20} configurations where the MVO achieved the best results according to Purity, Recall, Precision and F-measure.

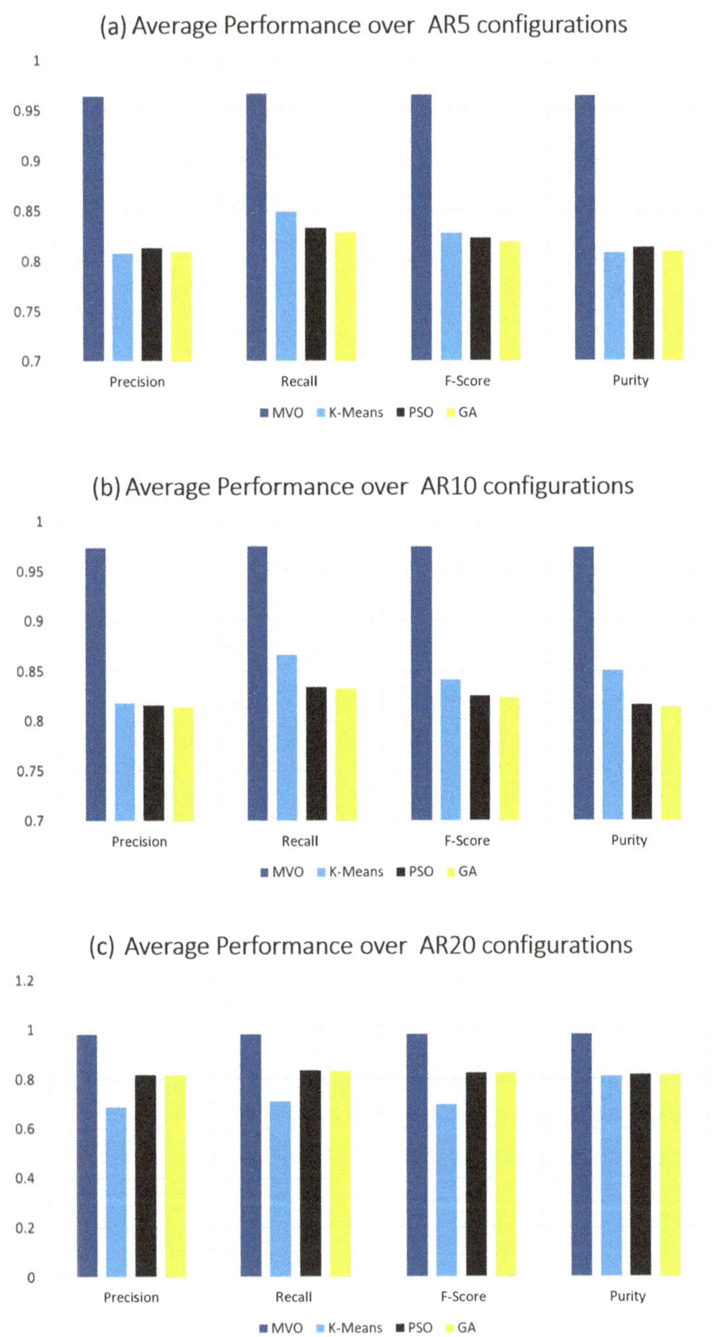

Fig. 10 Average Performance over: **a** AR5, **b** AR10, and **c** AR20 configurations

5 Conclusions and Future Work

In this chapter, a new method for unsupervised EEG-based user identification was proposed. The main purpose of the proposed approach is to find the optimal solution of the original brain EEG signal which can give unique features from auto-regressive models with three different coefficients (AR5, Ar10 and AR20). In general, the idea is to model the k-means working mechanism as an optimization process, where the samples are associated with their closest centroids. In this context, GA, PSO and MVO were considered to optimize k-means for EEG person identification. The proposed method was tested using EEG motor imagery which is a standard EEG dataset. The results of the proposed method are evaluated using four criteria, namely, precision, recall, F-score and purity.

The best results obtained by the MVO for all the evaluation measures are the PSO, GA and k-means, respectively.

For the future works, we intend to investigate another unsupervised techniques for EEG-based user identification with more challenging and more complex EEG dataset. Also, using multi-objective signal problem instances, such as user authentication or early detection of epilepsy based on EEG signals.

References

1. Abasi, Ammar Kamal, Ahamad Tajudin Khader, Mohammed Azmi Al-Betar, Syibrah Naim, Zaid Abdi Alkareem Alyasseri, and Sharif Naser Makhadmeh. 2020. An ensemble topic extraction approach based on optimization clusters using hybrid multi-verse optimizer for scientific publications. *Journal of Ambient Intelligence and Humanized Computing* 1–37.
2. Abasi, Ammar Kamal, Ahamad Tajudin Khader, Mohammed Azmi Al-Betar, Syibrah Naim, Zaid Abdi Alkareem Alyasseri, and Sharif Naser Makhadmeh. 2020. A novel hybrid multi-verse optimizer with k-means for text documents clustering. *Neural Computing & Applications*.
3. Abasi, Ammar Kamal, Ahamad Tajudin Khader, Mohammed Azmi Al-Betar, Syibrah Naim, Sharif Naser Makhadmeh, and Zaid Abdi Alkareem Alyasseri. 2019. An improved text feature selection for clustering using binary grey wolf optimizer. In *Proceedings of the 11th national technical seminar on unmanned system technology 2019*, 503–516. Springer.
4. Abasi, Ammar Kamal, Ahamad Tajudin Khader, Mohammed Azmi Al-Betar, Syibrah Naim, Sharif Naser Makhadmeh, and Zaid Abdi Alkareem Alyasseri. 2019. A text feature selection technique based on binary multi-verse optimizer for text clustering. In *2019 IEEE Jordan international joint conference on electrical engineering and information technology (JEEIT)*, 1–6. IEEE.
5. Abasi, Ammar Kamal, Ahamad Tajudin Khader, Mohammed Azmi Al-Betar, Syibrah Naim, Sharif Naser Makhadmeh, and Zaid Abdi Alkareem Alyasseri. 2020. Link-based multi-verse optimizer for text documents clustering. *Applied Soft Computing* 87: 106002.
6. Abdulkader, Sarah N., Ayman Atia, and Mostafa-Sami M. Mostafa. 2015. Brain computer interfacing: Applications and challenges. *Egyptian Informatics Journal* 16 (2): 213–230.
7. Al-Betar, Mohammed Azmi. 2017. β-hill climbing: An exploratory local search. *Neural Computing and Applications* 28 (1): 153–168.
8. Aljarah, Ibrahim, Al-Zoubi Ala'M, Hossam Faris, Mohammad A. Hassonah, Seyedali Mirjalili, and Heba Saadeh. 2018. Simultaneous feature selection and support vector machine

optimization using the grasshopper optimization algorithm. *Cognitive Computation* 10 (3): 478–495.
9. Alyasseri, Zaid Abdi Alkareem, Ahamad Tajudin Khadeer, Mohammed Azmi Al-Betar, Ammar Abasi, Sharif Makhadmeh, and Nabeel Salih Ali. 2019. The effects of EEG feature extraction using multi-wavelet decomposition for mental tasks classification. In *Proceedings of the international conference on information and communication technology*, 139–146.
10. Alyasseri, Zaid Abdi Alkareem, Ahamad Tajudin Khader, and Mohammed Azmi Al-Betar. 2017. Electroencephalogram signals denoising using various mother wavelet functions: A comparative analysis. In *Proceedings of the international conference on imaging, signal processing and communication*, 100–105. ACM.
11. Alyasseri, Zaid Abdi Alkareem, Ahamad Tajudin Khader, and Mohammed Azmi Al-Betar. 2017. Optimal electroencephalogram signals denoising using hybrid β-hill climbing algorithm and wavelet transform. In *Proceedings of the international conference on imaging, signal processing and communication*, 106–112. ACM.
12. Alyasseri, Zaid Abdi Alkareem, Ahamad Tajudin Khader, Mohammed Azmi Al-Betar, Ammar Kamal Abasi, and Sharif Naser Makhadmeh. 2019. EEG signal denoising using hybridizing method between wavelet transform with genetic algorithm. In *Proceedings of the 11th national technical seminar on unmanned system technology 2019*, 449–469. Springer.
13. Alyasseri, Zaid Abdi Alkareem, Ahamad Tajudin Khader, Mohammed Azmi Al-Betar, Ammar Kamal Abasi, and Sharif Naser Makhadmeh. 2019. EEG signals denoising using optimal wavelet transform hybridized with efficient metaheuristic methods. *IEEE Access* 8: 10584–10605.
14. Alyasseri, Zaid Abdi Alkareem, Ahamad Tajudin Khader, Mohammed Azmi Al-Betar, and Osama Ahmad Alomari. 2020. Person identification using EEG channel selection with hybrid flower pollination algorithm. *Pattern Recognition* 107393.
15. Alyasseri, Zaid Abdi Alkareem, Ahamad Tajudin Khader, Mohammed Azmi Al-Betar, and Mohammed A. Awadallah. 2018. Hybridizing β-hill climbing with wavelet transform for denoising ECG signals. *Information Sciences* 429: 229–246.
16. Alyasseri, Zaid Abdi Alkareem, Ahamad Tajudin Khader, Mohammed Azmi Al-Betar, J.P. Papa, Osama Ahmad Alomari, and Sharif Naser Makhadme. 2018. An efficient optimization technique of EEG decomposition for user authentication system. In *2018 2nd international conference on biosignal analysis, processing and systems (ICBAPS)*, 1–6. IEEE.
17. Alyasseri, Zaid Abdi Alkareem, Ahamad Tajudin Khader, Mohammed Azmi Al-Betar, João P. Papa, and Osama ahmad Alomari. EEG-based person authentication using multi-objective flower pollination algorithm. In *2018 IEEE congress on evolutionary computation (CEC)*, pages 1–8. IEEE, 2018.
18. Alyasseri, Zaid Abdi Alkareem, Ahamad Tajudin Khader, Mohammed Azmi Al-Betar, João P. Papa, and Osama Ahmad Alomari. 2018. EEG feature extraction for person identification using wavelet decomposition and multi-objective flower pollination algorithm. *IEEE Access*.
19. Alyasseri, Zaid Abdi Alkareem, Ahmad Tajudin Khader, Mohammed Azmi Al-Betar, Joao P. Papa, Osama Ahmad Alomari, and Sharif Naser Makhadmeh. 2018. Classification of EEG mental tasks using multi-objective flower pollination algorithm for person identification. *International Journal of Integrated Engineering* 10 (7).
20. Asanza, Víctor, Kerly Ochoa, Christian Sacarelo, Carlos Salazar, Francis Loayza, Carmen Vaca, and Enrique Peláez. 2016. Clustering of EEG occipital signals using k-means. In *Ecuador technical chapters meeting (ETCM), IEEE*, 1–5. IEEE.
21. Asanza, Víctor, Enrique Pelaez, and Francis Loayza. 2017. EEG signal clustering for motor and imaginary motor tasks on hands and feet. In *Ecuador technical chapters meeting (ETCM), 2017 IEEE*, 1–5. IEEE.
22. Bai, Liang, Xueqi Cheng, Jiye Liang, Huawei Shen, and Yike Guo. 2017. Fast density clustering strategies based on the k-means algorithm. *Pattern Recognition* 71: 375–386.
23. Barrow, John D., Paul C.W. Davies, and Charles L. Harper Jr. 2004. *Science and ultimate reality: Quantum theory, cosmology, and complexity*. Cambridge University Press.

24. Berger, Hans. 1929. Über das elektrenkephalogramm des menschen. *European Archives of Psychiatry and Clinical Neuroscience* 87 (1): 527–570.
25. Cura, Tunchan. 2012. A particle swarm optimization approach to clustering. *Expert Systems with Applications* 39 (1): 1582–1588.
26. Ewees, Ahmed A., Mohamed Abd El Aziz, and Aboul Ella Hassanien. 2017. Chaotic multi-verse optimizer-based feature selection. *Neural Computing and Applications* 1–16.
27. Faris, Hossam, Mohammad A. Hassonah, Al-Zoubi Ala'M, Seyedali Mirjalili, and Ibrahim Aljarah. 2017. A multi-verse optimizer approach for feature selection and optimizing SVM parameters based on a robust system architecture. *Neural Computing and Applications* 1–15.
28. Fathy, Ahmed, and Hegazy Rezk. 2018. Multi-verse optimizer for identifying the optimal parameters of PEMFC model. *Energy* 143: 634–644.
29. Fränti, Pasi, and Sami Sieranoja. 2017. K-means properties on six clustering benchmark datasets. *Applied Intelligence* 1–17.
30. Geem, Zong Woo, Joong Hoon Kim, and Gobichettipalayam Vasudevan Loganathan. 2001. A new heuristic optimization algorithm: Harmony search. *Simulation* 76 (2): 60–68.
31. Goldberger, Ary L., Luis A.N. Amaral, Leon Glass, Jeffrey M. Hausdorff, Plamen Ch. Ivanov, Roger G. Mark, Joseph E. Mietus, George B. Moody, Chung-Kang Peng, and H. Eugene Stanley. 2000. Physiobank, physiotoolkit, and physionet. *Circulation* 101 (23): e215–e220.
32. Hussain, Kashif, Mohd Najib Mohd Salleh, Shi Cheng, and Yuhui Shi. 2018. Metaheuristic research: A comprehensive survey. *Artificial Intelligence Review* 1–43.
33. Janiga, Damian, Robert Czarnota, Jerzy Stopa, Paweł Wojnarowski, and Piotr Kosowski. 2017. Performance of nature inspired optimization algorithms for polymer enhanced oil recovery process. *Journal of Petroleum Science and Engineering* 154: 354–366.
34. Jiang, Jian-Hui, Ji-Hong Wang, Xia Chu, and Ru-Qin Yu. 1997. Clustering data using a modified integer genetic algorithm (IGA). *Analytica Chimica Acta* 354 (1): 263–274.
35. Karaboga, Dervis, and Bahriye Basturk. 2008. On the performance of artificial bee colony (ABC) algorithm. *Applied Soft Computing* 8 (1): 687–697.
36. Katrawi, Anwar H., Rosni Abdullah, Mohammed Anbar, and Ammar Kamal Abasi. 2020. Earlier stage for straggler detection and handling using combined CPU test and late methodology. *International Journal of Electrical & Computer Engineering* 10. ISSN: 2088-8708.
37. Khoury, Justin, Burt A. Ovrut, Nathan Seiberg, Paul J. Steinhardt, and Neil Turok. 2002. From big crunch to big bang. *Physical Review D* 65 (8): 086007.
38. Kumari, Pinki, and Abhishek Vaish. 2014. Brainwave based authentication system: Research issues and challenges. *International Journal of Computer Engineering and Applications* 4 (1): 2.
39. Kumari, Pinki, and Abhishek Vaish. 2015. Brainwave based user identification system: A pilot study in robotics environment. *Robotics and Autonomous Systems* 65: 15–23.
40. Makhadmeh, Sharif Naser, Ahamad Tajudin Khader, Mohammed Azmi Al-Betar, and Syibrah Naim. 2018. Multi-objective power scheduling problem in smart homes using grey wolf optimiser. *Journal of Ambient Intelligence and Humanized Computing* 1–25.
41. Makhadmeh, Sharif Naser, Ahamad Tajudin Khader, Mohammed Azmi Al-Betar, Syibrah Naim, Ammar Kamal Abasi, and Zaid Abdi Alkareem Alyasseri. 2019. Optimization methods for power scheduling problems in smart home: Survey. *Renewable and Sustainable Energy Reviews* 115: 109362.
42. Makhadmeh, Sharif Naser, Ahamad Tajudin Khader, Mohammed Azmi Al-Betar, Syibrah Naim, Zaid Abdi Alkareem Alyasseri, and Ammar Kamal Abasi. 2019. A min-conflict algorithm for power scheduling problem in a smart home using battery. In *Proceedings of the 11th national technical seminar on unmanned system technology 2019*, 489–501. Springer.
43. Makhadmeh, Sharif Naser, Ahamad Tajudin Khader, Mohammed Azmi Al-Betar, Syibrah Naim, Zaid Abdi Alkareem Alyasseri, and Ammar Kamal Abasi. 2019. Particle swarm optimization algorithm for power scheduling problem using smart battery. In *2019 IEEE jordan international joint conference on electrical engineering and information technology (JEEIT)*, 672–677. IEEE.

44. Mavrovouniotis, Michalis, Changhe Li, and Shengxiang Yang. 2017. A survey of swarm intelligence for dynamic optimization: Algorithms and applications. *Swarm and Evolutionary Computation* 33: 1–17.
45. Mirjalili, Seyedali, Shahrzad Saremi, Seyed Mohammad Mirjalili, and Leandro dos S. Coelho. 2016. Multi-objective grey wolf optimizer: A novel algorithm for multi-criterion optimization. *Expert Systems with Applications* 47: 106–119.
46. Papa, J.P., L.P. Papa, R.R.J. Pisani, and D.R. Pereira. 2016. A hyper-heuristic approach for unsupervised land-cover classification. *IEEE Journal of Selected Topics in Applied Earth Observations and Remote Sensing* 9: 2333–2346.
47. Patel, Monika Raghuvanshi Rahul. 2017. An improved document clustering with multiview point similarity/dissimilarity measures. *International Journal of Engineering and Computer Science* 6 (2).
48. Ramadan, Rabie A., and Athanasios V. Vasilakos. 2017. Brain computer interface: Control signals review. *Neurocomputing* 223: 26–44.
49. Rao, Rajesh P.N. 2013. *Brain-computer interfacing: An introduction*. Cambridge University Press.
50. Rodrigues, Douglas, Gabriel F.A. Silva, J.P. Papa, Aparecido N. Marana, and Xin-She Yang. 2016. EEG-based person identification through binary flower pollination algorithm. *Expert Systems with Applications* 62: 81–90.
51. Sarier, Neyire Deniz. 2010. Improving the accuracy and storage cost in biometric remote authentication schemes. *Journal of Network and Computer Applications* 33 (3): 268–274.
52. Schalk, Gerwin, Dennis J. McFarland, Thilo Hinterberger, Niels Birbaumer, and Jonathan R. Wolpaw. 2004. BCI2000: A general-purpose brain-computer interface (BCI) system. *IEEE Transactions on Biomedical Engineering* 51 (6): 1034–1043.
53. Sharma, Pinki Kumari, and Abhishek Vaish. 2016. Individual identification based on neuro-signal using motor movement and imaginary cognitive process. *Optik-International Journal for Light and Electron Optics* 127 (4): 2143–2148.
54. Xiong, Hui, Junjie Wu, and Jian Chen. 2009. K-means clustering versus validation measures: A data-distribution perspective. *IEEE Transactions on Systems, Man, and Cybernetics, Part B (Cybernetics)* 39 (2): 318–331.
55. Zaw, Moe Moe, and Ei Ei Mon. 2013. Web document clustering using cuckoo search clustering algorithm based on Levy flight. *International Journal of Innovation and Applied Studies* 4 (1): 182–188.

Capacitated Vehicle Routing Problem—A New Clustering Approach Based on Hybridization of Adaptive Particle Swarm Optimization and Grey Wolf Optimization

Dao Vu Truong Son and Pham Nhat Tan

Abstract Capacitated vehicle routing problem (CVRP) is a challenging combinatorial optimization research and has drawn considerable interests from many scientists. However, no standard method has been established yet to obtain optimal solutions for all standard problems. In this research, we propose a two-phase approach: An improved k-Means algorithm for the clustering phase and a hybrid meta-heuristic based on Adaptive Particle Swarm—PSO and Grey Wolf Optimization—GWO (APGWO) for the routing phase. Our approach gives results very close to the exact solutions and better than the original k-Means algorithm. And for the routing phase, our experimental results show highly competitive solutions compared with recent approaches using PSO and GWO on many of the benchmark datasets.

Keywords Data clustering · Optimization · Evolutionary computation · Swarm intelligence · Meta-heuristics · Vehicle routing problem · GWO · PSO

1 Introduction and Related Works

CVRP is a constraint satisfaction problem to find minimum traveling distances of vehicles to serve all customers with varying demands. Previous researches proposed a lot of exact algorithms which can find the optimal solutions. These methods are branch and bound and cutting plane method by Lysgaard et al. [1], branch cut and price by Fukasawa et al. [2], algorithm based on set partitioning with extra cuts by Baldacci et al. [3] and so on. However, performance of these exact algorithms highly depends on the problem, and the computing time for large-scale issues grows

D. V. T. Son · P. N. Tan (✉)
International University, Vietnam National University - HCMC, Ho Chi Minh City, Vietnam
e-mail: pntan.iac@gmail.com

D. V. T. Son
e-mail: dvtson@hcmiu.edu.vn

I. Aljarah et al. (eds.), *Evolutionary Data Clustering: Algorithms and Applications*, Algorithms for Intelligent Systems,
https://doi.org/10.1007/978-981-33-4191-3_5

exponentially. The meta-heuristic approaches can obtain a "reasonably good" solution relatively quickly and can solve the large-scale problems, which make them suitable for practical situations.

A solution for a CVRP problem is a set of customer sequences. Verdú et al. [4] provided a comprehensive review of CVRP methods. One of the most popular methods is a cluster-first, route-second heuristic. In these approaches, the original problem is decomposed into smaller ones by grouping customers into clusters whose total demand is less than or equal to the vehicle capacity. After that, the customers in each group are routed using the traveling salesman problem (TSP). Clustering is the operation of segmenting a heterogeneous population into a number of clusters [5]. Verdú et al. [4] summarized some important results in clustering analysis for electrical customers. There are three key clustering approaches, namely, artificial neural networks, fuzzy logic, and statistical techniques. Another comprehensive review of clustering techniques from a data mining perspective could be found in Berkhin [6].

In statistical direction, multivariate statistics and inductive techniques are two main approaches. In multivariate statistics, a group of variables is considered together to classify a data set. MANOVA is a powerful tool for clustering customers by assessing multiple metric-dependent variables. K-Means algorithm is another popular statistical method that repeats two key steps of center determination and object allocation to clusters until total data variance to be minimized [7]. The K-Means algorithm is easily understood and implemented. However, K-Means algorithm has some shortcomings. Its results depend on initial value of centroids which cannot be determined in advance. In addition, clustering data is very sensitive to outliers that lead to unbalanced clusters. One popular inductive statistical method is decision tree. Decision tree classifies items in the form of a tree structure of attributes and the target value with decision nodes and leaf nodes. In these nodes, a single attribute-value is analyzed to choose which branch of the subtree will be selected to move further. A decision node has two or more branches. The leaf node indicates the value of the target attribute. Both categorical and numerical data can be solved by decision trees [8].

In fuzzy classification direction, there are some interesting approaches such as adaptive neuro-fuzzy inference system (ANFIS), fuzzy C-means (FCM), and fuzzy subtractive clustering. ANFIS can handle fuzzy aspects of the object data by using adaptive and learning control with fuzzy if-then rules. The FCM and related mountain methods effectively estimate cluster centroids. However, these techniques rely on "carefully" specified inputs such as number of clusters and initial partitions. Good estimation of these values proves challenging. Furthermore, these techniques require huge computational power. In this research, we choose the k-Means algorithm due to its simplicity and ease of calculation and improved it using a meta-heuristic. By combining the meta-heuristics in clustering, the disadvantages of K-means algorithm are reduced. Korayem et al. [9] previously chose to improve K-means performance for some of the CVRP dataset using GWO. Recently, swarm intelligence such as particle swarm optimization (PSO), ant colony optimization (ACO), artificial bee colony [10, 11] are proposed using swarm behaviors to solve complex problems.

In the PSO, solution particles try to move to better locations. PSO converges relatively fast using two mechanisms: remembering personal best and sharing global best experiences. Many have tried to use different versions of PSO to solve a CVRP, such as Chen et al. [12], Kao and Chen [13], Ai and Kachitvichyanukul [14], Marinakis et al. [15].

This study proposes a new hybrid algorithm for the CVRP, which is based on PSO and is hybridized with GWO. We choose swarm intelligence because they are population-based and do not rely on having a good initial solution. In this research, our main contributions are combining PSO, adaptive PSO, and a hybridized PSO-GWO with the traditional K-means clustering algorithm to generate the "K-PSO, K-APSO, and K-APSOGWO" algorithms. The algorithm is tested on a number of benchmark problems with promising results.

2 Mathematical Models of CVRP

Munari et al. [16] represented CVRP as a graph $G(N, V)$, where $N = C \cup \{0, n+1\}$ is the set of node associated to customers in C and to depot node 0 and $n + 1$. They use two nodes to represent the the same single depot and imposed that all routes start from 0 and return to $n + 1$. Set V contains the arcs (i, j) for each pair of node $(i, j) \in N$. The cost associated with an arc $(i, j) \in V$ is c_{ij}. Each node has a demand q_i such that $q_i > 0$ for each $i \in C$ and $q_0 = q_{n+1} = 0$. The objective of the problem is to determine a set of minimal cost routes that satisfies all the requirements defined above. They introduce binary decision variable x_{ij} that equals 1 if arc (i, j) is chosen and equals 0 otherwise; y_j is a continuous decision variable denoting the cumulated demand on the route that visits node $j \in N$ up to this visit. The flow-formulation is given as follows:

$$Minimize \sum_{i=0}^{n+1}\sum_{j=0}^{n+1} c_{ij}x_{ij}$$

subject to

$$\sum_{j=0, j\neq i}^{n+1} x_{ij} = 1, \quad \forall i = 1, \ldots, n \tag{1}$$

$$\sum_{i=0, i\neq h}^{n} x_{ih} - \sum_{j=0, j\neq h}^{n+1} x_{hj} = 0, \quad \forall h = 1, \ldots, n \tag{2}$$

$$\sum_{j=1}^{n} x_{0j} \leq K \tag{3}$$

$$y_j \leq y_i + q_j x_{ij} - Q(1 - x_{ij}), \forall i, j = 0, \ldots, n + 1 \tag{4}$$

$$d_i \leq y_i \leq Q, \forall i = 0, \ldots, n + 1 \tag{5}$$

$$x_{ij} \in \{0, 1\}, \forall i, j = 0, \ldots, n + 1 \tag{6}$$

Constraints (1) ensure that all customers are visited exactly once. Constraints (2) guarantee the correct flow of vehicles through the arcs, by stating that if a vehicle arrives to a node $h \in N$, then it must depart from this node. Constraint (3) limits the maximum number of routes to K, the number of vehicles. Constraints (4) and (5) ensure together that the vehicle capacity is not exceeded. The objective function imposes that the total travel cost of the routes is minimized. Constraints (4) also avoid subtours in the solution, i.e., cycling routes that do not pass through the depot.

3 Our Algorithm

3.1 Grey Wolf Optimization

GWO was first introduced by Mirjalili et al. [17] is an algorithm basing on the behavior of the wolves in catching their prey. This algorithm also focuses on the social dominant hierarchy of the wolf. The alphas are responsible for the leaders which take a role in making decisions about hunting, sleeping place, and so on. However, in some democratic behavior, the alpha follows the other wolves in the pack. The alphas are the best one in managing the pack. The following rank of wolf's hierarchy is beta. These wolves are the subordinate, who help the alpha in making decision. The main role of the beta is to support and give feedback to the alpha. The lowest level is omega. They have to submit to all the other dominant wolves. The omega also has the least right in the pack. Those wolves that do not belong to those groups above will be called delta. These wolves have variety of roles such as watching the boundaries of the territory and warning the pack and so on. To develop the mathematical model, the alpha, beta, and delta are the best, second best, and third best solutions, respectively. The next step of GWO is encircling prey, which is calculated by proposed equations (Fig. 1).

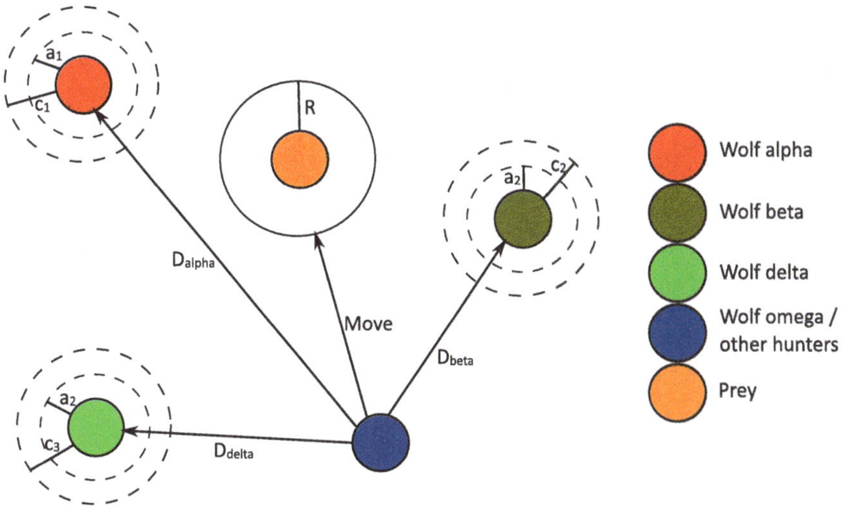

Fig. 1 Grey wolf optimization

To develop the mathematically model, first, the fittest solution is considered as alpha, the beta and delta are considered to the second and the third fittest solution, respectively. The next step of GWO is encircling prey, which is calculated by proposed equations:

$$\vec{D} = \left|\vec{C} \cdot \vec{X}_p(t) - \vec{X}(t)\right| \tag{7}$$

$$\vec{X}(t+1) = \vec{X}_p(t) - \vec{A} \cdot \vec{D} \tag{8}$$

where t indicates the current iteration, $\vec{A}$ and $\vec{C}$ are coefficient vectors, $\vec{X}_p$ is the position vector of the prey, and $\vec{X}$ is the position vector of a grey wolf. The coefficient vectors are calculated by equations:

$$\vec{A} = 2\vec{a} \cdot \vec{r_1} - \vec{a} \tag{9}$$

$$\vec{C} = 2\vec{r_1} \tag{10}$$

where $\vec{a}$ are linearly decreased from 2 to 0, $\vec{r_1}$ and $\vec{r_2}$ are random vector in [0, 1]. The following step is hunting, which defines the final position of the wolf $\vec{X}(t+1)$ using these equations:

$$\overrightarrow{D}_{\alpha} = \left|\overrightarrow{C}_1 \cdot \overrightarrow{X_{\alpha}} - \overrightarrow{X}\right| \tag{11}$$

$$\overrightarrow{D}_{\beta} = \left|\overrightarrow{C}_2 \cdot \overrightarrow{X_{\beta}} - \overrightarrow{X}\right| \tag{12}$$

$$\overrightarrow{D}_{\delta} = \left|\overrightarrow{C}_3 \cdot \overrightarrow{X_{\delta}} - \overrightarrow{X}\right| \tag{13}$$

$$\overrightarrow{X}_1 = \overrightarrow{X_{\alpha}} - \overrightarrow{A_1} \cdot \overrightarrow{D_{\alpha}} \tag{14}$$

$$\overrightarrow{X}_2 = \overrightarrow{X_{\beta}} - \overrightarrow{A_2} \cdot \overrightarrow{D_{\beta}} \tag{15}$$

$$\overrightarrow{X}_3 = \overrightarrow{X_{\delta}} - \overrightarrow{A_3} \cdot \overrightarrow{D_{\delta}} \tag{16}$$

$$\overrightarrow{X}(t+1) = \frac{\overrightarrow{X}_1 + \overrightarrow{X}_2 + \overrightarrow{X}_3}{3} \tag{17}$$

3.2 Particle Swarm Optimization

PSO was a swarm intelligence algorithm, proposed by Kennedy and Eberhart [18]. In PSO, each individual searches space with a velocity which is dynamically adjusted according to its own and its herb's experiences. The ith particle is represented as $X_i = (x_i^1, x_i^2, \ldots, x_i^k)$. The best previous position which giving the best fitness value of the ith particle is recorded and represented as in p_i^{best}. The index of the best particle in the population is noted as g_{best}^k. The velocity which defines the rate of the position changing for particle ith is represented as $V_i = (v_i^1, v_i^2, \ldots, v_i^k)$. The mathematically model of PSO is represented as the following equations:

$$v_k^{t+1} = w * v_k^t + c_1 * rand * \left(pbest_k^t - x_k^t\right) + c_2 * rand * \left(gBest - x_k^t\right) \tag{18}$$

$$x_k^{t+1} = x_k^t + v_k^t \tag{19}$$

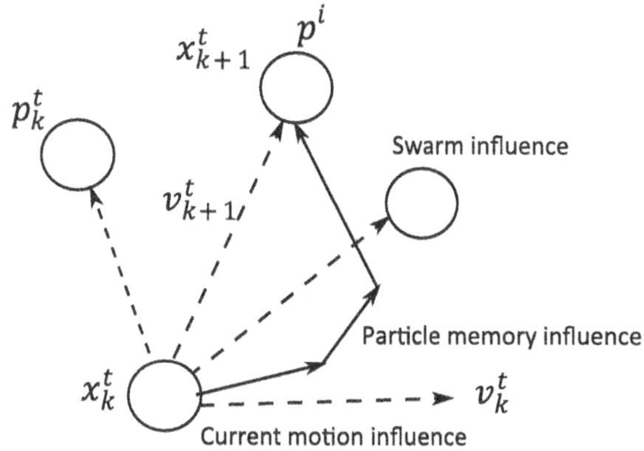

Fig. 2 Particle swarm optimization

where c_1 and c_2 are two constants, and rand is any random number between [0, 1]. The term $c_1 * rand * \left(pbest_k^t - x_k^t\right)$ is called "cognitive component" while $c_2 * rand * \left(gBest - x_k^t\right)$ is referred as "social component" (Fig. 2).

3.3 Adaptive Particle Swarm Optimization (APSO)

The value of c_1 and c_2, which are usually called "acceleration coefficients", are often set as constants, most likely $c_1 = c_2 = 1$ or $c_1 = c_2 = 2$. These values are found by empirical studies in order to balance the cognitive and social components, which also balance the exploration and exploration phases. In this study, we propose a formula to change the acceleration coefficients in each iteration. The new coefficients are calculated as followed:

$$c_1^t = 1.2 - \frac{f\left(x_k^t\right)}{f\left(gBest\right)} \tag{20}$$

$$c_2^t = 0.5 + \frac{f\left(x_k^t\right)}{f\left(gBest\right)} \tag{21}$$

where c_1^t and c_2^t stands for the coefficients at iteration t; $f(x_k^t)$ is the fitness of particle k at iteration t, and $f(gBest)$ is the swarm's global best fitness. The values of 1.2 and 0.5 are also found by empirical studies. We also modify the formula for inertia as followed:

$$w_t = (maxIter - t) * \frac{wMax - wMin}{maxIter} + wMin \tag{22}$$

Finally, we update the velocity and position of particles by the following equations:

$$v_k^{t+1} = w * v_k^t + c_1^t * rand * \left(pbest_k^t - x_k^t\right) + c_2^t * rand * \left(gBest - x_k^t\right) \quad (23)$$

$$x_k^{t+1} = x_k^t + v_k^t \quad (24)$$

The pseudocode of this algorithm is described as in Algorithm 1.

Algorithm 1 APSO algorithm

```
Initialize the particle population
Initialize parameters
while t < Max number of iteration do
    for each particle with position x_p do calculate fitness value f(x_p)
        if f(x_p) is better than pbest_p then
            pbest_p ← x_p
        end if
        if f(pbest_p) is better than gbest then
            gbest ← pbest_p
        end if
    end for
    update w according to Eq. (22)
    for each particle with position x_p do
        update c1, c2 according to Eq. (20) and Eq. (21)
        calculate velocity of each particle by Eq. (23)
        update position of each particle by Eq. (24)
    end for
    t = t + 1
end while
return gbest
```

3.4 *Adaptive Particle Swarm Optimization–Grey Wolf Optimization (APSOGWO)*

Şenel et al. [19] provided a novel hybrid PSO-GWO by replacing a particle of the PSO with a value being the mean of the three best wolves' positions. In this hybrid variant, we follow the same procedure of APSO and introduce a probability of mutation, which will trigger a small number of iterations of GWO within the APSO main loop. The probability of mutation is set at 0.1 in our case. The pseudocode for this is as in Algorithm 2.

Algorithm 2 APSOGWO Algorithm

Initialize the particle population
Initialize parameters
while $t < Max\ number\ of\ iteration$ **do**
    for each particle with position x_p **do** calculate fitness value $f(x_p)$
        if $f(x_p)$ is better than $pbest_p$ **then**
            $pbest_p \leftarrow x_p$
        end if
        if $f(pbest_p)$ is better than $gbest$ **then**
            $gbest \leftarrow pbest_p$
        end if
    end for
    update w according to Eq. (22)
    for each particle with position x_p **do**
        update c_1, c_2 according to Eq. (20) and Eq. (21)
        calculate velocity of each particle by Eq. (23)
        update position of each particle by Eq. (24)
    end for
    if $rand(0, 1) < prob$ **then**
        run GWO
        x_p = position of the best wolf
    end if
    $t = t + 1$
end while
return $gbest$

3.5 *Clustering with Evolutionary Algorithms*

This section presents the modified k-Means algorithm with evolutionary algorithms such as PSO, GWO, APSO, and APSOGWO. In a standard K-means algorithm, defining centroids is one important issue. Each centroid has a dimensional vector of location (cen_x^j, cen_y^j). The matrix of centroids has a size of $k \times 5$, where k is the number of vehicle, which are [index of centroid, x coordinate, y coordinate, current capacity of centroid C_j, total distance in this cluster]. The distance matrix of each set of centroids is $number\ of\ customer \times (k + 2)$ matrix in which the first column is the number of customer, the second to the $(k + 1)$ column is d_{ij}, which is distance from customer i to cluster j, the last column is the cluster that customer is assigned to.

3.5.1 Objective Function

Our objective is to minimize traveling distance from depot to all customers which is based on the distance and capacity constraints of each cluster. Originally, we aim to minimize the following function:

$$F_0 = \sum_{j=1}^{k} \sum_{i=1}^{n} \left| x_{ij} - cen_j \right|^2 \tag{25}$$

where k is the number of clusters, n is the number of customers, x_{ij} is the $i^{(}th)$ customer that belongs to cluster j, cen_j is the centroid of cluster j. As shown earlier, each cluster must satisfy a capacity constraint. We introduce a penalty into our objective function F_0 as shown in Eq. (24). This function penalizes infeasible solutions relative to the amount of constraint's violation [11]. Accordingly, the new fitness function can be defined as follows:

$$F = F_0 + p \times \left(\frac{\Delta capacity}{Q} \right)^c \tag{26}$$

where Q is the capacity of each vehicle, p and c are penalty parameters, $\Delta capacity$ is the amount of capacity violation. In our study, these parameters are determined experimentally.

3.5.2 k-Means with GWO

Korayem et al. [9] proposed a GWO-based clustering method to solve CVRP. In this approach, the solutions are represented by wolves. There are k numbers of clusters. Each wolf is a set of centroids corresponding to j cluster in k and is represented by a $k \times 2$ matrix. The population of wolves' hunts for the best possible solution which is defined as prey. The best solution is then the best position of the centroid. The positions of all wolves is represented by $(k \times number\ of\ population) \times 2$ matrix. The alpha position is the matrix of centroid that has the smallest objective function value, the beta position and delta position have the second and the third smallest objective function values, respectively. The distance and new position of all wolves is updated by Eqs. (11) to (17) by the parameter calculated by Eqs. (9) and (10) in Sect. 3.1.

3.5.3 k-Means with PSO

In K-means-PSO, similar to K-means-GWO, each population is defined by set of centroids. The best particle matrix of all centroids has the same size of the position of all population. The global best position is the set of centroids that has the minimum value of objective function. The velocity matrix is a $(k \times number\ of\ population) \times$ 2 matrix. The velocity and new position of each population is calculated by Eqs. (18) and (19) in Sect. 3.2.

3.5.4 k-Means with APSO and APSOGWO

In these two variants, the search procedures follow the principles of k-Means-PSO as stated before, with the algorithms following the pseudocode in Algorithm 1 and Algorithm 2, respectively.

3.5.5 Algorithm Approach

The clustering algorithm is developed as described below

- Initialize the position of population
- Calculate the fitness function for each population (a set of centroids)
  - Calculate the distance matrix from each customer to each centroids by $d_(j, i) = \sqrt{(x_x^i - cen_x^j)^2 + (x_y^i - cen_y^j)^2}$
  - Assign customer to the nearest centroid
  - Calculate the current capacity and total distance of from each centroid to the customers assign in this cluster
  - Calculate the violated capacity, if violated, assign the customer to the next nearest cluster.
  - Until all nearest capacity is violate, assign the customer to the cluster that has minimum violation
  - Calculate fitness function as Eq. (24)
- Define the alpha, beta, delta wolves for GWO or swarm of particles for PSO.
- Update the new position of population
- Loop until the maximum iteration
- Calculate total traveling distance by TSP code and violation of capacity.

4 Computational Results

All of the algorithms in this paper are coded using Matlab, and all experiments were run on a personal computer equipped with a Core i5 processor running at 2.2GHz.

Three different sets of benchmark problems were selected to evaluate the effectiveness of our proposed algorithm for the CVRP. The small set consists of instances having less than 100 customers. The medium set consists of instances having from 100 to 500 customers. The large set for those having more than 500 customers.

The benchmark set can be downloaded from the following website:

http://people.brunel.ac.uk/~mastjjb/jeb/orlib/vrpinfo.html. It has been widely used in previous studies.

Table 1 Computational results of best solutions

Instance	k-PSO	k-GWO	k-APSO	k-APSOGWO	BKS
A-n60-k9	**1392.7**	1415.2	1439	1399	1354
B-n31-k5	687	687	687	**684.7**	672
B-n39-k5	573.2	567.7	573.2	**564.7**	549
B-n41-k6	859.3	854.3	855.2	**851.4**	829
B-n50-k7	767.3	**765.5**	767.7	767.6	741
P-n20-k2	**210.5**	**210.5**	**210.5**	**210.5**	216
P-n76-k5	663	652.4	663	**648.9**	627
E-n76-k8	782.4	776.1	774.9	**759.4**	735
E-n101-k8	885.7	**871.5**	894	873.1	815
E-n101-k14	**1153.9**	1156.5	1180.2	1154.7	1067
P-n101-k4	716.3	720.8	**715.4**	718.2	681
X-n101-k25	29685	29031.7	30084.9	**28986.6**	27591
X-n209-k16	32884.1	32619	32839.6	**32597.7**	30656
X-n524-k153	168891	168514.3	169890.7	**167863.7**	154593
X-n716-k3	50051	48123.4	52028.1	**48034.4**	43414
X-n801-k40	81069.4	81322	81460.1	**80690**	73331
X-n1001-k43	**80468.7**	81464.4	83819.2	81853.6	72402
Leuven2-n4000	144107.2	158779.8	148483.7	**139227**	112998
Rank	2nd	3rd	4th	1st	

The parameters set for different algorithms are as follows:

- GWO: 30 search agents, 500 iterations
- PSO: 30 search agents, 500 iterations, $c_1 = c_2 = 1$
- APSO: 30 search agents, 500 iterations
- APSOGWO: 30 search agents (for PSO main loop), 20 search agents (for nested GWO loop), 500 iterations for PSO, 20 iterations for nested GWO, $wMax = 0.9$, $wMin = 0.2$.

The following tables list the computational results of our algorithm the test problems. The best solution, average solution, worst solution, and standard deviation (Std.) computed over 5 independent runs on each problem are summarized. The best results in each benchmark problems are in bold and underscored.

Table 1 reveals that APSOGWO is able to generate reasonable good solutions for most of CVRPs in terms of solution quality. Twelve out of eighteen test problems can be solved successfully by the proposed algorithm, and the result is much better than those using GWO (three best), PSO (four best), and APSO (two best).

Table 2 shows that APSOGWO placed second after GWO for providing best worst solutions and significantly better than PSO and APSO.

Table 2 Computational results of worst solutions

Instance	k-PSO	k-GWO	k-APSO	k-APSOGWO	BKS
A-n60-k9	1459.3	1444.4	1466.5	**1420.5**	1354
B-n31-k5	696.1	687.2	694	**687**	672
B-n39-k5	575.8	**567.9**	589.8	572.5	549
B-n41-k6	933.3	912.2	925.8	**864.8**	829
B-n50-k7	781.8	**770.8**	783.2	773.2	741
P-n20-k2	**210.5**	**210.5**	**210.5**	**210.5**	216
P-n76-k5	689.9	**660.6**	677.3	661.9	627
E-n76-k8	804	**792**	826.8	794.2	735
E-n101-k8	932.1	905.8	923.5	**882.1**	815
E-n101-k14	1198.9	**1180.1**	1204.9	1182.2	1067
P-n101-k4	730.8	724.4	**719.4**	726.4	681
X-n101-k25	31247	**30836.9**	31358.5	31503	27591
X-n209-k16	33852.4	**32894.9**	33283.3	33398.1	30656
X-n524-k153	172386.6	173178	171572.6	**169785.4**	154593
X-n716-k3	51912.8	**48724.5**	53123.4	52107.5	43414
X-n801-k40	**81296.6**	82214	82613.9	84729.8	73331
X-n1001-k43	83290.3	**82212.1**	85794.1	84964.2	72402
Leuven2-n4000	212129.2	235060.8	226410.7	**156047.8**	112998
Rank	3rd	1st	3rd	2nd	

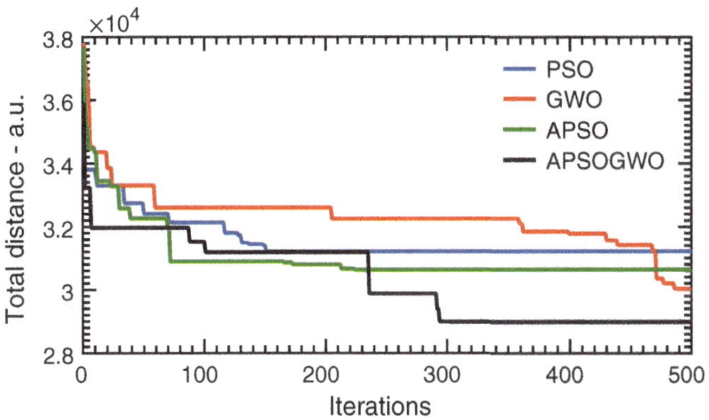

Fig. 3 Convergence of dataset X-n101-k25

Table 3 shows that APSOGWO and GWO provide the same number of best average results and significantly better than PSO and APSO.

Table 4 shows that APSOGWO placed second only after GWO and significantly better than PSO and APSO.

Table 3 Computational results of average solutions

Instance	k-PSO	k-GWO	k-APSO	k-APSOGWO	BKS
A-n60-k9	1434.7	1434.1	1445.9	**1410.3**	1354
B-n31-k5	693	687.1	689.7	**686.2**	672
B-n39-k5	571.6	**566**	573.3	**566**	549
B-n41-k6	897.4	872.8	889.2	**858.7**	829
B-n50-k7	777.1	**769.1**	774.6	770.5	741
P-n20-k2	**210.5**	**210.5**	**210.5**	**210.5**	216
P-n76-k5	675.5	**656.2**	668.7	657.1	627
E-n76-k8	792.5	786.6	798.5	**781.3**	735
E-n101-k8	901.6	890.2	909.1	**876.8**	815
E-n101-k14	1178.2	**1167.9**	1192.9	1169.1	1067
P-n101-k4	720.4	722.6	**718**	721.8	681
X-n101-k25	30469.1	**30069.9**	30614.3	30562.3	27591
X-n209-k16	33260.2	**32765.8**	32980.3	32920.5	30656
X-n524-k153	169997.7	170107	170571.1	**168958**	154593
X-n716-k3	51061.5	**48495.3**	52494.1	49655.2	43414
X-n801-k40	**81217.5**	81813.8	82162.1	82287.7	73331
X-n1001-k43	82151.1	**81960.9**	84482.4	83179.4	72402
Leuven2-n4000	144687.1	158221.9	148458.4	**139724.3**	112998
Rank	3rd	1st	4th	1st	

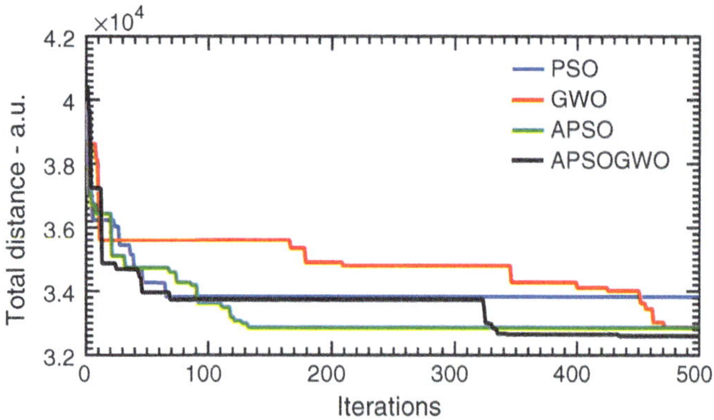

Fig. 4 Convergence of dataset X-n209-k16

Table 4 Computational results of standard deviations

Instance	k-PSO	k-GWO	k-APSO	k-APSOGWO
A-n60-k9	25.2	11.2	12.3	**8.1**
B-n31-k5	3.5	**0.2**	3.7	1
B-n39-k5	3.2	**1.5**	9.5	2.8
B-n41-k6	34.1	23.7	25.7	**4.9**
B-n50-k7	6	**2.1**	6.6	2.4
P-n20-k2	**0**	**0**	**0**	**0**
P-n76-k5	12.3	**3.5**	6.2	5.1
E-n76-k8	8.9	**6.4**	22.4	14.3
E-n101-k8	18.2	17.2	11	**3.4**
E-n101-k14	16.5	10.2	**9.9**	12.4
P-n101-k4	5.9	**1.3**	1.5	3.5
X-n101-k25	**723.1**	761.4	518.9	975.7
X-n209-k16	414.3	**124.6**	175.7	354.3
X-n524-k153	1433.2	1998.8	**686.5**	791.8
X-n716-k3	941	**325.0**	565.6	2160.2
X-n801-k40	**128.3**	453	616.2	2148.2
X-n1001-k43	1487.2	**430**	1136	1605.3
Leuven2-n4000	7116.4	6689.6	7513.5	**4683**
Rank	3rd	1st	4th	1st

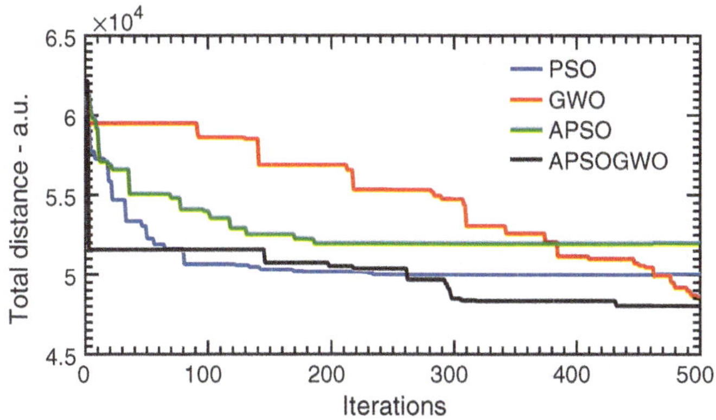

Fig. 5 Convergence of dataset X-n716-k3

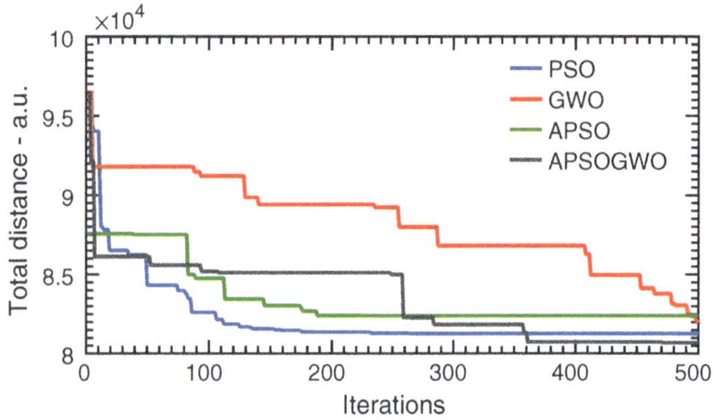

Fig. 6 Convergence of dataset X-n801-k40

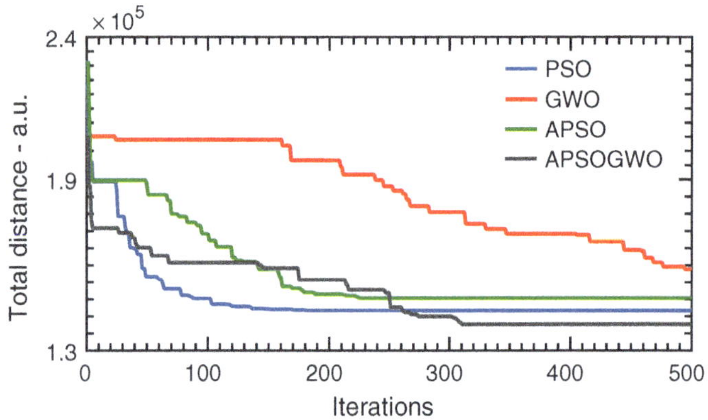

Fig. 7 Convergence of dataset Leuven4000

One thing worth noting is that with a very large problem size (with 4000 customers), APSOGWO provided the best solutions for best, worst, average, and even standard deviation results in 5 repetitions. Even though we did not compute for cases that have more than 4,000 customers, we believe that APSOGWO will be better as well.

We conducted a comparative study to compare our algorithm with a couple of swarm intelligence methods available for the CVRP. The smaller the objective function value, the better the solution. Benchmark sets X-n101-k25, X-n209-k16, X-n716-k3, X-n801-k40, and Leuven2-n4000-q150 are chosen to show the convergence of different iteration-best solutions for the K-APSOGWO. The curves reveal that K-PSO, K-APSO, and K-APSOGWO can converge faster than K-GWO. However, K-PSO and K-APSO cannot escape the local traps as well as the K-GWO. Since K-APSOGWO inherits the trap-escaping capability of the GWO, it can avoid local traps and provide the most accurate results (Figs. 3 and 4).

5 Conclusion

This chapter proposes a hybrid algorithm, APSOGWO, which takes advantage of particle swarm optimization and grey wolf algorithm for capacitated vehicle problems. Our contributions are two-fold. First, the "acceleration coefficients", c_1 and c_2 are modified in each iteration and dependent on the gBest value obtained so far. Second, we also introduce a probability of mutation, which will trigger a small number of iterations of GWO within the APSO main loop. The probability of mutation is set at 0.1 in our case. Computational results show that the performance of K-APSOGWO is competitive in terms of solution quality when compared with existing GWO- and PSO-based approaches. For future research, APSOGWO can be modified to extend its application to vehicle routing problems with time windows or multiple depots, among others (Figs. 5, 6, and 7).

References

1. Lysgaard, Jens, Adam Letchford, and R. Eglese. 2004. A new branch-and-cut algorithm for the capacitated vehicle problem. *Mathematical Programming*.
2. Fukasawa, Ricardo, Jens Lysgaard, Marcus Poggi, Marcelo Reis, Eduardo Uchoa, and Renato Werneck. 2004. Robust branch-and-cut-and-price for the capacitated vehicle routing problem. In *Integer programming and combinatorial optimization, 10th international IPCO conference*, New York, NY, USA, 7–11 June 2004.
3. Baldacci, R., E.A. Hadjiconstantinou, and A. Mingozzi. 2004. A new method for solving capacitated location problems based on a set partitioning approach. *Operation Research* 52: 723–738.
4. Verdú, S.V., M.O. Garcia, C. Senabre, A.G. Marin, and F.J.G. Franco. 2006. Classification, filtering, and identification of electrical customer load patterns through the use of self-organizing maps. *IEEE Transactions on Power Systems* 21: 1672–1682.
5. Linoff, G.S., and M.J. Berry. 2011. *Data mining techniques: For marketing, sales, and customer relationship management*. Wiley.
6. Berkhin, P. 2006. A survey of clustering data mining techniques. In *Grouping multidimensional data: Recent advances in clustering*, ed. J. Kogan, C. Nicholas, and M. Teboulle, 25–71. Springer.
7. Hammouda, K., and F. Karray. 2000. A comparative study of data clustering techniques. University of Waterloo, ON, Canada.
8. Amatriain, X., and J.M. Pujol. 2015. Data mining methods for recommender systems. In *Recommender systems handbook*, ed. F. Ricci, L. Rokach, and B. Shapira, 227–262. Springer US.
9. Korayem, Lamiaa, M. Khorsid, and Sally Kassem. 2015. Using grey wolf algorithm to solve the capacitated vehicle routing problem. *IOP Conference Series: Materials Science and Engineering* 83.
10. Lin, S.W., Z.J. Lee, K.C. Ying, and C.Y. Lee. 2009. Applying hybrid meta-heuristics for capacitated vehicle routing problem. *Expert Systems with Applications* 36 (2): 1505–1512.
11. Szeto, W.Y., Y. Wu, and S.C. Ho. 2011. An artificial bee colony algorithm for the capacitated vehicle routing problem. *European Journal of Operational Research* 215 (1): 126–135.
12. Chen, A.L., G.K. Yang, and Z.M. Wu. 2006. Hybrid discrete particle swarm optimization algorithm for capacitated vehicle routing problem. *Journal of Zhejiang University: Science* 7 (4): 607–614.

13. Kao, Y., and M. Chen. 2011. A hybrid PSO algorithm for the CVRP problem. In *Proceedings of the international conference on evolutionary computation theory and applications (ECTA '11)*, Paris, France, 539–543, Oct 2011.
14. Ai, T.J., and V. Kachitvichyanukul. 2009. Particle swarm optimization and two solution representations for solving the capacitated vehicle routing problem. *Computers and Industrial Engineering* 56 (1): 380–387.
15. Marinakis, Y., M. Marinaki, and G. Dounias. 2010. A hybrid particle swarm optimization algorithm for the vehicle routing problem. *Engineering Applications of Artificial Intelligence* 23 (4): 463–472.
16. Munari, Pedro, Twan Dollevoet, and Remy Spliet. 2017. A generalized formulation for vehicle routing problems.
17. Mirjalili, S., S.M. Mirjalili, and A. Lewis. 2014. Grey wolf optimizer. *Journal of Advances in Engineering Software* 69: 46–61.
18. Kennedy, J., and R. Eberhart. 1995. Particle swarm optimization. In *Proceedings IEEE international conference on neural networks (ICNN'95)*, Perth, WA, Australia, vol. 4, 1942–1948, Nov–Dec 1995.
19. Şenel, F.A., F. Gökçe, A.S. Yüksel, et al. 2019. A novel hybrid PSO–GWO algorithm for optimization problems. *Engineering with Computers* 35: 1359.

A Hybrid Salp Swarm Algorithm with β-Hill Climbing Algorithm for Text Documents Clustering

Ammar Kamal Abasi, Ahamad Tajudin Khader, Mohammed Azmi Al-Betar, Zaid Abdi Alkareem Alyasseri, Sharif Naser Makhadmeh, Mohamad Al-laham, and Syibrah Naim

Abstract Recently, the researchers' attention became more interest in partitioning particular sets of documents into various subsets, due to the massive number of documents that make pattern recognition, information retrieval, and text mining more complicated. This problem is known as a text clustering problem (TCD). Several metaheuristic optimization algorithms have been adapted to address TDC optimally. A new efficient metaheuristic optimization algorithm mimics the behavior of salps

---

A. K. Abasi (✉) · A. T. Khader · S. N. Makhadmeh
School of Computer Sciences, Universiti Sains Malaysia, George Town, Pulau Pinang, Malaysia
e-mail: ammar_abasi@student.usm.my

A. T. Khader
e-mail: tajudin@usm.my

S. N. Makhadmeh
e-mail: m_shareef_cs@yahoo.com

M. A. Al-Betar
Department of Information Technology - MSAI, College of Engineering and Information Technology, Ajman University, Ajman, United Arab Emirates
e-mail: mohbetar@bau.edu.jo

IT Department, Al-Huson University College, Al-Balqa Applied University, Irbid, Jordan

Z. A. A. Alyasseri
Faculty of Information Science and Technology, Center for Artificial Intelligence, Universiti Kebangsaan Malaysia, 43600 Bangi, Selangor, Malaysia
e-mail: zaid.alyasseri@uokufa.edu.iq

ECE Department, Faculty of Engineering, University of Kufa, Najaf, Iraq

M. Al-laham
Department of Management Information Systems, Amman University College, Al-Balqa Applied University (BAU), Amman, Jordan
e-mail: dr_laham@bau.edu.jo

S. Naim
Technology Department, Endicott College of International Studies (ECIS), Woosong University, Daejeon, South Korea
e-mail: syibrah@wsu.ac.kr

I. Aljarah et al. (eds.), *Evolutionary Data Clustering: Algorithms and Applications*, Algorithms for Intelligent Systems,
https://doi.org/10.1007/978-981-33-4191-3_6

in oceans known as the salp swarm algorithm (SSA) has been proposed and adapted to address different optimization problems. However, hybridizing optimization algorithms with another algorithm is becoming the focus of scholars to obtain a superior solution for the optimization problems. In this paper, a new hybrid optimization method of SSA algorithm and a well-known metaheuristic optimization algorithm called β-hill climbing algorithm (BHC), namely H-SSA, is proposed. The main aims of the proposed method to improve the quality of initial candidate solutions and enhance the SSA in terms of local search ability and convergence speed in attempting for optimal partitioning of the cluster. The proposed H-SSA performance is tested in the data cluster field using five standard datasets. In addition, the proposed method is tested using two scientific articles' datasets, and six standard text datasets in the text document clustering domain. The experiment results show that the proposed method boosted the solutions in terms of convergence rate, recall, precision, F-measure, accuracy, entropy, and purity criteria. For comparative evaluation, the proposed H-SSA compared with the pure SSA algorithm and well-known clustering techniques like DBSCAN, agglomerative, spectral, k-means++ k-means clustering techniques and the optimization algorithms like KHA, PSO, GA, HS, CMAES, COA, and MVO. The comparative results prove the efficiency of the proposed method, where it exhibited and yielded better performance than the compared algorithms and techniques.

Keywords Data clustering · Document clustering · Beta-hill climbing · Optimization · Evolutionary computation · Hill climbing · Swarm intelligence · Meta-heuristics

1 Introduction

Recently, one of the research topics became more interest due to its contributing grouping a particular set of text documents into subsets of clusters [34]. This topic is known as text document clustering (TDC). Several techniques have been structured to efficiently clusterize documents that have high intra-similarity by allocating the related documents at the same cluster [43]. This situation can be done according to several attributes that characterize the data [46].

TDC is considered the most prominent unsupervised technique due to its undiscovered area that increases the burden of finding the best cluster for documents [25]. The main aim of TDC algorithms is to cluster documents into related groups under the same topics. However, these topics are not titled in advanced. In this study, the procedures of clustering and partitioning the documents into specific subsets is the primary objective. These documents will be clustered based on an objective function.

Several algorithms have been built for TDC, where the K-means algorithm is the most prominent algorithm, due to its ability to deal with a massive number of data [4, 5, 34, 57]. However, the K-mean algorithm like other algorithms has disadvantages, where it could easy to fall in local optimal and not obtain the optimal cluster in some cases [62].

The metaheuristic optimization algorithms have been adapted instead of TDC algorithms, such as k-means, to efficiently address TDC, where this kind of optimization algorithms showed its robust performance in different fields [19]. The most prominent metaheuristic optimization algorithms are ray optimization algorithm (ROA) [47], genetic algorithm (GA) [44], grey wolf optimizer (GWO) [49–51], harmony search (HS) [13, 36], cuckoo search (CS) [68], binary artificial bee colony (ABC) [58], multi-verse optimizer (MVO) [4], fruit fly optimization algorithm (FFOA) [59], krill herd algorithm (KHA) [37], flower pollination algorithm [23], ant colony optimization (ACO) [66], particle swarm optimization (PSO) [28, 52], ant lion optimizer (ALO) [53], and dragonfly algorithm (DA) [54].

Salp swarm algorithm (SSA) is one of the recent and famous metaheuristic optimization algorithms due to its capability, flexibility, simplicity, and easy to be understood and utilized. In addition, SSA has a good ability to balancing exploration and exploitation through the searching mechanism [33]. SSA is one of the population-based metaheuristic algorithms that mimic the behavior of salps in nature that proposed by Mirjalili et al. [55]. In contrast, the SSA like other metaheuristic algorithms has some disadvantages that reduce its searching performance, such as stuck in local optima and slow convergence speed [38]. Therefore, an improvement in the SSA mechanism to speed up the convergence rate toward optimal solution without any stagnation in local optimal should be well prepared. This improvement can be done by enhancing the best solution at each iteration to find a better solution. Hybridizing a pure metaheuristic algorithm with another optimization algorithm is the most prominent technique to address algorithms drawbacks and improve their performance and results [19].

Normally, hybridizing a local search as a new operator in the SSA can improve the exploitation capability. The recent local search algorithm called β-hill climbing is proposed by [8]. It is utilized a stochastic operator with the neighborhood operator to enable the algorithm to escape the local optima. β-hill climbing optimizer is simple, adaptable,easy-to-use, and efficient. It has been adapted for a wide range of optimization problems such as ECG and EEG signal denoising [18, 20, 22], Generating Substitution-Boxes [24], gene selection [15], economic load dispatch problem [10, 12], mathematical optimization functions [6] multiple-reservoir scheduling [16], numerical optimization, [9], Sudoku game [11], classification problems [17], and cancer classification [14].

In this paper, a hybrid version of SSA and a local search optimization algorithm known as β-hill climbing (H-SSA) is proposed. The proposed method is prepared to enhance the SSA the best solution at each iteration and initial solutions. The proposed H-SSA performance is tested in the data cluster field using five standard datasets. In addition, the proposed method is tested using two scientific articles' datasets, and six standard text datasets in the text document clustering domain. The obtained results by the proposed method are compared several comparative techniques and algorithms, including optimization algorithms (e.g., PSO, GA, H-PSO, H-GA, MVO, KHA, HS) and clustering techniques (e.g., Spectral, Agglomerative, K-mean++, K-mean, DBSCAN).

The rest of the paper is divided into eight sections. The second section reviews previously conducted studies and related works. The third section presents and discusses the TDC problem in detail. The fourth section introduces the salp swarm algorithm's basic principles. The fifth section discusses the β-hill climbing principle, whereas the sixth section describes in detail the proposed method H-SSA. The seventh section provides the data set used, the experimental findings, and the proposed method's significance. The conclusions, as well as recommendations for further studies, are provided in the eighth section of the current paper.

2 Related Works

In the field of TDC, the researches use insistent and advanced techniques to address TDC. TDC can be addressed by partitioning documents into distinct clusters according to the similarity of the documents' content [65]. Several clustering methods were proposed to address the TDC problem, where the k-means method is one of the most popular methods. k-means method is only searching for local solutions in addressing the TDC problem. Therefore, it was hybridized with other algorithms to improve the poor local searching [45]. Also, β-hill climbing is one of the most prominent algorithms that only searching for local solutions. β-hill climbing is a recent single-based metaheuristic algorithm tailored for various optimisation problems [1, 8]. β-hill climbing has been hybridized with other algorithms to enhance the exploitation side of the searching process of particular problem. Using this algorithm, the results demonstrate its feasibility, significant improvement in performance while exploiting the search space [2, 15, 22].

As mentioned previously, several metaheuristic optimization algorithms have been adapted to address TDC problem. SSA is one of the recent swarm-based metaheuristic optimization algorithms due to its capability, flexibility, simplicity, and easy to be understood and utilized. SSA was adapted to address the choosing task for the best conductor in a radial distribution system [41]. The authors of [31] adapt SSA to tune a stabilizer in a power system. In the experiment results, SSA presented a high performance compared with other algorithms.

Lately, a few numbers of research have been hybridized the SSA with other algorithms to enhance the SSA searching mechanism and improve the solutions. The authors of [32] hybridize the SSA with the differential evolution algorithm to enhance the feature exploitation ability of the SSA due to the differential evolution algorithm performance in searching locally. SSA was hybridized with PSO in [40] to address the feature selection problem. The main purpose behind the hybridization is to improve the exploration and exploitation processes of SSA. The results prove the robust performance of the proposed hybrid version in addressing the problem.

3 Text Document Clustering Problem

TDC can be characterized as the problem of the NP-complete, which involves finding clusters in documents that are heterogeneous through the minimization of objective in the present paper(min $f(\boldsymbol{x})$ is equal to minimize the Euclidean distance). This section describes the formulation of the TDC problem, the Preprocessing of TDC, and the TDC algorithm, in addition to the similarity measures, as well as the objective function.

3.1 Problem Formulation

The TDC problem can be characterized by considering a specific set $(Docs)$ of the t documents at a high level. It aims to partition the documents into a specific number of predetermined clusters (k) subsets, whereby $(Docs)$ represents a documents' vector $(Docs = (Doc_1, Doc_2, \ldots, Doc_i, \ldots, Doc_d))$. Doc_i denotes the number of the document i, and Doc_d stands for the entire documents' number in $Docs$. Each cluster holds a specific cluster's centroid (K_{cent}), which represents a specific vector of terms' weights of length f $(k_{cnt} = (k_{cnt1}, k_{cnt2}, \ldots, k_{cntj}, \ldots, k_{cntf}))$, whereby k_{cnt} signifies a centroid of k_{th} cluster, k_{cnt1} stands for a position value 1 in the centroid of the cluster k, and k_{cntf} stands for the number of the entire special features of the centroid (terms) [39, 60].

To determine a partition $k_{cnt} = (k_{cnt1}, k_{cnt2}, \ldots, k_{cntj}, \ldots, k_{cntf}$ that satisfies certain conditions:

- $k_{cnt} \neq \varnothing$
- $k_{cnt} \bigcap k_{cnt'} = \varnothing if K \neq K'$
- $\bigcup_{k=1}^{K} k_{cnt} = 0$
- Objects that are in the same cluster are very similar to one another. However, objects that are in diverse clusters are different from each other.

3.2 Preprocessing of Text Document

Before the application of the clustering algorithm, standard preprocessing steps are implemented to preprocess text documents, including tokenization and stop word removal, as well as stemming, in addition to steps of term weighting. Text documents can be converted to a format that is numerical or matrix via the preprocessing steps. In the present work, the term frequency-inverse document frequency (TFIDF) can be generally used as a scheme of weighting.

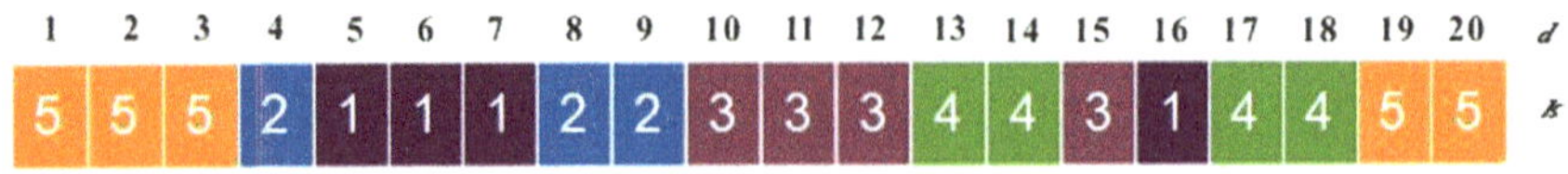

Fig. 1 Solution representation

3.3 Clustering Algorithm and Similarity Measures

The clustering algorithm implementation for documents, as well as the features of the document to produce clusters based on the similarity measures, establishes a substantial phase in the clustering of documents. The TDC algorithm represents a method of unsupervised learning. This method aims to find the best solution to partition a set of documents [36]. The TDC algorithm functions in accordance with certain evaluation criteria, including objective function, in addition to fitness [27]. Often, similarity measures can be represented in relation to distance or dissimilarity measures. The Euclidean distance measure [61] is a standard measure, and it is used as an objective function for various algorithms of text clustering so that the resemblance between the documents, as well as the cluster's centroid, can be computed, whereby the aim of the cost function is to minimize distance between the documents within each cluster.

3.3.1 Solution Representation

Every solution within the population suggests one candidate solution to the problem; each solution can be indicated as a specific vector $\boldsymbol{x} = (x_1, x_2, \ldots, x_d)$, whereby d signifies the number of documents, and each variable's value x_i can take the value of k decision $k \in \{1, \ldots, K\}$, whereby k signifies a random cluster's number, and K indicates the clusters' number as shown in Fig. 1. Each dimension can be treated as one document. As shown in Fig. 1, the solution X possesses dim (here dim = 20), twenty documents can be distributed between five clusters, e.g., document 5 (namely, x_5) can be in the cluster one; cluster five possesses four documents $\{1, 2, 3, 19, 20\}$.

3.3.2 Objective Function

The measure of the Euclidean distance has been used in this work. It is a specific standard measure and it has been widely used for similar purposes [27, 36]. This measure has been widely applied with the aim of computing resemblance between two documents (namely, the document, as well as the cluster's centroid) [61], e.g., when considering two documents $d_1 = (t_{11}, t_{12}, \ldots, t_{1n})$ and $d_2 = (t_{21}, t_{22}, \ldots, t_{2n})$, the Euclidean distance can be calculated according to:

$$D(d_1, d_2) = \left(\sum_{i=1}^{n} |t_{1,i} - t_{2,i}|^2 \right)^{1/2}, \tag{1}$$

where $t_{1,i}$ indicates the term's weight i in document one, and $t_{2,i}$ indicates the term's weight i in document two (namely, the feature's weight i in the cluster's centroid). The distance values range between (0, 1), whereby the 0 value denotes the most appropriate value; the 1 value denotes the worst value. Euclidean distance signifies a distinct case of the Minkowski distances [26] that can be expressed in Eq. (2), except for the p value that equals 2 in the Euclidean distance, which represents the key difference.

$$D(d_1, d_2) = \left(\sum_{i=1}^{n} |t_{1,i} - t_{2,i}|^p \right)^{1/p} \tag{2}$$

For the text document's solution $\boldsymbol{x}$ to be assessed, the resemblance between each cluster of the documents, as well as the cluster's centroid can be computed. In this work, the measure of distance between documents in a similar cluster can be utilized by means of computing the resemblance between each document, as well as each cluster's centroid for k_{cnt}, where $k_{cnt} = (k_{cnt1}, k_{cnt2}, \ldots, k_{cntj}, \ldots, k_{cntf})$ is a specific cluster centroid vector of the length f.

To distribute each of the documents in its appropriate cluster's centroid, each document cluster's centroid is recomputed using Eq. (3). That is, a documents' group can be placed in a similar cluster with a closer clustering centroid.

$$k_{cntj} = \frac{\sum_{i=1}^{d} (a_{ki}) d_{ij}}{\sum_{i=1}^{d} a_{ki}}, \tag{3}$$

whereby k_{cntj} represents the cluster's centroid j, d signifies the overall number of documents, whereas d_{ij} denotes to j_{th} feature weight for document i. a_{ki} signifies a binary matrix size $d \times k$ based on:

$$a_{ki} = \begin{cases} 1 & \text{document } i \text{ assigned to the cluster } j, \\ 0 & \text{otherwise} \end{cases} \tag{4}$$

Finally, each solution's objective function $\boldsymbol{x}$ of the introduced algorithm should be formulated. Thus, the distance of an average document to a cluster's centroid (ADDC) is used based on Eq. 5.

$$\min f(\boldsymbol{x}) = \frac{\sum_{i=1}^{k} \left(\frac{1}{d_i} \sum_{j=1}^{d_i} D(k_{cnti}, d_j) \right)}{k}, \tag{5}$$

where $f(\boldsymbol{x})$ signifies an objective function (namely, minimizing the distance), d_i indicates the number of documents in cluster i, k signifies the number of clusters, in addition to $D(k_{cnti}, d_j)$, which indicates distance between cluster's centroid j, as well as document i.

4 Salp Swarm Algorithm (SSA)

SSA is a swarm-intelligence algorithm which introduced by Mirjalili in [55]. It is inspired from swarming behavior of salps when navigating and foraging in oceans. The main procedure of the standard algorithm of SSA can be summarized by the following phases:

Phase 1: ***Parameters Initialization***. Both parameters of the SSA and the text document problem should be initialized within possible parameters range value s. In this case, the general formulation of the SSA initialization can be introduced as follows:

$$\max \text{ or } \min\{f(s) \mid s \in \mathbf{S}\},$$

where $f(s)$ is the objective function; $s = \{s_i \mid i = 1, \ldots, N\}$ is the set of decision variables. $\mathbf{s} = \{s_i \mid i = 1, \ldots, d\}$ is the possible value range for each decision variable, where $s_i \in [LB_i, UB_i]$, where LB_i and UB_i are the lower and upper bounds for the decision variable s_i, respectively, and N is the number of decision variables.
Also, for the rest SSA parameters must be initialized in this phase too, these parameters can summarize as follows:

- $Salp_size$: is a SSA population size.
- F^*_{best}: refers to the best current solution (best agent).

The next phases will provide a full explanation of these parameters.

Phase 2: ***Initialize SSA population memory***. The SSA can be represented as a two-dimensional matrix with size $Salp_size \times N$ which have sets of slap location vectors as many as $Salp_size$ (see Eq. (6)). Where these slaps are randomly generated as follows: $s_i^j = LB_i + (UB_i - LB_i) \times U(0, 1)$, $\forall i = 1, 2, \ldots, z$ and $\forall j = 1, 2, \ldots, Salp_size$, and $U(0, 1)$ generates a uniform random number between 0 and 1. The generated solutions are stored in the SSAM in ascending order according to their objective function values where $f(s^1) \leq f(s^2) \leq \ldots \leq f(s^{SSA_s})$.

$$\text{SSAM} = \begin{bmatrix} s_1^1 & s_2^1 & \cdots & s_N^1 \\ s_1^2 & s_2^2 & \cdots & s_N^2 \\ \vdots & \vdots & \cdots & \vdots \\ s_1^{SSA_s} & s_2^{SSA_s} & \cdots & s_N^{SSA_s} \end{bmatrix}. \tag{6}$$

Also in this step, the global best slap location S_{best} is memorized where $S_{best} = s^1$.

Phase 3: ***Intensification of the current slap population***. To model the SSA chains, the population divided into two sets which are leader and followers. The leader is the slap at the front of the chain, whereas the rest of the slaps are considered as slaps. Similarly to other swarm-based techniques, the position of slaps is defined in an n-dimensional search space where n is the number of variables of a given problem. Therefore, the position of all slaps are stored in a two-dimensional matrix called s. It is also assumed that there is a food source called F in the search space as the swarm's target. To update the position of the leader the following equation is proposed:

$$s_i^j = \begin{cases} F(j) + c_1 \times ((ub_j - lb_j) \times c_2 + lb_j)Bigr\rfloor & c_3 \geq 0, \\ F(j) - c_1 \times ((ub_j - lb_j) \times c_2 + lb_j)Bigr\rfloor & c_3 < 0 \end{cases} \quad (7)$$

where s_i^j shows the position of the first slap (leader) in the jth dimension, F_j is the position of the food source in the jth dimension, ub_j indicates the upper bound of jth dimension, lb_j indicates the lower bound of jth dimension, c_1, c_2, and c_3 are random numbers.

Phase 4: ***Updating the position of the leading slap (best solution (S_{best}))***. During for each iteration in SSA procedure, the global best slap location F_{best} will be updated if $f(s'^j) < f(S_{best})$.

Phase 5: ***Stop condition***. SSA repeats phase 3 and phase 4 till the stop criterion is met. The stop criterion is normally met depends on two criteria, such as the quality of the final outcomes or the number of iterations.

5 β-Hill Climbing Algorithm

The idea of β-hill climbing algorithm extended from a simple trajectory-based algorithm called hill climbing algorithm, where the idea of the hill climbing algorithm for finding the optimal solution for any problem. Firstly it starts to generate a random solution then step by step try to improve this solution by finding the better parameters from the search space. This trajectory approach will be repeated until reaches the optimal solution. The main problem with a trajectory-based algorithm such as hill climbing it is always looking for uphill movements to finding the optimal solutions. For that, it leads to getting easily stuck in the local optima [8]. For solving the stuck in the local optima problem and accept the downhill movements Al-Betar in 2017 introduced a new extension of the hill climbing algorithm which called β-hill climbing algorithm [8]. The main idea behind the β-hill climbing algorithm is using a simple operator to make a balance between both exploration and exploitation during the search.

As aforementioned, the β-hill climbing algorithm is a trajectory search algorithm which starts with single random solution, $\boldsymbol{S} = (s_1, s_2, \ldots, s_N)$. After number of iteration in the searching space, a new solution, $\mathbf{s}' = (s'_1, s'_2, \ldots, s'_N)$, will be generated by improving the current solution $\boldsymbol{s}$. The improvement will happen based on two operators, namely, $\mathcal{N}$-operator and β-operator, where these operators represent the sources for exploitation and exploration, respectively. Specifically, the $\mathcal{N}$-operator works as neighborhood search, while β-operator works as similar to mutation operator. At each iteration, the new solution can be improved by $\mathcal{N}$-operator stage or β-operator stage until the optimal solution is reached.

The algorithm begins to generate the solution randomly, then the solution is evaluated using the objective function $f(\boldsymbol{s})$. The solution is then modified using $\mathcal{N}$-operator, which employs the $improve(\mathcal{N}(\boldsymbol{s}))$ function within a random range of its neighbors. The solution $\boldsymbol{s}$ is as follows:

$$s'_i = s_i \pm U(0, 1) \times bw \qquad \exists i \in [1, N]$$

where i is randomly selected from the space range, $i \in [1, 2, \ldots, N]$. The parameter bw representees the bandwidth between the current value and the new value.

In β-operator, within the β range where $\beta \in [0, 1]$, variables of new solution will be assigned based on selected randomly from available range or from the existing values of the current solution as follows:

$$s'_i \leftarrow \begin{cases} s_r & rnd \leq \beta hc \\ s_i & otherwise, \end{cases}$$

where rnd generates a uniform random number between 0 and 1 and $s_r \in \boldsymbol{S}_i$ is the possible range for the decision variable s'_i.

Finally, the β-hill climbing has successfully achieved optimal results in many global problems such as EEG and ECG signal denoising, EEG channel selection, feature selection, and sudoku problem [7, 11, 18, 20–22].

6 The Proposed Hybrid Salp Swarm Algorithm for TDC

This section provides the introduced method (H-SSA) in detail. The introduced SSA algorithm's hybrid strategy consists of two main stages. They include improving the quality of the initial candidate solutions in stage one and improving the best solution in stage two that SSA provides at each iteration. Figure 2 displays the steps of the introduced hybrid strategy and every single phase is described as follows.

6.1 Improving the Initial Candidate Solutions Quality Using β-Hill Climbing

The SSA algorithm is characterized by being too sensitive towards early condition solutions. Therefore, the starting point of the algorithm may have a major impact on the efficiency of the ultimate optimum solution's [64].

Two or even more algorithms have been hybridized in recently conducted studies to obtain the optimal solutions for tackling the problems of optimization [42, 63]. The main hybridization drivers avoid local optimization and enhance initial solutions. This results in enhancing global (diversification) searchability and it can be more essential in problems of multi-dimension. In accordance with this perspective, the introduced hybrid algorithm can combine the local search algorithm's main features to improve the initial solution. Thus, the result is directed as an SSA initial solution.

Following the step of preprocessing, the H-SSA can start initializing a few parameters; it can randomly generate solutions. Every single solution can be associated with one salp. In other words, the number of solutions, as well as the number of salps is equal. It is important to mention that the original SSA's initialization phase (Sect. 4 can be adopted to cluster with specific modifications, which are associated with the problem's variables nature, considering that the problem of clustering is discrete; the SSA algorithm was initially utilized to tackle continuous problems of structural optimization [56]. The SSA has been introduced with the aim of dealing with the decision variables' discrete values of every single TDC solution via utilizing a rounding function so that continuous values are converted to discrete values. These changes can be applied as follows:

1. Generation of initial solution function Eq. (6) is modified to:

$$x_i^j = \lfloor rand()\%(ub_j - lb_j))Bigr\rfloor \quad \forall i \in (1, 2, \ldots, n) \wedge \forall j \in (1, 2, \ldots, d), \tag{8}$$

   whereby n signifies the candidate number of solutions and d signifies the number of documents, $lb_j = 1$, ub_j signifies the number of clusters, and $rand()$ signifies a specific function, through which an arbitrary distribution number is generated within the range [0.0, 1.0]. $Bigr$ is a function that is approximately the value of x_i^j to the closest integer; it is less than or it is equal to x_i^j,

2. The updates' equations of the leader's position and the follower's position is changed in Eq. 9 to

$$x_i^j = \begin{cases} \lfloor F(j) + c1 \times ((ub_j - lb_j) \times c2 + lb_j)Bigr\rfloor & c3 < 0.5, \\ \lfloor F(j) - c1 \times ((ub_j - lb_j) \times c2 + lb_j)Bigr\rfloor & c3 \geq 0.5, \end{cases} \tag{9}$$

   whereby x_i^j demonstrates the first salp position (leader) in jth dimension and F_j demonstrates the food source position in jth dimension, as well as ub_j, which demonstrates the jth dimension's upper bound, whereas lb_j demonstrates the jth dimension's lower bound, in addition to c_1, c_2 , and with c_3, which represent arbitrary numbers.

$$x_i^j = \lfloor \frac{1}{2}(x_i^j + x_i^{j-1}) \rfloor, \tag{10}$$

whereby $i \geq 2$ and x_i^j demonstrates the ith follower salp position in the dimension of jth.

A solutions' set can be randomly produced in the phase of initialization utilizing Eq. 8. In this phase, the BHC algorithm can consider solutions in the form of an input to enhance each solution and recalculate the solution's quality using Eq. 5. In case the generated quality of the solution can outperform the quality of the old solution, it can memorize the solution that is generated; otherwise, it will continue with an old solution. This process can be repeated considering the entire solutions in a population (see, Algorithm 1).

Algorithm 1 Enhancing the quality of initial candidate solutions algorithm

Input: Population matrix U of size n solutions
Output: Improving Population matrix U of size n solutions by β-hill climbing algorithm
for *Each solution* x_i *in Population matrix U* **do**
  Step1: calculate the solution x_i quality using Eq. 5
  Step2: make the solution x_i instead of the first step in β-hill climbing algorithm.
  Step3: run β-hill climbing algorithm
  Step4: calculate the solution x_i quality using Eq. 5

  if *The solution* x_i *improved* **then**
    Step5a: replace the solution x_i with the old solution.
  end
  else
    Step5b: the solution x_i remains unchanged.
  end
end

6.2 *Improving the Best Solution of Salp Swarm Algorithm with β-Hill Climbing*

Similar to the remaining metaheuristic algorithms, SSA can be designed to accomplish two phases, including exploration and exploitation phases. In these phases, the algorithm must be equipped with specific mechanisms so that an extensive search can be carried out for a search space. The promising regions of the search space can be identified in the exploration phase. On the other hand, the local search is highlighted in the exploitation phase, as well as convergence to reach in an exploration phase promising areas. The key adequate solution in both phases is the leading solution. In the exploration phase, it is highly possible that the solutions learn from the lead-

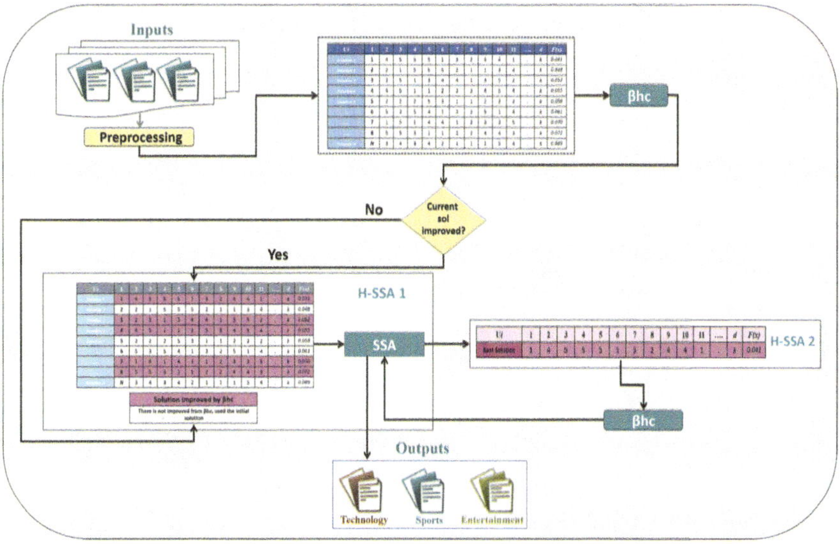

Fig. 2 The proposed methodology

ing solution. Furthermore, In the exploitation phase, it focuses only on the leading solution. A liberal bias for improving the leading solution may share good attributes along the searching path for the global optima, as well as lead the search so that better solutions are found. At every single iteration, the leading solution, which SSA produces, is the β-hill climbing initial state. The stage intensifies the probability of the existence of useful attributes in the optimal solution. Therefore, the optimal solution can send these to the existing solutions and assist them in improving the fitness value. The enhancement process of the optimal solution is provided in detail in Algorithm 2. The optimization framework of the proposed method is demonstrated in Fig. 2, as well as pseudocode in Algorithm 3.

Algorithm 2 Enhance the best solution of SSA

Input: The best solution $best_{SI}$
Output: Improving solution $best_{SI}$ by β-hill climbing algorithm
Step1: make the solution $best_{SI}$ instead of the first step in β-hill climbing algorithm.
Step2: run β-hill climbing algorithm
Step3: calculate the solution $best_{SI}$ quality using Eq. 5
if *The solution* $best_{SI}$ *improved* **then**
| Step4a: replace the solution $best_{SI}$ with the old solution.
end
else
| Step4b: the solution $best_{SI}$ remains unchanged.
end

Algorithm 3 Pseudocodes and general steps of SSA algorithm

```
Create SSA population considering ub and lb.
run Algorithm 1
while the end criterion is not satisfied do
    Evaluate the objective function for all solutions.
    Update the Best solution vector.
    run Algorithm 2
    Update c1
    for each solutions indexed by i do
        if i == 1 then
            Update the position of the leading salpby Eq. 9
        else
            Update the position of the leading salpby Eq. 10
        end if
    end for
    Amend the solutions based on the upper and lower bounds of variables.
end while
Produce the best Solution.
```

7 Results and Discussion

This section describes the experimental architecture, experimental datasets, state-of-the-art algorithms, evaluation measures of the system, and parameter descriptions. The convergence rate is also given in this study, and the findings are addressed and analyzed. The datasets are described in the next section.

7.1 The Experimental Design

This section discusses the proposed method and involves a comparison of the introduced H-SSA method with the remaining techniques of the state of the art. To provide better insights into this work, the performance of the introduced algorithm is evaluated utilizing three artificial datasets types, involving scientific articles datasets (SAD), data clustering datasets (DCD), in addition to text clustering datasets (TCD). To achieve the results' uniformity, the introduced algorithm has been conducted 30 times along with all the datasets, utilizing similar initial solutions for the meta-heuristic algorithms. This specific number has been selected based on the literature. Therefore, the introduced method is appropriately validated. Also, an almost reasonable comparison that involved the entire competing algorithms has been conducted. The algorithms that are local-based for the clustering technique can run 100 iterations each run time, and 100 iterations are experimentally appropriate for convergence of the intensification search algorithm. Also, 1000 iterations can be appropriate for convergence of the diversification search algorithm of the population-based algorithms. Seven measures of an external evaluation have been implemented as has always been conventionally done, including accuracy and recall, as well as precision and F-measure, in addition to purity and entropy criteria, along with the intra-cluster distances' sum, which is a measure of internal quality. For conducting a comparative assessment, the obtained findings by utilizing these evaluation measures were compared as opposed to the obtained findings by utilizing ten state-of-the-art algorithms, which include the original SSA K-mean and K-mean++, as well as DBSCAN and Agglomerative, in addition to Spectral, KHA, HS, PSO, GA, CMAES, COA, along with MVO by using a similar objective function.

7.2 Experimental Benchmark Datasets and Parameter Setting

To validate the efficacy of the introduced H-SSA algorithm, three artificial datasets types have been selected to test the clustering problem. Five DCD and two SAD, as well as six TCD based on the literature, have been chosen as testbeds for the conducted comparison. The datasets are benchmark datasets in the field of clustering. These were commonly applied to examine the recently developed algorithms' performance including: CMC and Iris, as well as Seeds and Glass, in addition to Wine. TCD is obtainable at the computational intelligence laboratory (LABIC) as a numerical form after the term extraction. The characteristics of the TCD are described as in the following:

CSTR. The Centre for Speech Technology Research (CSTR) is a research center that is interdisciplinary. In CSTR, informatics and linguistics as well as the English language are linked. In 1984, this center was established to foster research in various areas, including information access, as well as speech recognition. The selected dataset comprises 299 documents. These documents are included in four classes, including theory and artificial intelligence, as well as robotics, in addition to systems.

20Newsgroups. It comprises 19,997 articles. They are included in 20 classes and collected from various Usenet newsgroups. Thus, the initial 100 documents for the experiment have been chosen from the superior three classes within the dataset as follows: comp_windows_x, talk_politics_misc, and rec_autos classes.

Tr12 , Tr41, and Wap (Datasets from Karypis Lab). These have been selected from different sources to ensure diversity. Tr12 and Tr41, as well as Wap, comprise 313, 878, and 1560 documents. They are included in 8 and 10, as well as 20 classes, respectively. The details can be accessible in [69].

Classic4. It contains 2000 documents along with four classes: CACM and CRAN, as well as CISI, in addition to MED (i.e., 500 documents in each class). This dataset originally comprises 7,095 documents. They are included in any of four previous categories, whereas Classic3, as well as Classic4, represent versions of a similar dataset.

SAD includes scientific articles that have been published in internationally recognized conferences such as (NIPS 2015, as well as AAAI 2013). The SAD's features are discussed herein:

NIPS 2015. Kaggle site has provided this dataset, which comprises 403 articles. These articles were published in the conference of Neural Information Processing Systems (NIPS). In the machine learning domain, this conference is a fundamentally ranked conference. The topics involved deep learning and computer vision, as well as cognitive science, in addition to reinforcement learning. The NIPS 2015 dataset contains the paper ID along with the paper title, as well as the type of event (poster and oral or spotlight presentation). It also contains the pdf file name, the abstract, and the paper text. However, the title of the paper, the abstract, and the text are utilized in the experimentation of this paper. Most articles are associated with machine learning, as well as natural language processing.

AAAI 2013. The UCI has repository has provided this dataset. There are 150 articles in this dataset. These articles have been accepted by another conference, which is a fundamentally ranked conference. It is the AI domain (known as AAAI 2013). Each of the papers contains information that includes the title of the paper, topics (i.e., keywords that are author-selected low-level from the provided list of the conference), keywords (i.e., keywords that are author-generated), the paper abstract, as well as keywords of high-level (i.e., keywords that are author-selected high-level from the provided list of the conference). Many articles revolved around artificial intelligence like multi-agent systems and reasoning, as well as machine learning like data mining, in addition to knowledge discovery, and so on.

Table 1 shows related information given in each dataset. This information involves the number of the dataset, the dataset's name, objects or the number of documents, the clusters' number, in addition to the number of features. Thus, it is important that the introduced H-SSA's parameter values are identified, in addition to additional comparative algorithms. Table 2 shows a comparison between all the algorithmic parameters of algorithms.

7.3 Evaluation Measures

Considering the methods of text clustering,based, seven measures of the external evaluation were normally utilized. These measures included accuracy and recall, as well as precision, in addition to F-measure, purity, along with entropy [29]. The utilized measures have been calculated after the findings have been achieved. The calculation of these measures is provided in detail herein.

Table 1 Datasets description

Type	ID	Datasets	Number of objects/Documents	Number of clusters (K)	Number of features (t)
DCD					
	DS1	CMC	1473	3	9
	DS2	Iris	150	3	4
	DS3	Seeds	210	3	7
	DS4	Glass	214	7	9
	DS5	Wine	178	3	13
TCD					
	DS6	CSTR	299	4	1725
	DS7	20Newsgroups	300	3	2275
	DS8	tr12	313	8	5329
	DS9	tr41	878	10	6743
	DS10	Wap	1560	20	7512
	DS11	Classic4	2000	4	6500
SAD					
	DS12	NIPS 2015	403	2	22888
	DS13	AAAI 2013	150	4	1897

Table 2 Parametric values for all algorithms being compared

Parameters	Algorithm	Value
Population size	All Optimization algorithms	20
Maximum number of iteration	All Optimization algorithms	1000
runs	All Optimization algorithms	30
WEP Max	proposed method (H-MVO)	1.00
WEP Min	proposed method (H-MVO)	0.20
p	proposed method (H-MVO)	6.00
Crossover probability	GA	0.80
Mutation probability	GA	0.02
Maximum inertia weight	PSO	0.90
Minimum inertia weight	PSO	0.20
C1	PSO	2.00
C2	PSO	2.00
Vf	KHA	0.02
$Dmax$	KHA	0.002
$NMax$	KHA	0.05
$PARMin$	HS	0.45
$PARMax$	HS	0.90
$bwMin$	HS	0.10
$bwMax$	HS	1.00
HMCR	HS	0.90

7.3.1 Accuracy

This measure has been used to compute objects or the percentage of text documents assigned to correct clusters [30, 48] and calculation can be done utilizing Eq. (11).

$$Ac = \frac{1}{n}\sum_{j=1}^{k} n_{ij}, \tag{11}$$

whereby n refers to objects or the number of text documents, k refers to the number of clusters, whereas n_{ij} represents the correct number of i text documents in the j cluster.

7.3.2 Precision

The assignment ratio of objects or the text documents is assigned by using this measure of the best clusters for the total number of cluster documents [35]. By using Eq. (12), the precision of the class i in the cluster j can be calculated.

$$P(i, j) = \frac{n_{i,j}}{n_j}, \tag{12}$$

whereby n_j signifies the number of the total documents in the cluster j, and ni, j represents the number of the correctly assigned objects or text documents of the class i in the cluster j.

7.3.3 Recall

Recall provides the correctly assigned objects' ratio or text documents' ratio to the overall number of related documents (namely, the overall number of documents in the specific class). By formulating Eq. (13), the recall measure can be calculated.

$$R(i, j) = \frac{n_{i,j}}{n_i}, \tag{13}$$

whereby ni, j signifies the number of the correctly assigned objects or text documents of the class i in the cluster j, and n_i signifies the total number of documents in the class i.

7.3.4 F-Measure

This measure is external, which represents a harmonic combination between precision, as well as recall. The closer the F-measure value to 1, the stronger the clustering solution will be [3]. For computing the F-measure, Eq. (14) is used.

$$F(i, j) = \frac{2 \times P(i, j) \times R(i, j)}{P(i, j) + R(i, j)}, \tag{14}$$

whereby $P(i, j)$ signifies the class i precision in the cluster j, and $R(i, j)$ signifies the class i recall in the cluster j. Equation (15) shows the F-measure calculation of the entire clusters.

$$F = \sum_{i=1}^{k} \frac{n_j}{n} \max F(i, j) \tag{15}$$

7.3.5 Purity Measure

This measure computes the percentage of each cluster in a large class via assigning each cluster to the most recurrent class [67]. Accordingly, the best-obtained value of purity is close to 1, given that a large-size rate of class in each cluster is computed for agreement with a specific estimated cluster's size. By using Eq. (16), the purity measure of the entire clusters can be calculated.

$$purity = \frac{1}{n} \sum_{i=1}^{k} \max(i, j), \tag{16}$$

whereby $\max(i, j)$ refers to a large class i size in the cluster j, k signifies the number of clusters, as well as n, which refers to the entire number of documents in this dataset.

7.3.6 Entropy Measure

The objects or text documents' distribution for a single class to clusters is analyzed by this measure. For every single class, this measure assesses the distribution of documents to the correct clusters. To calculate the cluster's entropy, a couple of steps are followed: (i) the distribution of documents for clusters for every single class can be calculated, (ii) all the entropies, which are obtained by the first step, are used to calculate the clusters' entropy. The cluster j's entropy can be computed by utilizing Eq. (17).

$$E(kj) = -\sum_{i=1} p(i, j) log(p_i, j), \tag{17}$$

whereby $E(kj)$ refers to the cluster j's entropy, as well as $p(i, j)$, which refers to the objects or text documents' probability in the cluster i that belongs to the class j. Equation (18) is used to calculate entropy for the entire clusters.

$$Entropy = -\sum_{i=1}^{K} \frac{n_i}{n} E(kj), \tag{18}$$

whereby K refers to the number of clusters, n_i refers to the number of documents in the cluster i, whereas n refers to the total number of documents in datasets.

7.4 Analysis of Results

To measure the obtained clusters via the introduced H-SSA algorithm, as well as other algorithms, six measures of the evaluation were utilized. The obtained experimental results showed that the introduced H-SSA2 algorithm has effectively worked to tackle the TDC problem. The evaluation has been performed based on the performance of the algorithm, being compared with current comparative algorithms. Thus, the results of Five DCD and six TCD, as well as two SAD, were achieved according to accuracy and recall, as well as precision, in addition to F-measure and purity, along with entropy as shown in Table 3, Table 4, and Table 5, respectively. The optimal results were highlighted with bold font. These results were reached by the introduced H-SSA2 in all datasets according to the measure of accuracy, that is, (DS1 and DS2), then by the H-SSA1 in DCD, as well as TCD, along with H-PSO in SAD. In accordance with the measure of evaluation, which represents one of the most commonly utilized standard measures in the text clustering domain, the improvement of the initial solutions' starting point and the optimal solution at every single iteration were found to be effective in comparison with the original SSA algorithm and the remaining comparative algorithms.

7.5 Convergence Analysis

The convergence behavior shows the effectiveness of various versions of H-SSA (namely, H-SSA1, as well as H-SSA2) compared with current state-of-the-art methods. The convergence rate of the clustering algorithm indicates an evaluation criterion, which aims to find the best solution. Figures 3 and 4, as well as Fig. 5 demonstrate the behaviors of convergence of HS, GA, PSO, KHA, SSA, H-PSO, H-GA, H-SSA1, as well as H-SSA2 on specific datasets; 30 runs were conducted for each dataset. An average value has been computed depending on the convergence behavior of each algorithm. Also, the documents' average distance to the values of the cluster's centroid (ADDC) was plotted in comparison with 1000 iterations on 13 datasets.

Table 3 Results of entropy, purity, F-measure, recall, precision, and accuracy for five data clustering datasets

Dataset	Measure	Clustering techniques					Optimization algorithms									
		Spectral	K-mean	K-mean++	Agglomerative	DBSCAN	BHC	MVO	KHA	PSO	H-PSO	HS	GA	H-GA	H-SSA1	H-SS2
DS1	Accuracy	0.5956	0.6412	0.5937	0.6464	0.6167	0.6434	0.6500	0.6244	0.6190	0.6774	0.5904	0.5990	0.6064	0.6695	**0.6854**
	Precision	0.6226	0.7281	0.6596	0.6934	0.6266	0.7296	0.6929	0.6984	0.6709	0.6514	0.6193	0.6599	0.6673	0.7210	**0.7309**
	Recall	0.5797	0.6593	0.6178	0.6855	0.5848	0.6747	0.6750	**0.6905**	0.6381	0.6315	0.5865	0.6411	0.6025	0.6531	0.6769
	F-measure	0.5909	0.6789	0.6213	0.6739	0.5893	**0.6876**	0.6678	0.6847	0.6440	0.6687	0.5864	0.6314	0.6674	0.6499	0.6620
	Purity	0.6054	0.7004	0.6454	0.7017	0.6124	**0.7158**	0.6917	0.7049	0.6631	0.6529	0.6138	0.6631	0.6228	0.6891	0.7087
	Entropy	0.5029	0.5087	0.5040	0.5170	0.4723	0.4998	0.4553	0.5119	0.4237	0.4549	0.5276	0.5094	0.4216	0.4162	**0.4100**
	Rank	14	5	12	7	13	3	4	6	9	8	15	11	10	2	1
DS2	Accuracy	0.6768	0.6738	0.6677	0.7242	0.6866	0.6708	0.7204	0.6993	0.6722	0.7203	0.6908	0.7190	0.7278	0.7221	**0.7314**
	Precision	0.7353	0.6940	0.6919	0.7703	0.7055	0.6949	0.7624	0.7747	0.7343	0.7417	0.7481	**0.8056**	0.7741	0.7876	0.8013
	Recall	0.7140	0.6501	0.6639	0.7558	0.6584	0.6699	0.7456	0.7676	0.6921	0.7476	0.7154	0.7372	0.7624	0.7630	**0.7797**
	F-measure	0.7094	0.6613	0.6562	0.7476	0.6659	0.6749	0.7447	0.7549	0.6946	0.7629	0.7215	0.7530	**0.7965**	0.7776	0.7513
	Purity	0.7410	0.6813	0.6835	0.7768	0.6854	0.6915	0.7681	**0.7772**	0.7153	0.7442	0.7395	0.7723	0.7065	0.7595	0.7668
	Entropy	0.5284	0.4188	0.4163	0.5573	0.3918	0.4144	0.4916	0.5467	0.4628	0.5337	0.5280	0.4778	0.4630	0.4430	**0.4570**
	Rank	10	15	14	7	12	13	5	6	11	8	9	3	4	2	1
DS3	Accuracy	0.6975	0.7247	0.7020	0.7779	0.6026	0.7038	0.7577	0.7328	0.7231	0.7858	0.6298	0.7021	0.7588	0.7722	**0.7916**
	Precision	0.6786	0.7290	0.7604	0.8146	0.6973	0.6851	0.7921	0.7976	0.7721	0.7996	0.7196	0.7658	0.7816	0.8038	**0.8249**
	Recall	0.6860	0.6922	0.7461	**0.8267**	0.6927	0.6934	0.7783	0.7958	0.7420	0.7518	0.6876	0.7180	0.7516	0.7861	0.8069
	F-measure	0.6557	0.6884	0.7373	0.8145	0.6738	0.6692	0.7719	0.7842	0.7501	0.7722	0.6881	0.7215	0.7713	0.6303	**0.8179**
	Purity	0.7138	0.7152	0.7682	0.8339	0.7013	0.7287	0.7916	0.8101	0.7697	0.8001	0.7171	0.7521	0.7081	0.7749	**0.8374**
	Entropy	0.5053	0.4680	0.4818	0.4119	0.4474	0.4638	0.5326	0.4757	0.4148	0.3697	0.4638	0.4328	0.4028	0.3853	**0.3779**
	Rank	14	11	9	2	15	12	7	4	8	3	13	10	6	5	1

(continued)

Table 3 (continued)

Dataset	Measure	Clustering techniques					Optimization algorithms									
		Spectral	K-mean	K-mean++	Agglomerative	DBSCAN	BHC	MVO	KHA	PSO	H-PSO	HS	GA	H-GA	H-SSA1	H-SS2
DS4	Accuracy	0.4966	0.5191	0.5424	0.5826	0.5499	**0.5937**	0.5345	0.5230	0.5066	0.5490	0.5056	0.4964	0.5446	0.5699	0.5876
	Precision	0.5620	0.5366	0.6310	0.6108	0.6028	0.6293	0.6024	0.5812	0.5689	0.6032	0.5296	0.5220	**0.6716**	0.6463	0.6600
	Recall	0.5231	0.4901	0.5652	0.6341	0.5938	0.6442	0.5922	0.5438	0.5432	0.6228	0.4820	0.4970	0.6340	0.6300	**0.6752**
	F-measure	0.5302	0.4912	0.5859	0.6149	0.5827	**0.6196**	0.5865	0.5472	0.5392	0.5902	0.4967	0.4914	0.5337	0.5929	0.5971
	Purity	0.5554	0.5179	0.6072	0.6392	0.6105	0.6524	0.6090	0.5664	0.5659	0.6214	0.5122	0.5198	0.6562	0.6101	**0.6603**
	Entropy	0.4114	0.4189	0.5202	0.4592	0.5036	0.5095	0.5661	0.4704	0.4389	0.4364	0.3782	0.4148	0.4262	0.4255	**0.4100**
	Rank	12	14	8	4	7	2	9	10	11	6	13	15	5	3	1
DS5	Accuracy	0.4496	0.4755	0.5016	0.4877	0.5107	0.5204	0.5345	0.4782	0.4638	0.5672	0.4645	0.4481	0.5055	0.5287	**0.5770**
	Precision	0.5169	0.4899	0.5912	0.5633	0.5570	0.5597	0.5665	0.5415	0.5245	0.5415	0.4829	0.4799	0.5629	0.5777	**0.6127**
	Recall	0.4744	0.4463	0.5194	0.5504	0.5452	0.5455	**0.5897**	0.5014	0.4957	0.5734	0.4355	0.4550	0.5125	0.5479	0.5749
	F-measure	0.4823	0.4512	0.5445	0.5465	0.5388	0.5510	0.5662	0.5015	0.4925	0.5545	0.4573	0.4444	0.5393	0.5673	**0.5917**
	Purity	0.5122	0.4786	0.5667	0.5644	0.5681	0.5682	0.5998	0.5267	0.5230	0.5667	0.4719	0.4752	0.5679	0.5638	**0.6161**
	Entropy	0.5222	0.5186	0.6067	0.5664	0.6071	0.5978	0.6188	0.5877	0.5970	0.5607	0.5109	0.5032	0.5269	0.5087	**0.4943**
	Rank	11	13	8	7	9	6	4	10	12	3	14	15	5	2	1
Average		12.2	11.6	10.2	5.4	11.2	7.2	5.8	7.2	10.2	5.6	12.8	10.8	6.0	2.8	1.0
Final Rank		14	13	9	3	12	7	5	7	9	4	15	11	6	2	1

Note The lowest ranked algorithm is the best one

Table 4 Results of entropy, purity, F-measure, recall, precision, and accuracy for six text clustering datasets

Dataset	Measure	Clustering techniques					Optimization algorithms									
		Spectral	K-mean	K-mean++	Agglomerative	DBSCAN	BHC	MVO	KHA	PSO	H-PSO	HS	GA	H-GA	H-SSA1	H-SS2
DS6	Accuracy	0.4319	0.3573	0.4355	0.4360	0.4005	0.4365	0.4593	0.3649	0.4356	0.5494	0.4464	0.3399	0.5056	0.5648	**0.5717**
	Precision	0.3597	0.4091	0.3953	0.4423	0.3393	0.3766	0.5715	0.4213	0.5340	0.6065	0.4235	0.4417	0.6130	**0.6385**	0.6317
	Recall	0.4925	0.3092	0.4076	0.4666	0.4256	0.5063	0.4829	0.5355	0.4360	**0.5600**	0.5060	0.3418	0.4950	0.5026	0.5439
	F-measure	0.3971	0.3460	0.3546	0.3266	0.3046	0.4133	0.5244	0.4139	0.4819	0.5577	0.3377	0.3886	0.5859	0.5668	**0.5864**
	Purity	0.4485	0.3525	0.4096	0.4816	0.4076	0.4524	0.5685	0.3874	0.4953	**0.6135**	0.4355	0.4050	0.5743	0.5012	0.5827
	Entropy	0.4893	0.8201	0.5246	0.5076	0.4586	0.8018	0.5207	0.4344	0.6199	0.5076	0.4786	0.7170	0.5104	0.4967	**0.5016**
	Rank	10.0000	15.0000	11.0000	9.0000	12.0000	13.0000	5.0000	7.0000	6.0000	2.0000	8.0000	14.0000	4.0000	3.0000	1.0000
DS7	Accuracy	0.3633	0.3180	0.3784	0.4055	0.3038	0.3695	0.4044	0.3216	0.3498	0.4442	0.3122	0.3676	0.4308	0.5070	**0.5185**
	Precision	0.3424	0.3121	0.3652	0.3990	0.3094	0.3526	0.4392	0.3829	0.4134	0.4821	0.3601	0.4209	0.4814	0.4915	**0.5456**
	Recall	0.3280	0.3100	0.3662	0.3576	0.3017	0.3340	0.3842	0.3136	0.3497	0.4220	0.3170	0.3676	**0.4887**	0.4292	0.4753
	F-measure	0.3136	0.3406	0.3619	0.3548	0.3193	0.3164	0.4109	0.2996	0.3803	0.4454	0.3214	0.3936	0.4433	0.5127	**0.5180**
	Purity	0.3110	0.3741	0.4134	0.4417	0.3027	0.3281	0.4344	0.3421	0.4097	0.4725	0.3355	0.4081	0.4837	**0.5077**	0.4426
	Entropy	0.6125	0.8028	0.6611	0.5953	0.7473	0.7940	0.7121	0.6767	0.7723	0.6631	0.6481	0.7547	0.6833	0.6456	**0.6475**
	Rank	*10*	*14*	7	5	*15*	*13*	6	*12*	*9*	4	*11*	8	3	2	1
DS8	Accuracy	0.3373	0.2971	0.3795	0.4481	0.3045	0.3881	0.4485	0.3357	0.4075	0.4796	0.3776	0.3677	0.5205	0.5267	**0.5448**
	Precision	0.4508	0.3522	0.4215	0.4923	0.3218	0.4379	0.5075	0.3748	0.4298	0.5355	0.4385	0.4128	0.4798	0.5895	**0.6234**
	Recall	0.3055	0.2944	0.3778	0.4341	0.3234	0.3812	0.4398	0.3099	0.4264	0.3893	0.3453	0.3549	0.4974	**0.5123**	0.5095
	F-measure	0.3781	0.3222	0.4176	0.4090	0.4048	0.4258	0.4706	0.2916	0.4278	**0.5860**	0.4470	0.3826	0.5428	0.5671	0.5797
	Purity	0.4132	0.3908	0.4808	0.5553	0.4728	0.4926	0.5448	0.3803	0.4878	0.5876	0.4986	0.4513	**0.6038**	0.5929	0.5894
	Entropy	0.4932	0.7138	0.5094	0.3978	0.5010	0.7116	0.5224	0.5183	0.5720	0.5757	0.4517	0.6233	0.6380	0.5697	**0.5618**
	Rank	11	15	9	5	13	10	6	14	8	4	7	12	3	2	1

(continued)

Table 4 (continued)

Dataset	Measure	Clustering techniques					Optimization algorithms									
		Spectral	K-mean	K-mean++	Agglomerative	DBSCAN	BHC	MVO	KHA	PSO	H-PSO	HS	GA	H-GA	H-SSA1	H-SS2
DS9	Accuracy	0.3787	0.4126	0.4260	0.4610	0.3704	0.3728	0.4630	0.4054	0.4870	0.4740	0.3590	0.4320	0.5940	0.5524	**0.6130**
	Precision	0.3630	0.3945	0.3769	0.3445	0.4242	0.4325	0.4569	0.4191	0.4505	0.5010	0.3710	0.4140	0.5245	0.5361	**0.5557**
	Recall	0.3480	0.3813	0.3559	0.3487	0.4291	0.4391	0.4419	0.4579	0.4497	0.4768	0.3888	0.4008	**0.5127**	0.5019	0.5020
	F-measure	0.3691	0.3876	0.4299	0.4187	0.4152	0.4342	0.4569	0.4137	0.4497	0.4701	0.3701	0.4071	**0.5647**	0.5074	0.5164
	Purity	0.4866	0.4108	0.5471	0.4760	0.6025	0.6208	0.6081	0.6063	0.5790	0.5273	0.4863	0.5603	0.6420	0.6361	**0.6476**
	Entropy	0.5386	0.5874	0.4745	0.4821	0.4776	0.5850	0.5355	0.4041	0.5391	0.5029	0.4539	0.5469	0.5371	0.4890	**0.4808**
	Rank	14	15	11	12	8	9	6	5	7	4	13	10	2	3	1
DS10	Accuracy	0.4905	0.5012	0.4937	0.4933	0.4846	0.5038	0.5291	0.5191	0.5623	0.5532	0.5032	0.5316	0.6523	0.5812	**0.6674**
	Precision	0.4454	0.4626	0.4913	0.4479	0.5004	0.4571	0.5213	0.4773	0.5249	0.5450	0.4630	0.5314	0.5749	0.5900	**0.6213**
	Recall	0.3423	0.4011	0.3508	0.4471	0.4176	0.3601	0.4496	0.4106	0.4811	0.4907	0.4447	0.4706	0.5671	**0.5807**	0.5758
	F-measure	0.4141	0.4315	0.4462	0.4487	0.4878	0.4292	0.4831	0.4131	0.5017	0.5521	0.4611	0.4998	**0.6187**	0.5888	0.5940
	Purity	0.5142	0.4759	0.5797	0.5445	0.4007	0.5146	0.6069	0.5939	0.6125	0.6939	0.5589	0.4917	**0.7315**	0.6795	0.6938
	Entropy	0.5468	0.7044	0.5961	0.4875	0.5706	0.6858	0.6625	0.5595	0.5765	0.6310	0.5470	0.6216	0.6585	0.6216	**0.6200**
	Rank	13	15	11	8	12	14	6	10	5	4	9	7	2	3	1
DS11	Accuracy	0.6326	0.5858	0.6799	0.6683	0.6206	0.6992	0.7043	0.5761	0.6363	0.6853	0.5963	0.6621	**0.7273**	0.7126	0.7184
	Precision	0.6620	0.5699	0.6880	0.6449	0.5708	0.6929	0.6919	0.6246	0.6604	0.7354	0.6064	0.6726	**0.7774**	0.7066	0.7049
	Recall	0.6528	0.5259	0.7028	0.6224	0.5796	**0.7109**	0.6844	0.5550	0.6164	0.6754	0.5674	0.6320	0.6944	0.6721	0.7032
	F-measure	0.6764	0.5472	0.6920	0.6281	0.6173	**0.7074**	0.6881	0.5529	0.6377	0.6297	0.5787	0.6519	0.6887	0.6816	0.6987
	Purity	0.6352	0.5938	0.6227	0.5812	0.6089	0.6272	0.6742	0.5550	0.6243	0.6413	0.5983	0.6320	0.6733	0.6943	**0.7302**
	Entropy	0.4002	0.5601	0.3540	0.4033	0.4854	0.5504	0.5113	0.5321	0.5306	0.5656	0.4946	0.5781	0.6166	0.5624	**0.4347**
	Rank	7	15	2	9	12	6	4	14	11	8	13	10	3	5	1
Average		10.8	14.8	8.5	8.0	12.0	10.8	5.5	10.3	7.7	4.3	10.2	10.2	2.8	3.0	1.0
Final Rank		12	15	8	7	14	12	5	11	6	4	9	9	2	3	1

Note The lowest ranked algorithm is the best one

Table 5 Results of entropy, purity, F-measure, recall, precision, and accuracy for two scientific articles datasets

Dataset	Measure	Clustering techniques					Optimization algorithms									
		Spectral	K-mean	K-mean++	Agglomerative	DBSCAN	BHC	MVO	KHA	PSO	H-PSO	HS	GA	H-GA	H-SSA1	H-SS2
DS12	Accuracy	0.5593	0.5433	0.5545	0.5472	0.5281	0.5646	0.6081	0.6306	0.5802	0.6556	0.6211	0.5503	0.6272	0.6356	**0.6576**
	Precision	0.4929	0.4489	0.4994	0.5340	0.5443	0.4962	0.6113	0.5914	0.6440	0.6384	0.5413	0.6209	0.6246	0.6585	**0.6675**
	Recall	0.3911	0.5211	0.3913	0.4797	0.4706	0.3937	0.5756	0.4986	0.5717	0.6075	0.5156	0.5571	0.5901	0.6119	**0.6661**
	F-measure	0.4547	0.5197	0.4701	0.4661	0.4251	0.4681	0.5941	0.5298	0.6461	0.6195	0.4991	0.5797	0.5965	0.6114	**0.6604**
	Purity	0.5685	0.5735	0.5262	0.5839	0.6459	0.5689	0.6899	0.5127	0.7089	**0.7877**	0.6919	0.5015	0.7589	0.6325	0.7279
	Entropy	0.5555	0.5415	0.5218	0.5030	0.5565	0.4876	0.5005	0.5556	0.5190	0.4536	0.5915	0.5085	0.4544	0.4534	**0.4260**
	Rank	15	11	14	10	12	13	6	9	5	2	8	7	3	4	1
DS13	Accuracy	0.4188	0.4887	0.4896	0.5635	0.4892	0.5039	0.6238	0.5811	0.6242	0.6811	0.5945	0.5395	0.6685	0.6766	**0.6874**
	Precision	0.4865	0.6323	0.5459	0.5219	0.6175	0.6455	0.6625	0.6139	0.5635	**0.7289**	0.6125	0.5429	0.6439	0.6699	0.6878
	Recall	0.3977	0.6304	0.5734	0.4415	0.5324	0.6381	0.6087	0.5860	0.6414	0.6100	0.6447	0.5255	0.6083	0.6310	**0.6478**
	F-measure	0.4805	0.6295	0.5043	0.4983	0.5615	0.6376	0.6395	0.6084	0.5875	**0.6554**	0.6172	0.5103	0.6342	0.6393	0.6529
	Purity	0.4280	0.6464	0.5512	0.5299	0.6127	0.6593	0.6380	0.6502	0.6187	0.6572	0.6758	0.5629	0.6456	0.6749	**0.6794**
	Entropy	0.6190	0.6054	0.5402	0.5579	0.6267	0.5506	0.5780	0.5632	0.5917	0.5532	0.5268	0.5859	**0.5016**	0.5504	0.5340
	Rank	15	10	12	14	11	7	6	8	9	2	5	13	4	3	1
Average		15.0	10.5	13.0	12.0	11.5	10.0	6.0	8.5	7.0	2.0	6.5	10.0	3.5	3.5	1.0
Final Rank		15	11	14	13	12	9	5	8	7	2	6	9	3	3	1

Note The lowest ranked algorithm is the best one

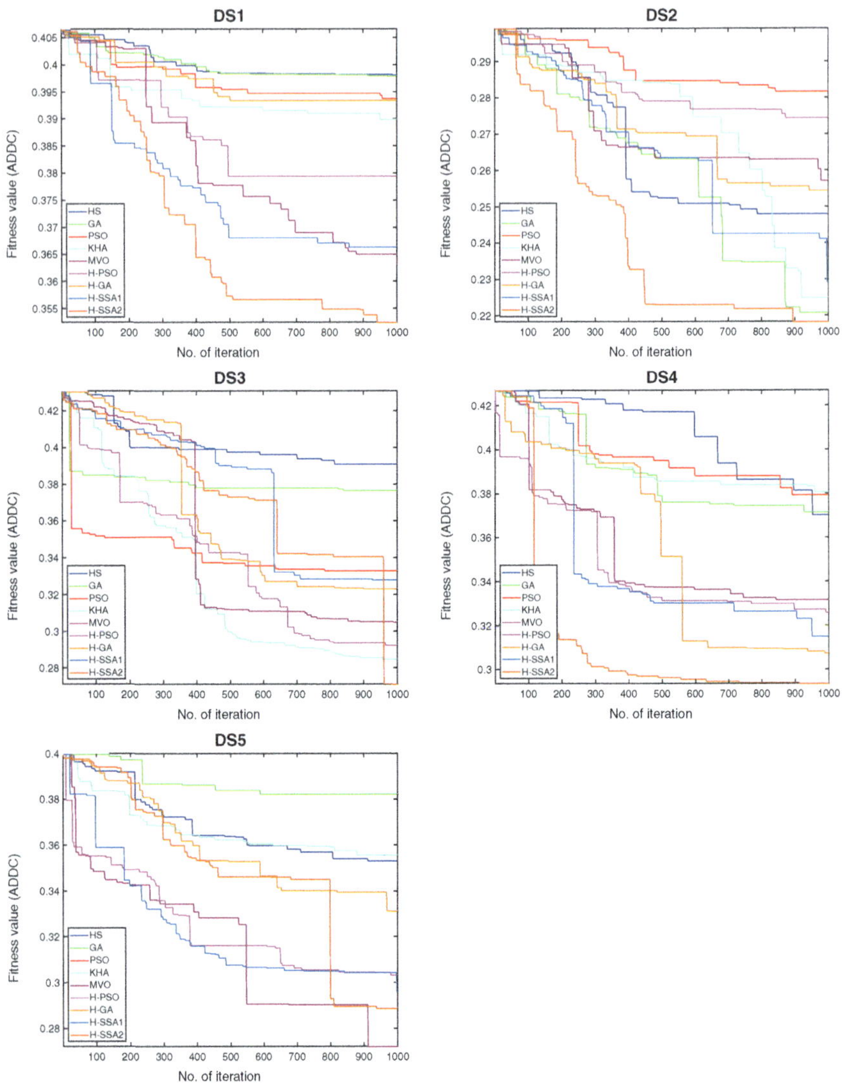

Fig. 3 Optimization algorithms convergence rate for five data clustering datasets

The convergence methods of state of the art were faster compared to H-SSA1, as well as H-SSA2. Nevertheless, H-SSA2 was found to be more effective in comparison with the original SSA regarding the performance of the algorithm and the time of execution. Thus, a better clustering quality has been generated compared to the algorithms that are most popular. Even though the basic SSA has slowly approached local optima compared to hybrid versions, namely, H-SSA1, as well as H-SSA2, it has more efficiently converged to the best solution. H-SSA2 has achieved the optimal

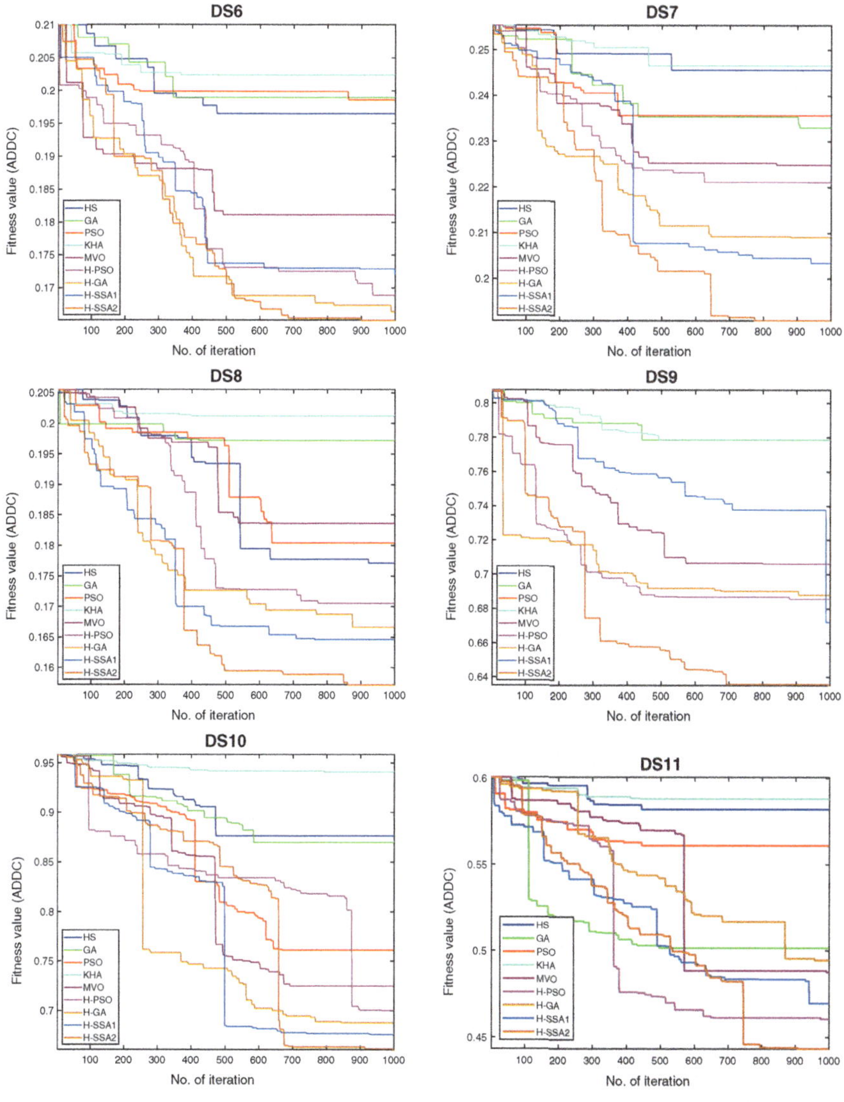

Fig. 4 Optimization algorithms convergence rate for six text clustering datasets

solution compared to the comparative algorithms. Also, high-quality clusters were achieved compared to different comparative algorithms. By examining the convergence behavior of the introduced versions, it was found that the proposed H-SSA2 has achieved the most optimal performance and has quickly achieved optimal results as well. However, its convergence was found to be quicker than the convergence of the versions of the SSA algorithm along with known clustering algorithms.

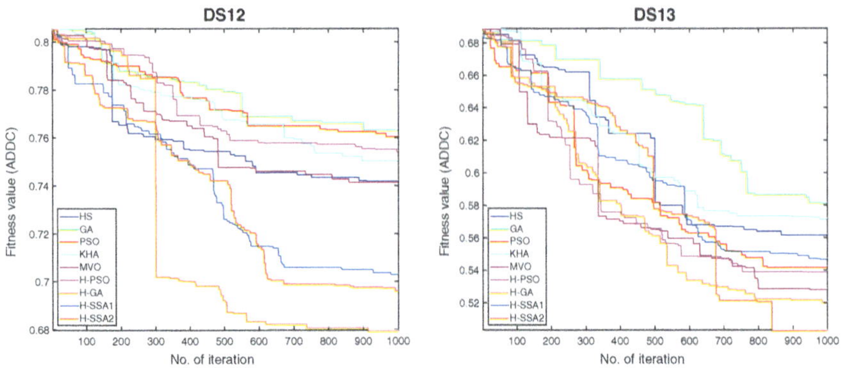

Fig. 5 Optimization algorithms convergence rate for two scientific articles datasets

8 Conclusion and Future Works

The problem of text clustering is a major concern. Many researchers have therefore tried to solve this problem. The population-based SSA algorithm is a novel optimization algorithm. It aims at solving numerous global optimization problems. SSA aims to simultaneously provide good exploration for the search space's various regions to find the best solution at the exploitation cost. The aim of this work is to tackle two critical issues related to the SSA algorithm, involving the initial value of the objective function for the candidate solutions, along with the most optimal solution that SSA can produce at every single iteration.

Therefore, a new hybrid algorithm has been developed in this work with the aim of solving the text clustering problem based on the k-means clustering combination along with the SSA algorithms to tackle critical issues. Thus, an enhanced SSA version has been created to enhance the quality of the initial candidate solutions and enhance the optimal solution as well. By utilizing such a hybridization, H-SSA has carried out a more efficient and more effective search; it has converged quickly to optimal solutions.

Considering the assessment of the novel H-SSA, seven evaluation measures were applied, involving error rate and accuracy, as well as precision, in addition to recall and F-measure, along with purity, entropy, and the behavior of convergence, along with statistical analysis. These measures are the most commonly utilized evaluation criteria in the data and text mining domain, which aim to evaluate the novel clustering method. Thus, the novel H-SSA can produce the best-recorded findings for the entire benchmark datasets that are used compared with current versions along with several clustering methods and techniques that are successfully implemented in the literature. Therefore, the SSA algorithm with the k-means is an efficient approach for clustering techniques. Consequently, many success stories are expected in the domain of data and text clustering. The findings of this work showed that the newly introduced hybridization can be active, which is an effective method for tackling the problems of clustering. The experimental findings were compared with current comparative algorithms. The novel SSA (H-SSA) hybridization has properly solved the clustering problems regarding data and text. H-SSA can, therefore, contribute to the domain of clustering. Also, different clustering problems are expected to be explored in future research to confirm the capability of the introduced algorithm in this domain. Accordingly, a more powerful local search can be hybridized to achieve further enhancements of the SSA exploitation capability. In conclusion, the introduced algorithm can be further investigated regarding the benchmark function datasets in future works.

References

1. Abasi, Ammar Kamal, Ahamad Tajudin Khader, Mohammed Azmi Al-Betar, Syibrah Naim, Zaid Abdi Alkareem Alyasseri, and Sharif Naser Makhadmeh. 2020. An ensemble topic extraction approach based on optimization clusters using hybrid multi-verse optimizer for scientific publications. *Journal of Ambient Intelligence and Humanized Computing* 1–37.
2. Abasi, Ammar Kamal, Ahamad Tajudin Khader, Mohammed Azmi Al-Betar, Syibrah Naim, Zaid Abdi Alkareem Alyasseri, and Sharif Naser Makhadmeh. 2020. A novel hybrid multi-verse optimizer with k-means for text documents clustering. *Neural Computing & Applications*.
3. Abasi, Ammar Kamal, Ahamad Tajudin Khader, Mohammed Azmi Al-Betar, Syibrah Naim, Sharif Naser Makhadmeh, and Zaid Abdi Alkareem Alyasseri. 2019. An improved text feature selection for clustering using binary grey wolf optimizer. In *Proceedings of the 11th national technical seminar on unmanned system technology 2019*, 503–516. Springer.
4. Abasi, Ammar Kamal, Ahamad Tajudin Khader, Mohammed Azmi Al-Betar, Syibrah Naim, Sharif Naser Makhadmeh, and Zaid Abdi Alkareem Alyasseri. 2019. A text feature selection technique based on binary multi-verse optimizer for text clustering. In *2019 IEEE Jordan international joint conference on electrical engineering and information technology (JEEIT)*, 1–6. IEEE.
5. Abasi, Ammar Kamal, Ahamad Tajudin Khader, Mohammed Azmi Al-Betar, Syibrah Naim, Sharif Naser Makhadmeh, and Zaid Abdi Alkareem Alyasseri. 2020. Link-based multi-verse optimizer for text documents clustering. *Applied Soft Computing* 87: 106002.
6. Abed-alguni, Bilal H., and Faisal Alkhateeb. 2018. Intelligent hybrid cuckoo search and β-hill climbing algorithm. *Journal of King Saud University-Computer and Information Sciences*.
7. Abualigah, Laith Mohammad, Ahamad Tajudin Khader, Mohammed Azmi Al-Betar, and Zaid Abdi Alkareem Alyasseri. 2017. β-hill climbing technique for the text document clustering. In *New trends in information technology NTIT2017 conference*, Amman, Jordan, 1–6. IEEE.
8. Al-Betar, Mohammed Azmi. 2017. β-hill climbing: An exploratory local search. *Neural Computing and Applications* 28 (1): 153–168.
9. Al-Betar, Mohammed Azmi, Ibrahim Aljarah, Mohammed A Awadallah, Hossam Faris, and Seyedali Mirjalili. 2019. Adaptive β-hill climbing for optimization. *Soft Computing* 1–24.
10. Al-Betar, Mohammed Azmi, Mohammed A. Awadallah, Iyad Abu Doush, Emad Alsukhni, and Habes ALkhraisat. 2018. A non-convex economic dispatch problem with valve loading effect using a new modified β-hill climbing local search algorithm. *Arabian Journal for Science and Engineering*.
11. Al-Betar, Mohammed Azmi, Mohammed A. Awadallah, Asaju Laaro Bolaji, and Basem O. Alijla. 2017. β-hill climbing algorithm for sudoku game. In *Second Palestinian international conference on information and communication technology (PICICT 2017)*, Gaza, Palestine, 1–5. IEEE.

12. Al-Betar, Mohammed Azmi, Mohammed A. Awadallah, Ahamad Tajudin Khader, Asaju Laaro Bolaji, and Ammar Almomani. 2018. Economic load dispatch problems with valve-point loading using natural updated harmony search. *Neural Computing and Applications* 29 (10): 767–781.
13. Al-Betar, Mohammed Azmi, and Ahamad Tajudin Khader. 2012. A harmony search algorithm for university course timetabling. *Annals of Operations Research* 194 (1): 3–31.
14. Alomari, Osama Ahmad, Ahamad Tajudin Khader, Mohammed Azmi Al-Betar, and Zaid Abdi Alkareem Alyasseri. 2018. A hybrid filter-wrapper gene selection method for cancer classification. In *2018 2nd international conference on biosignal analysis, processing and systems (ICBAPS)*, 113–118. IEEE.
15. Alomari, Osama Ahmad, Ahamad Tajudin Khader, Mohammed Azmi Al-Betar, and Mohammed A. Awadallah. 2018. A novel gene selection method using modified MRMR and hybrid bat-inspired algorithm with β-hill climbing. *Applied Intelligence* 48 (11): 4429–4447.
16. Alsukni, Emad, Omar Suleiman Arabeyyat, Mohammed A. Awadallah, Laaly Alsamarraie, Iyad Abu-Doush, and Mohammed Azmi Al-Betar. 2019. Multiple-reservoir scheduling using β-hill climbing algorithm. *Journal of Intelligent Systems* 28 (4): 559–570.
17. Alweshah, Mohammed, Aram Al-Daradkeh, Mohammed Azmi Al-Betar, Ammar Almomani, and Saleh Oqeili. 2019. β-hill climbing algorithm with probabilistic neural network for classification problems. *Journal of Ambient Intelligence and Humanized Computing* 1–12.
18. Alyasseri, Zaid Abdi Alkareem, Ahamad Tajudin Khader, and Mohammed Azmi Al-Betar. 2017. Optimal electroencephalogram signals denoising using hybrid β-hill climbing algorithm and wavelet transform. In *Proceedings of the international conference on imaging, signal processing and communication*, 106–112. ACM.
19. Alyasseri, Zaid Abdi Alkareem, Ahamad Tajudin Khader, Mohammed Azmi Al-Betar, Ammar Kamal Abasi, and Sharif Naser Makhadmeh. 2019. EEG signals denoising using optimal wavelet transform hybridized with efficient metaheuristic methods. *IEEE Access* 8: 10584–10605.
20. Alyasseri, Zaid Abdi Alkareem, Ahamad Tajudin Khader, Mohammed Azmi Al-Betar, and Laith Mohammad Abualigah. 2017. ECG signal denoising using β-hill climbing algorithm and wavelet transform. In *ICIT 2017, the 8th international conference on information technology*, 1–7.
21. Alyasseri, Zaid Abdi Alkareem, Ahamad Tajudin Khader, Mohammed Azmi Al-Betar, and Osama Ahmad Alomari. 2020. Person identification using EEG channel selection with hybrid flower pollination algorithm. *Pattern Recognition* 107393.
22. Alyasseri, Zaid Abdi Alkareem, Ahamad Tajudin Khader, Mohammed Azmi Al-Betar, and Mohammed A Awadallah. 2018. Hybridizing β-hill climbing with wavelet transform for denoising ECG signals. *Information Sciences* 429: 229–246.
23. Alyasseri, Zaid Abdi Alkareem, Ahamad Tajudin Khader, Mohammed Azmi Al-Betar, Mohammed A. Awadallah, and Xin-She Yang. 2018. Variants of the flower pollination algorithm: A review. In *Nature-inspired algorithms and applied optimization*, 91–118. Springer.
24. Alzaidi, Amer Awad, Musheer Ahmad, Mohammad Najam Doja, Eesa Al Solami, and M.M. Sufyan Beg. 2018. A new 1D chaotic map and $beta$-hill climbing for generating substitution-boxes. *IEEE Access* 6: 55405–55418.
25. Bharti, Kusum Kumari, and Pramod Kumar Singh. 2016. Chaotic gradient artificial bee colony for text clustering. *Soft Computing* 20 (3): 1113–1126.
26. Boley, Daniel, Maria Gini, Robert Gross, Eui-Hong Sam Han, Kyle Hastings, George Karypis, Vipin Kumar, Bamshad Mobasher, and Jerome Moore. 1999. Document categorization and query generation on the world wide web using webace. *Artificial Intelligence Review* 13 (5–6): 365–391.
27. Bouras, Christos, and Vassilis Tsogkas. 2012. A clustering technique for news articles using wordnet. *Knowledge-Based Systems* 36: 115–128.
28. Cura, Tunchan. 2012. A particle swarm optimization approach to clustering. *Expert Systems with Applications* 39 (1): 1582–1588.

29. Deepa, M., P. Revathy, and P. Student. 2012. Validation of document clustering based on purity and entropy measures. *International Journal of Advanced Research in Computer and Communication Engineering* 1 (3): 147–152.
30. Del Buono, Nicoletta, and Gianvito Pio. 2015. Non-negative matrix tri-factorization for co-clustering: An analysis of the block matrix. *Information Sciences* 301: 13–26.
31. Ekinci, Serdar, and Baran Hekimoglu. 2018. Parameter optimization of power system stabilizer via salp swarm algorithm. In *2018 5th international conference on electrical and electronic engineering (ICEEE)*, 143–147. IEEE.
32. Elaziz, Mohamed Abd, Lin Li, KPN Jayasena, and Shengwu Xiong. 2019. Multiobjective big data optimization based on a hybrid salp swarm algorithm and differential evolution. *Applied Mathematical Modelling*.
33. Faris, Hossam, Seyedali Mirjalili, Ibrahim Aljarah, Majdi Mafarja, and Ali Asghar Heidari. 2020. Salp swarm algorithm: Theory, literature review, and application in extreme learning machines. In *Nature-inspired optimizers*, 185–199. Springer.
34. Figueiredo, Elliackin, Mariana Macedo, Hugo Valadares Siqueira, Clodomir J. Santana Jr, Anu Gokhale, and Carmelo J.A. Bastos-Filho. 2019. Swarm intelligence for clustering a systematic review with new perspectives on data mining. *Engineering Applications of Artificial Intelligence* 82: 313–329.
35. Forsati, Rana, Andisheh Keikha, and Mehrnoush Shamsfard. 2015. An improved bee colony optimization algorithm with an application to document clustering. *Neurocomputing* 159: 9–26.
36. Forsati, Rana, Mehrdad Mahdavi, Mehrnoush Shamsfard, and Mohammad Reza Meybodi. 2013. Efficient stochastic algorithms for document clustering. *Information Sciences* 220: 269–291.
37. Hossein, Gandomi Amir, and Amir Hossein Alavi. 2012. Krill herd: A new bio-inspired optimization algorithm. *Communications in Nonlinear Science and Numerical Simulation* 17 (12): 4831–4845.
38. Hegazy, AhE, M.A. Makhlouf, and GhS El-Tawel. 2019. Feature selection using chaotic salp swarm algorithm for data classification. *Arabian Journal for Science and Engineering* 44 (4): 3801–3816.
39. Huang, Anna. 2008. Similarity measures for text document clustering. In *Proceedings of the sixth New Zealand computer science research student conference (NZCSRSC2008)*, Christchurch, New Zealand, 49–56.
40. Ibrahim, Rehab Ali, Ahmed A. Ewees, Diego Oliva, Mohamed Abd Elaziz, and Lu Songfeng. 2019. Improved salp swarm algorithm based on particle swarm optimization for feature selection. *Journal of Ambient Intelligence and Humanized Computing* 10 (8): 3155–3169.
41. Ismael, Sherif M., Shady H.E. Abdel Aleem, Almoataz Y. Abdelaziz, and Ahmed Faheem Zobaa. 2018. Practical considerations for optimal conductor reinforcement and hosting capacity enhancement in radial distribution systems. *IEEE Access* 6: 27268–27277.
42. Jangir, Pradeep, Siddharth A. Parmar, Indrajit N. Trivedi, and R.H. Bhesdadiya. 2017. A novel hybrid particle swarm optimizer with multi verse optimizer for global numerical optimization and optimal reactive power dispatch problem. *Engineering Science and Technology, an International Journal* 20 (2): 570–586.
43. Jensi, R., and G. Wiselin Jiji. 2014. A survey on optimization approaches to text document clustering. arXiv:1401.2229.
44. Karaa, Wahiba Ben Abdessalem, Amira S. Ashour, Dhekra Ben Sassi, Payel Roy, Noreen Kausar, and Nilanjan Dey. 2016. Medline text mining: An enhancement genetic algorithm based approach for document clustering. In *Applications of intelligent optimization in biology and medicine*, 267–287. Springer.
45. Karaboga, Dervis, and Celal Ozturk. 2011. A novel clustering approach: Artificial bee colony (ABC) algorithm. *Applied Soft Computing* 11 (1): 652–657.
46. Katrawi, Anwar H., Rosni Abdullah, Mohammed Anbar, and Ammar Kamal Abasi. 2020. Earlier stage for straggler detection and handling using combined CPU test and LATE methodology. *International Journal of Electrical & Computer Engineering* 10. ISSN: 2088-8708.

47. Kaveh, A., and M. Khayatazad. 2012. A new meta-heuristic method: Ray optimization. *Computers & Structures* 112: 283–294.
48. Lin, Yung-Shen, Jung-Yi Jiang, and Shie-Jue Lee. 2014. A similarity measure for text classification and clustering. *IEEE Transactions on Knowledge and Data Engineering* 26 (7): 1575–1590.
49. Makhadmeh, Sharif Naser, Ahamad Tajudin Khader, Mohammed Azmi Al-Betar, and Syibrah Naim. 2018. An optimal power scheduling for smart home appliances with smart battery using grey wolf optimizer. In *2018 8th IEEE international conference on control system, computing and engineering (ICCSCE)*, 76–81. IEEE.
50. Makhadmeh, Sharif Naser, Ahamad Tajudin Khader, Mohammed Azmi Al-Betar, and Syibrah Naim. 2019. Multi-objective power scheduling problem in smart homes using grey wolf optimiser. *Journal of Ambient Intelligence and Humanized Computing* 10 (9): 3643–3667.
51. Makhadmeh, Sharif Naser, Ahamad Tajudin Khader, Mohammed Azmi Al-Betar, Syibrah Naim, Ammar Kamal Abasi, and Zaid Abdi Alkareem Alyasseri. 2019. Optimization methods for power scheduling problems in smart home: Survey. *Renewable and Sustainable Energy Reviews* 115: 109362.
52. Makhadmeh, Sharif Naser, Ahamad Tajudin Khader, Mohammed Azmi Al-Betar, Syibrah Naim, Zaid Abdi Alkareem Alyasseri, and Ammar Kamal Abasi. 2019. Particle swarm optimization algorithm for power scheduling problem using smart battery. In *2019 IEEE Jordan international joint conference on electrical engineering and information technology (JEEIT)*, 672–677. IEEE.
53. Mirjalili, Seyedali. 2015. The ant lion optimizer. *Advances in Engineering Software* 83: 80–98.
54. Mirjalili, Seyedali. 2016. Dragonfly algorithm: A new meta-heuristic optimization technique for solving single-objective, discrete, and multi-objective problems. *Neural Computing and Applications* 27 (4): 1053–1073.
55. Mirjalili, Seyedali, Amir H. Gandomi, Seyedeh Zahra Mirjalili, Shahrzad Saremi, Hossam Faris, and Seyed Mohammad Mirjalili. 2017. Salp swarm algorithm: A bio-inspired optimizer for engineering design problems. *Advances in Engineering Software* 114: 163–191.
56. Mirjalili, Seyedali, Seyed Mohammad Mirjalili, and Abdolreza Hatamlou. 2016. Multi-verse optimizer: A nature-inspired algorithm for global optimization. *Neural Computing and Applications* 27 (2): 495–513.
57. Niknam, Taher, and Babak Amiri. 2010. An efficient hybrid approach based on PSO, ACO and k-means for cluster analysis. *Applied Soft Computing* 10 (1): 183–197.
58. Ozturk, Celal, Emrah Hancer, and Dervis Karaboga. 2015. Dynamic clustering with improved binary artificial bee colony algorithm. *Applied Soft Computing* 28: 69–80.
59. Pan, Wen-Tsao. 2012. A new fruit fly optimization algorithm: Taking the financial distress model as an example. *Knowledge-Based Systems* 26: 69–74.
60. Park, Hae-Sang, and Chi-Hyuck Jun. 2009. A simple and fast algorithm for k-medoids clustering. *Expert Systems with Applications* 36 (2): 3336–3341.
61. Patel, Monika Raghuvanshi Rahul. 2017. An improved document clustering with multiview point similarity/dissimilarity measures. *International Journal of Engineering and Computer Science* 6 (2).
62. Sahoo, G., et al. 2017. A two-step artificial bee colony algorithm for clustering. *Neural Computing and Applications* 28 (3): 537–551.
63. Sayed, Gehad Ismail, Ashraf Darwish, and Aboul Ella Hassanien. 2017. Quantum multiverse optimization algorithm for optimization problems. *Neural Computing and Applications* 1–18.
64. Sayed, Gehad Ismail, Ashraf Darwish, and Aboul Ella Hassanien. 2018. A new chaotic multiverse optimization algorithm for solving engineering optimization problems. *Journal of Experimental & Theoretical Artificial Intelligence* 30 (2): 293–317.
65. Shahnaz, Farial, Michael W. Berry, V. Paul Pauca, and Robert J. Plemmons. 2006. Document clustering using nonnegative matrix factorization. *Information Processing & Management* 42 (2): 373–386.
66. Shelokar, P.S., Valadi K. Jayaraman, and Bhaskar D. Kulkarni. 2004. An ant colony approach for clustering. *Analytica Chimica Acta* 509 (2): 187–195.

67. Wei, Tingting, Lu Yonghe, Huiyou Chang, Qiang Zhou, and Xianyu Bao. 2015. A semantic approach for text clustering using wordnet and lexical chains. *Expert Systems with Applications* 42 (4): 2264–2275.
68. Zaw, Moe Moe, and Ei Ei Mon. 2015. Web document clustering by using PSO-based cuckoo search clustering algorithm. In *Recent advances in swarm intelligence and evolutionary computation*, 263–281. Springer.
69. Zhao, Ying, and George Karypis. 2001. Criterion functions for document clustering: Experiments and analysis.

Controlling Population Diversity of Harris Hawks Optimization Algorithm Using Self-adaptive Clustering Approach

Hamza Turabieh and Majdi Mafarja

Abstract Harris Hawks optimization (HHO) algorithm is a new adaptive search method that can be employed as approximation heuristic for complex optimization problems. HHO algorithm processes a population of search space with two operations: Soft besiege and Hard besiege. One of main problems in the use of population-based algorithms is premature convergence. A premature stagnation of the search creates a shortage of diversity, which affects the relationship between exploration and exploitation processes. Here, we propose a self-adaptive clustering approach based on k-means and density for all solutions in the population pool. The k-means clustering method will examine the distribution (i.e., locations) of the solutions over the search space. Once most of solutions (more than 90%) are related to one cluster, an exploitation process will gain a higher priority compared to exploration one. Redistributing the solutions over several clusters by a redistribution process will keep the ratio between exploration and exploitation stable over all iterations. The proposed approach will be tested over several mathematical test functions. The obtained results show that the proposed modification can enhance the performance of HHO algorithm and prevent premature convergence.

Keywords HHO · Harris Hawks Optimization · Data clustering · Optimization · Evolutionary computation · Nature-inspired algorithms · Swarm intelligence · Meta-heuristics · Unsupervised learning

---

H. Turabieh
Information Technology Department, CIT College, Taif University, Taif, Saudi Arabia
e-mail: h.turabieh@tu.edu.sa

M. Mafarja (✉)
Department of Computer Science, Birzeit University, Birzeit, West Bank, Palestine
e-mail: mmafarja@birzeit.edu

I. Aljarah et al. (eds.), *Evolutionary Data Clustering: Algorithms and Applications*, Algorithms for Intelligent Systems,
https://doi.org/10.1007/978-981-33-4191-3_7

1 Introduction

Population-based algorithms usually have two criticisms while optimizing complex problems which are trapped in local optima, and computationally inefficient while searching without contributing to the best solution found [3]. Low population diversity is a major factor for premature convergence of population-based algorithms [5, 11]. Finding a certain level of population diversity is an acceptable solution that will help curb premature convergence. In general, every optimization algorithm should address two criteria: exploration and exploitation of a search space. Exploration refers to visit new areas of the search space, while exploitation refers to visit those areas within neighborhood points [5, 20]. As a result, the population-based search algorithm should find a balanced ratio between exploration and exploitation to prevent the premature convergence state and low population diversity.

In general, the overall behavior of population-based algorithms such as genetic algorithms (GAs) [6], ant colony optimization (ACO) [8], evolutionary strategies (ESs) [2], evolutionary programming (EP) [9], particle swarm optimization (PSO) [14],brain storm optimization (BSO) [17], Harris Hawks optimization (HHO) [10], to name recent will-known algorithms, is determined by the relationship between exploitation and exploration throughout the run [16]. Several researchers believe that population-based algorithms are effective due to their excellent ratio between both operations (i.e., exploration and exploitation).

Up to date, researchers are still working to enhance the performance of population-based algorithms. Recent development and enhancements can be found in [4, 12, 21]. Kelly et al. [11] proposed a novel selection method called *knobelty*, to dynamic control the exploration and exploitation for GP algorithm by generating mixed population of parents. The proposed approach overcomes the low population diversity. Asiain et al. [1] proposed a controller exploitation-exploration (CEE) with reinforcement learning (RL) approach. The proposed CEE consists of three parts: controller, fast-tracked learning, and the actor-critic. The proposed CEE is able to find a good ratio between exploitation and exploration and prevent premature convergence. Nguyen et al. [13] proposed a dynamic sticky binary PSO, which controls the PSO parameters to improve the exploration and exploitation processes for feature selection and knapsack problems. Zhang et al. [24] highlighted the importance of exploitation and exploration for evolutionary algorithms with a different adaptive variation for local fine-tuning while searching for global optima. Interested readers about recently published papers related to exploration and exploitation for population-based algorithms can read the following cited papers [15, 18, 19, 23].

In 2019, Asghar et al. [10] proposed HHO algorithm, that mimics the Harris Hawks hunting nature. This chapter presents a self-adaptive learning method based on k-means clustering method to control the population diversity of HHO algorithm. The motivation of this chapter is to prevent the premature convergence of HHO once the majority of solutions belong to one cluster throughout the run and no improvements occur on the best so far solution. The proposed approach will redistribute half of the current population to enhance the ratio for exploration process and discover new areas.

The rest of this chapter is organized as follows: Sect. 2 presents the proposed enhanced of HHO algorithm by hybridizing it with k-means method. Section 3 presents the obtained results and comparison with the original HHO algorithm. Section 4 presents the conclusion and future works.

2 Proposed Approach

A self-adaptive controlling method to control the population diversity of Harris Hawks Optimization Algorithm is proposed in this chapter. In the beginning, HHO generates randomly a set of initial solutions (Rabbit positions) distributed on the search space as shown in Fig. 1. The original HHO algorithm comes with two main phases for exploration and exploitation processes. The exploration process that simulates the nature of Harris' hawks tracks the expected prey (solution) with their powerful eyes. Harris' hawks use two strategies to perch a prey. The first strategy depends on the location of other Harris' hawks and the expected prey location. While the second strategy depends on monitoring random locations in their eyes range, both strategies have an equal chance. Equation (1) demonstrates the exploration process for HHO algorithm.

$$X(t+1) = \begin{cases} X_{rand}(t) - r_1 \left| X_{rand}(t) - 2r_2 X(t) \right| & q \geq 0.5 \\ (X_{rabbit}(t) - X_m(t)) - r_3(LB + r_4(UB - LB)) & q < 0.5 \end{cases} \tag{1}$$

where X refers to a position vector, the variable $X(t+1)$ presents the hawks position in the following iteration t, while $X_{rabbit}(t)$ presents the location of the prey (i.e., rabbit) in the current iteration t. $X_{rand}(t)$ is a hawk that is selected randomly from the current population. X_m is the average position of the current population of hawks. r_1, r_2, r_3, r_4, and q are random numbers between 0 and 1, which are generated randomly at each iteration. LB and UB are the upper and lower bounds of variables. The average position for all hawks is calculated based on Eq. (2), where $X_i(t)$ presents the location of each hawk in iteration t, and N refers to the total number of hawks in the population.

$$X_m(t) = \frac{1}{N} \sum_{i=1}^{N} X_i(t) \tag{2}$$

One of the main advantages of HHO is the switching process between exploration to exploitation. This switching process depends on the escaping energy of the prey. In general, the energy decreases in escaping behavior. Equation (3) presents the decreasing value of the prey energy, where E is the escaping energy of the prey, T denotes the maximum number of iterations, and E_0 is a random initial energy of the prey within the range $(-1, 1)$ that is generated at each iteration. The switching

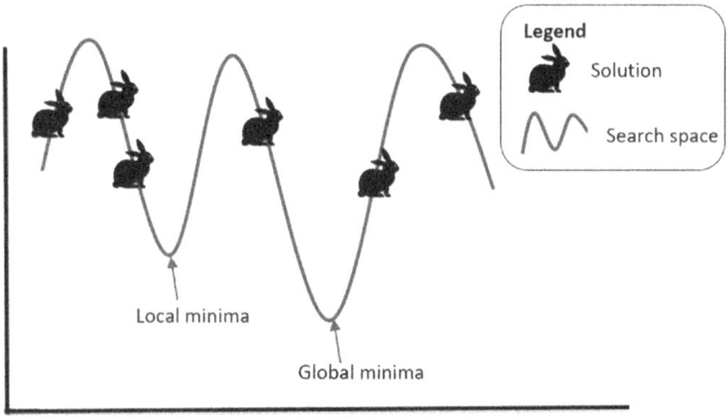

Fig. 1 Initial distribution of solutions over the search space

process between exploration to exploitation is controlled by the value of E. The exploration process is executed when $|E| \geq 1$, while exploitation process is executed when $|E| < 1$.

$$E = 2E_0\left(1 - \frac{t}{T}\right) \tag{3}$$

The exploitation process occurs based on the prey energy status E. The exploration process inside HHO has two strategies called soft and hard besiege processes. Let a variable r presents the chance of rabbit in successfully escaping ($r < 0.5$) or not successfully escaping ($r \geq 0.5$) before surprise pounce. The soft besiege occurs when $r \geq 0.5$ and $|E| \geq 0.5$. Since $|E| \geq 0.5$, this indicates that the prey still have a good enough energy to escape. The hawk keeps flying until the energy of rabbit (i.e., prey) starts decrease. Equation (4) demonstrates the soft besiege process, where $\Delta X(t)$ refers to the difference between the current location in the search space and the position vector at iteration t. $J = 2(1 - r_5)$ indicates a simulation of random jump of rabbit while escaping process, where r_5 is a random value between 0 and 1.

$$X(t+1) = \Delta X(t) - E\left|J X_{rabbit}(t) - X(t)\right| \tag{4}$$

$$\Delta X(t) = X_{rabbit}(t) - X(t) \tag{5}$$

Equation (6) presents the hard besiege process, when both $|E|$ and $r < 0.5$. The hawks reduce the average positions distance with the escaping prey where both variables Y and Z are obtained based on Eqs. (8) and (9), where $X_m(t)$ is achieved based on Eq. (2). The hawks update their position while executing soft besiege process is based on Eq. (6), where both variables Y and Z are evaluated based on two Equations that are Eqs. (8) and (9)), respectively. Algorithm 1 shows the pseudocode of the original HHO algorithm.

$$X(t+1) = \begin{cases} Y \; if F(Y) < F(X(t)) \\ Z \; if F(Z) < F(X(t)) \end{cases} \tag{6}$$

$$X(t+1) = X_{rabbit}(t) - E\,|\Delta X(t)| \tag{7}$$

$$Y = X_{rabbit}(t) - E\,|J X_{rabbit}(t) - X(t)| \tag{8}$$

$$Z = Y + S \times LF(D) \tag{9}$$

Algorithm 1 The original HHO.

Defines:
Set population size = N
Set number of iterations T
Set Iter =1
Output: Best solution (i.e., Rabbit location)
Generate population randomly $X_i(1, 2, \ldots, N)$
while (Iter $< T$) **do**
  Calculate the fitness values for each hawk
  Set X_{rabbit} as the location of rabbit (best location)
  for (each hawk (X_i)) **do**
    Update the initial energy E_0
    Update the jump strength J
    Update the value of E using Eq. (3)
    if ($|E| \geq 1$) **then**
      Update the location vector using Eq. (1)
    if ($|E| < 1$) **then**
      if ($r \geq 0.5$ and $|E| \geq 0.5$) **then**
        Update the location vector using Eq. (4)
      else if ($r \geq 0.5$ and $|E| < 0.5$) **then**
        Update the location vector using Eq. (7)
      else if ($r < 0.5$ and $|E| \geq 0.5$) **then**
        Update the location vector using Eq. (6)
      else if ($r < 0.5$ and $|E| < 0.5$) **then**
        Update the location vector using Eq. (6)
  Iter=Iter+1
Return X_{rabbit}

In general, HHO and other population-based algorithms may suffer from premature convergence after a set of iterations. Figure 2 demonstrates a pictorial presentation of premature convergence case, where all solutions may have high similarity (i.e., cluster) and stuck in local optima. To overcome this problem, an enhanced version of HHO has been proposed by hybridizing it with k-means clustering approach. The proposed idea will check the distribution (i.e., location) of the solutions (preys)

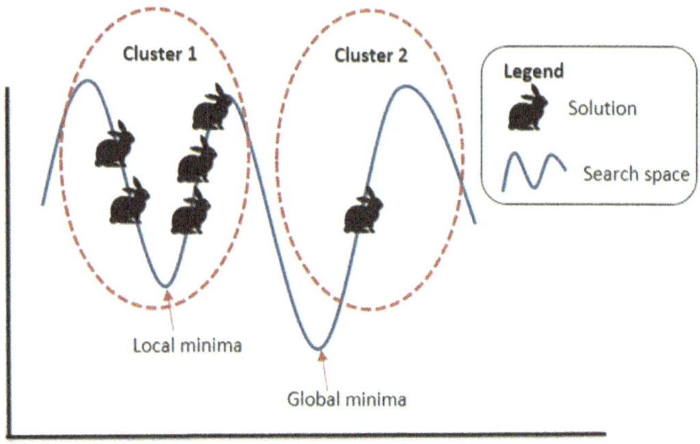

Fig. 2 Premature convergence case

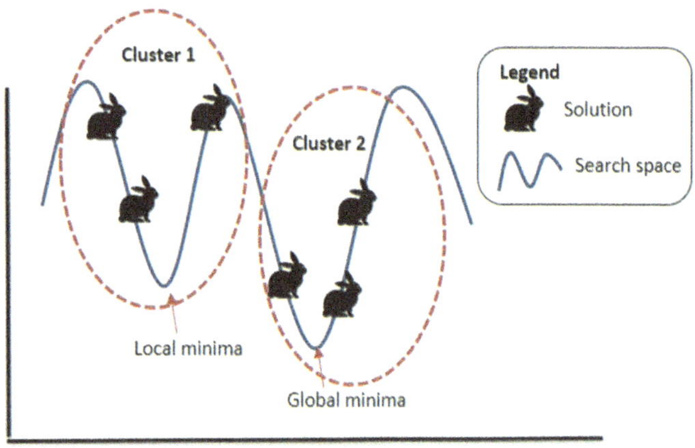

Fig. 3 Redistribution of the solutions after numbers of iterations

over the search space. If the majority of preys at certain iterations starts belong to one cluster and no improvement occurs on the best solution for M iterations, a redistribution process will generate random solutions and update the current population. Figure 3 shows the redistribution process for solutions. This process will keep the diversity of the population at higher levels and prevent premature convergence. Figure 2 demonstrates the modified HHO algorithm.

The proposed approach will distribute half of the solutions in the current population randomly over the search space. This mechanism will keep population diversity high, which enhance the exploration process and prevent the premature convergence. Adding the k-means clustering method will increase the execution time for the original HHO algorithm. However, it will enhance the overall performance of HHO algorithm.

Algorithm 2 The modified HHO.

```
Defines:
Set population size = N
Set number of iterations = T
Set Iter =1
Output: Best solution (i.e., Rabbit location)
Generate population randomly X_i(1, 2, ..., N)
while (Iter < T) do
   Calculate the fitness values for each hawk
   Define X_rabbit as the location best location.
   for (each hawk (X_i)) do
      Update the initial energy E_0 and jump strength J
      Update the energy (E) based on Eq. (3)
      if (|E| ≥ 1) then
         Update the location vector based on Eq. (1)
      if (|E| < 1) then
         if (r ≥0.5 and |E| ≥ 0.5 ) then
            Update the location vector based on Eq. (4)
         else if (r ≥0.5 and |E| < 0.5 ) then
            Update the location vector based on Eq. (7)
         else if (r <0.5 and |E| ≥ 0.5 ) then
            Update the location vector based on Eq. (6)
         else if (r <0.5 and |E| < 0.5 ) then
            Update the location vector based on Eq. (6)
   Cluster= k-means(X_rabbit,2)                    ▷ only two clusters will be used
   if (the majority of X_rabbit belong to one cluster and no improvement for M iterations) then
      generate N/2 new random positions X_rabbit.
      Update the current population.
   Iter=Iter+1
Return X_rabbit
```

3 Results and Analysis

To evaluate the enhanced HHO, we employed the modified HHO on different examples of mathematical test functions with known global optima. We employed 13 mathematics benchmark functions and compared the obtained results with the original HHO [7, 10, 22]. These mathematical test functions are categorized into three sets: *unimodal* (UM) test functions, multimodal (MM) test functions, and composition (CM) test functions. Functions with a single global optimum from (F1–F7) are UM, while functions with multi global optimum from (F8–F13). Tables 1 and 2 demonstrate each mathematical functions for two categories (i.e., UM and MM). Each mathematical function was simulated for 21 independent runs with dimension 30 and 100. All experiments are evaluated using Matlab 2019b, 2.81 GHz, and 16 GB RAM. Table 3 presents the parameters setting used in this chapter.

Table 4 demonstrates the obtained average and standard deviation for each function. In all experiments, we employed 500 iterations for original and modified HHO algorithm. It is clear that for UM functions, the proposed modified approach outper-

Table 1 Details of unimodal (UM) benchmark test functions

Test function	Dimensions	Range	f_{min}
$f_1(x) = \sum_{i=1}^{n} x_i^2$	30	[−100, 100]	0
$f_2(x) = \sum_{i=1}^{n} \|x_i\| + \prod_{i=1}^{n} \|x_i\|$	30	[−10, 10]	0
$f_3(x) = \sum_{i=1}^{n} \left(\sum_{j-1}^{i} x_j\right)^2$	30	[−100, 100]	0
$f_4(x) = \max_i \{\|x_i\|, 1 \leq i \leq n\}$	30	[−100, 100]	0
$f_5(x) = \sum_{i=1}^{n-1} \left[100\left(x_{i+1} - x_i^2\right)^2 + (x_i - 1)^2\right]$	30	[−30, 30]	0
$f_6(x) = \sum_{i=1}^{n} ([x_i + 0.5])^2$	30	[−100, 100]	0
$f_7(x) = \sum_{i=1}^{n} ix_i^4 + random[0, 1)$	30	[−128, 128]	0

form the original HHO in five functions based on the average (i.e., f_1, f_2, f_3, f_4, and f_7) under the same settings. While for MM functions, the proposed approach outperform in one function (i.e., f_8) and gain same results in 3 mathematical functions (i.e., f_9, f_{10}, and f_{11}). Moreover, the proposed modification can enhance the overall performance of the original HHO and prevent the premature convergence. Table 5 shows the number of calling time for population redistribution method for the modified HHO. It is clear that the original HHO has a fast premature convergence since all solutions (preys) belong to one cluster at the early iterations. Moreover, the original HHO exploitation ratio is more higher than exploration one.

Figure 4 shows the convergence performance of f_3 for the original and modified HHO algorithm with Dim equals 30. It is obvious that the original HHO has a fast convergence compared to the modified one. From Table 5, the average redistribution function called 372.95 times out of 500 iterations. Moreover, the reported results for average value in Table 4 for f_3 show that the modified HHO can outperform the original one.

Figure 5 shows the performance of f_{10} with Dim equals 100. It is clear that the original HHO has a fast convergence, while the modified approach tries to explore more search space areas by updating the population once no improvement occurs. Moreover, from Table 5, the updating function is called 432 times out of 500, it is clear that the original HHO has a high exploitation process compared to exploration one.

In general, HHO algorithm exploitation ratio is higher than exploration one. Form Table 5 more than 70% of the total iterations, the redistribution method is called. A premature convergence occurs once most of the solutions belong to one cluster and HHO can not enhance the current best solution. The reported results show that controlling the population diversity and having a good ratio between exploration and exploitation process will reduce the probability of premature convergence of HHO algorithm.

Table 2 Details of multimodal (MM) benchmark test functions

Test function	Dimensions	Range	$f_{\min}$
$f_8(x) = \sum_{i=1}^{n} -x_i \sin\left(\sqrt{\lvert x_i \rvert}\right)$	30	[−500, 500]	−418.9829 × n
$f_9(x) = \sum_{i=1}^{n} \left[x_i^2 - 10\cos(2\pi x_i) + 10\right]$	30	[−5.12, 5.12]	0
$f_{10}(x) = -20\exp(-0.2\sqrt{\frac{1}{n}\sum_{i=1}^{n} x_i^2}) - \exp\left(\frac{1}{n}\sum_{i=1}^{n} cos\,(2\pi x_i)\right) + 20 + e$	30	[−32, 32]	0
$f_{11}(x) = \frac{1}{4000}\sum_{i=1}^{n} x_i^2 - \prod_{i=1}^{n}\cos\left(\frac{x_i}{\sqrt{i}}\right) + 1$	30	[−600, 600]	0
$f_{12}(x) = \frac{\pi}{n}\left\{10\sin(\pi y_1) + \sum_{i=1}^{n-1}(y_i - 1)^2\left[1 + 10\sin^2(\pi y_{i+1})\right] + (y_n - 1)^2\right\}$ $+\sum_{i=1}^{n} u(x_i, 10, 100, 4)$ $y_i = 1 + \frac{x_i+1}{4} u(x_i, a, k, m) = \begin{cases} k(x_i - a)^m & x_i > a \\ 0 - a & < x_i < a \\ k(-x_i - a)^m & x_i < -a \end{cases}$	30	[−50, 50]	0
$f_{13}(x) = 0.1\left\{\sin^2(3\pi x_1) + \sum_{i=1}^{n}(x_i - 1)^2\left[1 + \sin^2(3\pi x_i + 1)\right] + (x_n - 1)^2\left[1 + \sin^2(2\pi x_n)\right]\right\} +$ $\sum_{i=1}^{n} u(x_i, 5, 100, 4)$	30	[−50, 50]	0

Table 3 Parameters setting for original and modified HHO

Parameter	Value
Number of iterations (T)	500
Population size (N)	50

Table 4 Comparison between original and modified HHO algorithm

	Original HHO		Modified HHO		Original HHO		Modified HHO	
	Dim 30		Dim 30		Dim 100		Dim 100	
	Avg.	Std.	Avg.	Std.	Avg.	Std.	Avg.	Std.
f_1	2.93E−95	1.22E−94	**0.00E+00**	0.00E+00	4.02E−94	1.82E−93	**0.00E+00**	0.00E+00
f_2	9.11E−52	3.09E−51	**4.73E−94**	2.13E−93	3.94E−50	1.79E−49	**6.64E−95**	2.51E−94
f_3	1.41E−78	6.44E−78	**0.00E+00**	0.00E+00	2.19E−66	9.79E−66	**0.00E+00**	0.00E+00
f_4	3.20E−49	7.47E−49	**3.75E−94**	1.14E−93	5.20E−49	1.44E−48	**1.95E−92**	5.01E−92
f_5	**0.006925**	0.007514	0.195261	0.294316	**0.03484**	0.044785	0.003743	0.04682
f_6	**0.000172**	0.000303	0.000679	0.000859	**0.000457**	0.000577	0.016174	0.015386
f_7	0.000213	0.000336	**9.07E−05**	7.85E−05	0.000213	0.000178	**1.42E−04**	1.48E−04
f_8	−12569	0.687329	**−12568.4**	1.09969	**−41851.6**	205.618	−41889.4	12.77664
f_9	**0**	0	**0**	0	0	0	**0**	0
f_{10}	**8.88E−16**	0	**8.88E−16**	0	**8.88E−16**	0	**8.88E−16**	0
f_{11}	**0**	0	**0**	0	**0**	0	**0**	0
f_{12}	**6.98E−06**	1.07E−05	7.85E−05	7.14E−05	**4.86E−06**	9.50E−06	1.80E−04	1.40E−04
f_{13}	**0.000128**	0.000206	0.000662	0.000593	**0.000146**	0.000213	0.013219	0.020858

Table 5 Number of calling times for updating method

	Dim 30	Dim100
	Avg. call	Avg. call
f_1	368.57	436
f_2	373.19	448
f_3	372.95	457
f_4	371.33	426
f_5	372.67	433
f_6	374.43	437
f_7	369.76	447
f_8	370.67	429
f_9	371.81	444
f_{10}	369.24	432
f_{11}	371.33	441
f_{12}	371.19	436
f_{13}	373.05	430

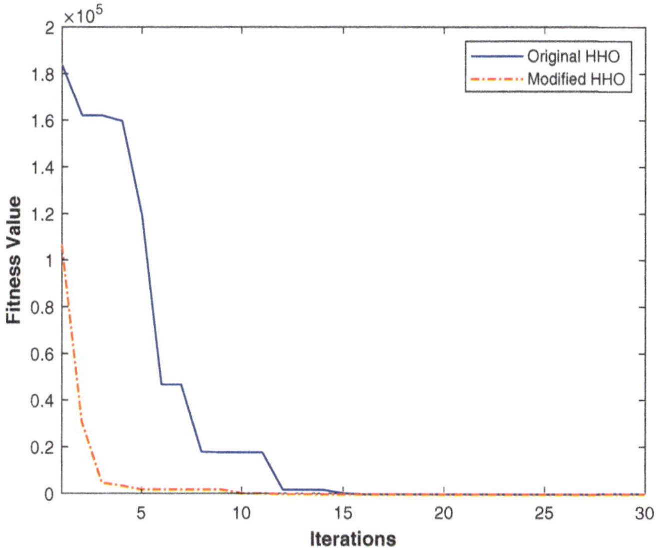

Fig. 4 Convergence diagram for f_3 with Dim = 30

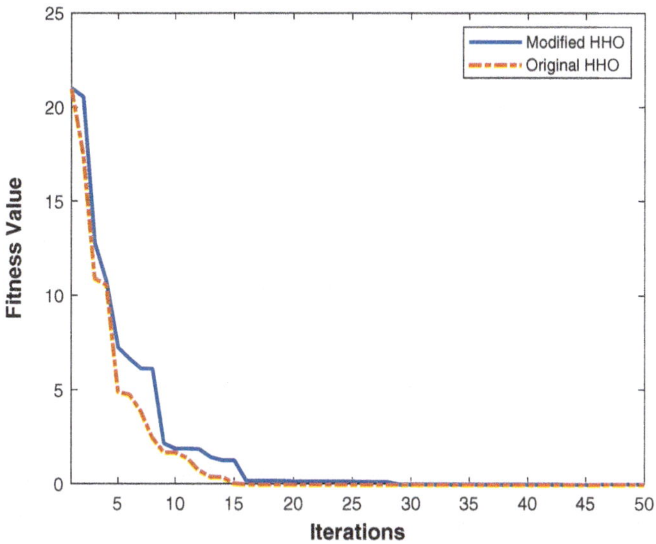

Fig. 5 Convergence diagram for f_{10} with Dim = 100

4 Conclusion and Future Works

In this chapter, we introduced a modified version of HHO algorithm by hybridizing it with k-means method. The k-means is employed to examine the population diversity of the population. Once the majority of solutions belong to one cluster and HHO can find any improvement for predetermined iterations (i.e., premature convergence case), we update the current population by replacing half of the current population with new solutions randomly. This updating process will redistribute the solutions over the search space, which will increase the exploration process by discovering new locations for the HHO. The obtained results show that the proposed redistribution approach has been called more than 70% of the overall number of iterations. Our future works will examine the proposed approach on several real NP-hard optimization problems.

References

1. Asiain, Erick, Julio B. Clempner, and Alexander S. Poznyak. 2019. Controller exploitation-exploration reinforcement learning architecture for computing near-optimal policies. *Soft Computing* 23 (11): 3591–3604.
2. Bäck, Thomas. 1996. *Evolutionary algorithms in theory and practice: Evolution strategies, evolutionary programming, genetic algorithms*. New York, NY, USA: Oxford University Press, Inc.
3. Bongard, Josh C., and Gregory S. Hornby. 2010. Guarding against premature convergence while accelerating evolutionary search. In *Proceedings of the 12th annual conference on genetic and evolutionary computation, GECCO '10*, 111–118. New York, NY, USA: ACM.
4. Bujok, Petr, Josef Tvrdk, and Radka Polkovi. 2019. Comparison of nature-inspired population-based algorithms on continuous optimisation problems. *Swarm and Evolutionary Computation* 50: 100490.
5. Črepinšek, Matej, Shih-Hsi Liu, and Marjan Mernik. 2013. Exploration and exploitation in evolutionary algorithms: A survey. *ACM Computing Surveys* 45 (3): 35:1–35:33
6. DeJong, Kenneth A., and Kenneth A. De Jong. 2002. *Evolutionary computation*. The MIT Press.
7. Digalakis, J.G., and K.G. Margaritis. 2001. On benchmarking functions for genetic algorithms. *International Journal of Computer Mathematics* 77 (4): 481–506.
8. Dorigo, Marco, and Gianni Di Caro. 1999. Ant colony optimization: A new meta-heuristic. In *Proceedings of the 1999 congress on evolutionary computation-CEC99 (Cat. No. 99TH8406)*, vol. 2, 1470–1477. IEEE.
9. Fogel, Lawrence J. 1999. *Intelligence through simulated evolution: Forty years of evolutionary programming*. New York, NY, USA: Wiley, Inc.
10. Heidari, Ali Asghar, Seyedali Mirjalili, Hossam Faris, Ibrahim Aljarah, Majdi Mafarja, and Huiling Chen. 2019. Harris hawks optimization: Algorithm and applications. *Future Generation Computer Systems* 97: 849–872.
11. Kelly, Jonathan, Erik Hemberg, and Una-May O'Reilly. 2019. Improving genetic programming with novel exploration—Exploitation control. In *Genetic programming*, ed. Lukas Sekanina, Hu Ting, Nuno Lourenço, Hendrik Richter, and Pablo García-Sánchez, 64–80. Cham: Springer International Publishing.
12. Mahdavi, Sedigheh, Shahryar Rahnamayan, and Abbas Mahdavi. 2019. Majority voting for discrete population-based optimization algorithms. *Soft Computing* 23 (1): 1–18.

13. Nguyen, B.H., B. Xue, P. Andreae, and M. Zhang. 2019. A new binary particle swarm optimization approach: Momentum and dynamic balance between exploration and exploitation. *IEEE Transactions on Cybernetics* 1–15.
14. Olorunda, O., and Engelbrecht, A.P. 2008. Measuring exploration/exploitation in particle swarms using swarm diversity. In *2008 IEEE congress on evolutionary computation (IEEE world congress on computational intelligence)*, 1128–1134, June 2008.
15. Rezapoor Mirsaleh, Mehdi, and Mohammad Reza Meybodi. 2018. Balancing exploration and exploitation in memetic algorithms: A learning automata approach. *Computational Intelligence* 34 (1): 282–309.
16. Sharma, Harish, Jagdish Chand Bansal, and K.V. Arya. 2014. Self balanced differential evolution. *Journal of Computational Science* 5 (2): 312–323 (Empowering Science through Computing + BioInspired Computing).
17. Shi, Yuhui. 2011. Brain storm optimization algorithm. In *International conference in swarm intelligence*, 303–309. Springer.
18. Shojaedini, Ehsan, Mahshid Majd, and Reza Safabakhsh. 2019. Novel adaptive genetic algorithm sample consensus. *Applied Soft Computing* 77: 635–642.
19. Stanovov, Vladimir, Shakhnaz Akhmedova, and Eugene Semenkin. 2019. Selective pressure strategy in differential evolution: Exploitation improvement in solving global optimization problems. *Swarm and Evolutionary Computation* 50: 100463.
20. Tan, K.C., S.C. Chiam, A.A. Mamun, and C.K. Goh. 2009. Balancing exploration and exploitation with adaptive variation for evolutionary multi-objective optimization. *European Journal of Operational Research* 197 (2): 701–713.
21. Wu, Guohua, Rammohan Mallipeddi, and Ponnuthurai Nagaratnam Suganthan. 2019. Ensemble strategies for population-based optimization algorithms—A survey. *Swarm and Evolutionary Computation* 44: 695–711.
22. Yao, Xin, Yong Liu, and Guangming Lin. 1999. Evolutionary programming made faster. *IEEE Transactions on Evolutionary Computation* 3 (2): 82–102.
23. Yelghi, Aref, and Cemal Kse. 2018. A modified firefly algorithm for global minimum optimization. *Applied Soft Computing* 62: 29–44.
24. Zhang, Hu, Jianyong Sun, Tonglin Liu, Ke Zhang, and Qingfu Zhang. 2019. Balancing exploration and exploitation in multiobjective evolutionary optimization. *Information Sciences* 497: 129–148.

A Review of Multiobjective Evolutionary Algorithms for Data Clustering Problems

Ruba Abu Khurma and Ibrahim Aljarah

Abstract Clustering is a challenging machine learning problem that deals with unlabeled data instances. It divides them into homogeneous groups according to similarity/dissimilarity between these instances. This process is called unsupervised learning. Detecting the data structure is not an easy task, especially if there are no previous assumptions to guide the clustering process. Multi-objective evolutionary algorithms have been used as alternative solutions in place of traditional clustering and single-objective evolutionary methods. By using multi-objective evolutionary algorithms, multiple objectives can be optimized simultaneously. This chapter introduces the clustering problem, multi-objective optimization, and reviews several multi-objective clustering techniques. It focuses on five subjects: chromosome encoding methods, fitness functions for optimization, evolutionary operators, storing current non-dominated solutions, and selection of the final solution from the non-dominated front. Finally, various applications of multi-objective clustering algorithms are discussed in the areas of image processing, bioinformatics and social network, etc. The chapter concludes by highlighting some of the open issues that could be investigated in the future.

Keywords Data clustering · Multi-objective optimization · Evolutionary computation · Nature-inspired algorithms · Swarm intelligence · Meta-heuristics unsupervised learning · Pareto optimality

R. Abu Khurma · I. Aljarah (✉)
King Abdullah II School for Information Technology, The University of Jordan, Amman, Jordan
e-mail: i.aljarah@ju.edu.jo

R. Abu Khurma
e-mail: RBA9150539@fgs.ju.edu.jo

I. Aljarah et al. (eds.), *Evolutionary Data Clustering: Algorithms and Applications*, Algorithms for Intelligent Systems,
https://doi.org/10.1007/978-981-33-4191-3_8

1 Introduction

Recently, the amount of data stored in data repositories such as databases, data warehouses, and metadata has increased the need for specialized data analysis tools [28]. One of the new technologies is unsupervised techniques. The merit of these methods is that they are data driven. This means that the learning process is not guided by labeled training data. However, unsupervised methods are inductive methods that, unlike supervised methods, do not rely on examples of training. Therefore, unsupervised learning is based on the distinct structure of the data. It uses some criteria to detect the hidden pattern of data points in the data space [35].

Clustering is a popular unsupervised method of learning from a massive amount of unlabeled data items. It has drawn the attention of researchers in the fields of bioinformatics, medicine, and pattern recognition. The clustering principle is to obtain homogeneity within clusters by grouping similar data items in the same cluster. Furthermore, it posses heterogeneity between clusters by distributing dissimilar data items into different clusters [5–7].

Some aspects considered challenges for the clustering process [67]. First, the same dataset may lead to different clustering solutions. This is because each clustering algorithm has its method of grouping data elements and each one may use different properties of these elements to determine their corresponding clusters. Thus, it is very difficult to obtain stable clustering results for the dataset using different clustering algorithms. For example, the K-means algorithm and graph-based partitioning algorithm formulate the clusters and determines the elements membership using different methods.

The second challenge for clustering algorithms is to optimize the cluster parameters like the number of clusters and cluster centers [40]. The clustering algorithm may fix this dilemma after spending a long time determining these numbers but still have a chance of failing because it is considering a single evaluation function to estimate the quality of a cluster such as the spatial separation, compactness, and connectedness.

The final challenge is that the traditional clustering algorithms such as K-means have a local minima [37]. This is due to the high sensitivity towards the initially selected centroids. The other major problem is that these methods apply a greedy search to optimize a single criterion. In the case of K-means, the algorithm tries to achieve the largest similarity between data instances in the cluster. This objective is called cluster compactness (intra-cluster cohesion).

In the literature, various evolutionary algorithms such as Genetic Algorithms (GAs) have been used as alternative solutions to address the problem of local minima [64]. Although these methods can mitigate the local minima problem and approach near global best, they are unable to deal with the different characteristics of the dataset simultaneously [1]. This is because the single-objective evolutionary algorithm has a fitness function that tries to optimize a single validity index [2]. Therefore, there has been a growing trend towards using multi-objective algorithms to simultaneously process different data characteristics. Considering different characteristics of the dataset will bring more benefits to the process of clustering optimization [9, 10, 58, 66].

The Multi-Objective Optimization Problem (MOOP) depends on a number of conflicting objectives. At least, two competing criteria must be optimized. Unlike scalar-objective optimization, the MOOP does not have a single solution to make optimization for all the evaluation functions simultaneously. However, there will be trade-off solutions generated from the set of all possible solutions [45].

Evolutionary Algorithms (EAs) have a major role in tackling the MOOPs and have been proposed in the literature to address various MOOPs in many applications. Due to the promising results achieved by Multi-Objective Evolutionary Algorithms (MOEAs), they are useful for use in the clustering problem [45]. Dealing with multiple objectives simultaneously may help consider different data properties that may improve the robustness of the clustering method and enhance clustering results.

In general, the MOEAs have two different methods for multi-objective optimization: the first is to convert the multi-objective problem into a single-objective problem based on a formula that gives a rank to each objective based on its importance. The second is the Pareto-based method of generating a group of non-dominated solutions by simultaneous optimization of the objectives. The obtained solutions are trade-off solutions that can't be further optimized based on one objective without adversely affecting other competing objectives.

In the clustering problem, there is no advance information about available objectives so that their relative importance can't be detected in advance [3]. Thus, the first method that relies on the order of objectives in a single-objective function based on their weights, then improving this function may degrade the clustering results [4, 7, 24, 25, 59].

The second multi-objective optimization is more convenient for clustering because it optimizes objectives simultaneously instead of merging them into a single-objective function. The obtained trade-off solutions can then be compared based on user needs and problems at hand.

This chapter reviews the subject of evolutionary multi-objective clustering. Due to the low coverage of this research area, we have directed our efforts to systematically study many of the previous works in this area. The chapter discusses the clustering problem, multi-objective optimization, and poses a framework for evolutionary multi-objective clustering. Besides, a detailed discussion is provided about the different multi-objective clustering methods and finally, the applications of multi-objective clustering are described.

The chapter is structured as follows: Sect. 2 presents the clustering problem. Section 3 identifies clustering as an optimization problem. Section 4 explains the concepts of multi-objective optimization. Section 5 discusses the importance of multi-objective clustering. Section 6 poses a framework of evolutionary multiple objectives clustering. Section 7 discusses various multi-objective clustering techniques. Section 8 discusses the applications of multi-objective evolutionary clustering. Finally, Sect. 9 contains the conclusion of the chapter and paves the way for future studies.

2 Clustering Problem

Clustering is a main machine learning process to determine the hidden structure of a group of unlabeled data instances [8]. The major challenge of clustering is that the learning process is unsupervised because there is no knowledge available about the structural characteristics of the dataset such as distribution, volume, shapes, etc. A simple definition for Clustering is that it is a process of finding the intrinsic grouping of a set of instances and distribute them into homogeneous sets based on certain grouping criteria. Accordingly, there are two clustering methods: distance clustering and conceptual clustering. The former one uses the similarity measure (distance) to collect close instances in the same cluster. The conceptual clustering collects instances in clusters based on a common descriptive concept.

In general, no best criterion can be determined in an absolute way. The goal of clustering is user-dependent which means that the user can define the clustering criterion based on his/her needs. For example, the user may be interested in finding similar behavior or unusual behavior (outliers detection) in a customer database.

Formally speaking, clustering is an unsupervised learning method for dividing space into K areas $\{C_1, C_2, \ldots, C_K\}$ based on some criteria. K is the number of clusters that maybe not given in advance. Given a set of data points $\{X_1, X_2, \ldots, X_n\}$ that are distributed in a multi-dimensional space, the purpose of clustering is to produce a partition matrix of size $K \times n$. The partition matrix is named by $U = [u_{ij}]$ where $i \in [1, K]$ and $j \in [1, n]$. The element of the partition matrix denoted by u_{ij} means that an instance x_j is a member of the cluster C_i. Based on this definition, there two types of clustering: partitional and overlapping [36]. In partitional clustering, there are hard partition of objects into mutually disjoint groups so that $C_{r1} \cap C_{r2} = \phi$ where $r1 \neq r2$ and $C_1 \cup C_2 \cup \ldots \cup C_K = X$ where $C_i \neq \phi$. In other words, hard partition clustering has a condition that $u_{ij} = 1$ iff $x_j \in Ci$ and $u_{ij} = 0$ iff $x_j \notin Ci$. The other clustering method is the overlapping clustering. It relaxes the partitional clustering constraints. Hence, the data instance may belong entirely to one or more clusters such as in soft clustering or may belong to different clusters with different degrees such as in fuzzy clustering. Thus, the condition in fuzzy clustering is $0 \leq u_{ij} \leq 1$.

Figure 1 illustrates the results of clustering for some clustering algorithms on toy datasets. The used clustering algorithms are MiniBatchKMeans, AffinityPropagation, MeanShift, SpectralClustering, Ward, AgglomerativeClustering, DBSCAN, OPTICS', Birch and GaussianMixture. The used datasets are noisy-circles, noisy-moons, varied, aniso and no-structure. This example illustrates the diverse behavior of different clustering algorithms and gives some intuition on their performance across 2D datasets. The distribution of points shows that in most cases the clustering results were good. On the other hand, the last row in Fig. 1 shows that the dataset is fairly homogeneous so the clustering results were not good.

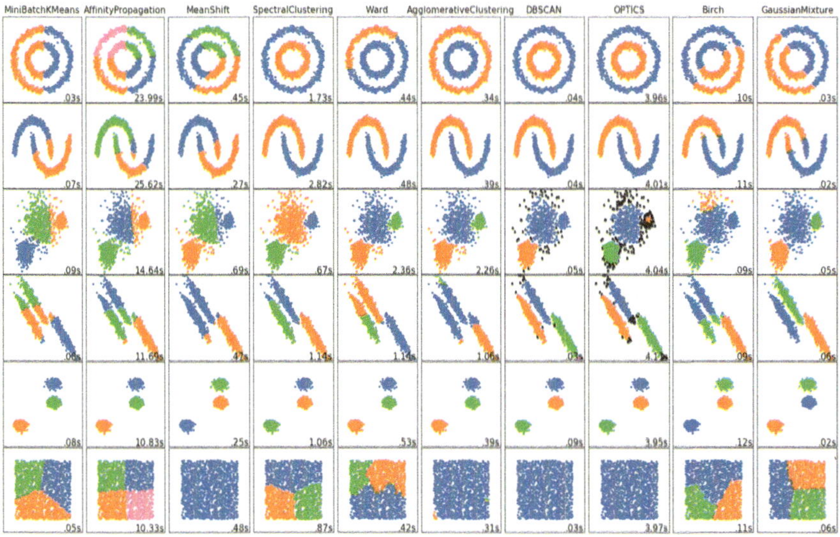

Fig. 1 Clustering example

3 Formulating Clustering as an Optimization Problem

Clustering is formalized as an optimization problem because it searches to determine the best grouping of data points in such a way the grouping criteria are optimized. These grouping criteria are called validity indices. Consequently, the objective function of an optimization algorithm may rely on optimizing one cluster validity index (i.e., single-objective algorithm) or a set of validity indices (i.e., multi-objective algorithm). The search space of the clustering problem represents the set of all possible groupings of data points. Thus, searching to find the best grouping of data points in this large search space makes the clustering problem an NP-hard problem. Formally, the clustering problem is formulated as an optimization problem [32].

Let X a finite set of data points, a distance measure $\delta(i, j) \in \Re_0^+$ for $i, j \in X$, a positive integer K and a criterion function $J(C, \delta(.,.))$ on a K-partition $C = C_1, \ldots, C_K$ of X and measure $\delta(.,.)$. An optimization clustering problem seeks to find the partition of X into disjoint sets $\{C_1 \ldots C_K\}$ that maximizes the expression $J(C, \delta(.,.))$. The number $R(K, N)$ that represents the possible solutions for the division of a dataset of size N into K partitions is given by Eq. 1:

$$R(K, N) = \frac{1}{K!} \sum_{i=1}^{K} (-1)^{K-i} \binom{K}{i} (i)^N \approx \frac{K^N}{K!} \qquad (1)$$

The search space of the clustering problem increases in exponential manner even if K number of partitions is fixed. This means that exhaustive search methods will not be able to traverse all the possible solutions even for medium-sized problems.

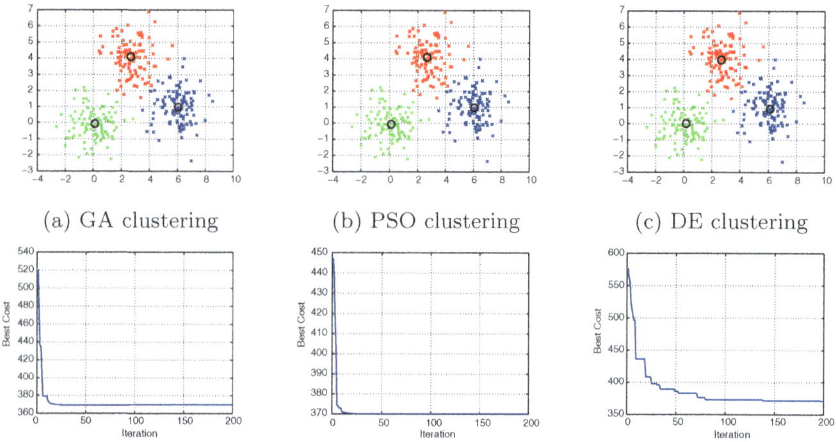

(a) GA clustering (b) PSO clustering (c) DE clustering

(d) GA convergence (e) PSO convergence (f) DE convergence

Fig. 2 Evolutionary clustering

Figure 2 shows an example for evolutionary clustering. In this example, Genetic algorithm (GA), Particle Swarm Optimization (PSO), and Differential Evolution (DE) are used to optimize the clustering problem. Clusters generated using these evolutionary algorithms differ in their convergence behavior. As seen that PSO achieved the best cost then GA and finally DE.

4 Multi-objective Optimization Problem

The Multi-Objective Optimization Problem (MOPs) optimizes more than one objective. These objectives are usually conflicting which indicates that no solution can be found to optimize all these objectives simultaneously [45]. Instead, trade-off solutions called Pareto optimal solutions are found among different objectives which are called Pareto Front (PF) in the objective space and Pareto Set (PS) in the solution space. Many Pareto optimal solutions can be generated [69]. However, the task is to find those solutions that converge closer to PF and distribute uniformly among them [72]. Equation2 formally describes the multi-objective optimization problem.

$$\begin{aligned} min\ F(x) &= (f_1(x), f_2(x), \ldots, f_k(x))^T \\ subject\ to\ x &= (x_1, x_2, \ldots, x_n)^T \in \Omega \end{aligned} \tag{2}$$

where $F(x)$ consists of k objective functions and $f_k(x)$ represents the k_{th} objective function. $x \in \Omega$ is a decision variable vector, Ω is the decision space, and n is the dimension number of the decision variable x. It is said that the decision vector $x_A \in \Omega$ dominate the decision vector $x_B \in \Omega$ If and only if the Eq. 3 is satisfied.

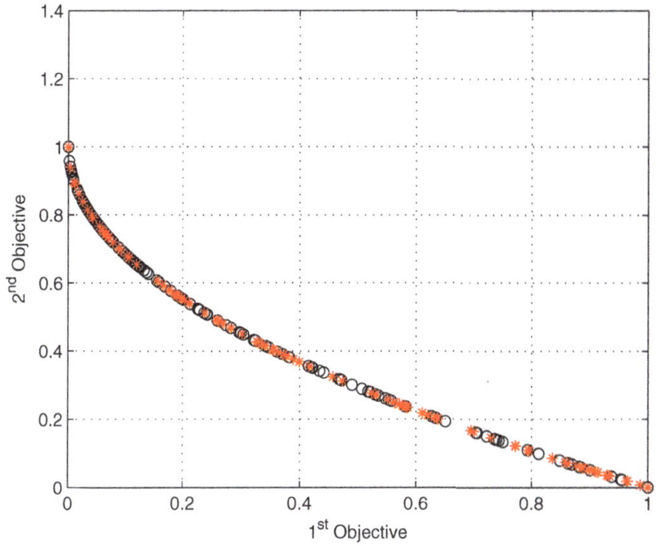

Fig. 3 Near-optimal Pareto solutions generated by MOPSO

$$\forall i = 1, 2, \ldots, k,\ f_i(x_A) \preceq fi(x_B) \\ \wedge \exists j = 1, 2, \ldots, kf_j(x_A) \prec f_j(x_B) \tag{3}$$

A vector $x^* \in \Omega$ is Pareto optimal if no vector $x \in \Omega$ can dominate it. In other words, x^* is a non-dominated solution. The Pareto Optimal Set (PS) consists of all non-dominated vectors. The set composed of all the Pareto optimal objective vectors is called the Pareto Front (PF). The definition of PF can be written as in Eq. 4:

$$PF = F(x^\star) = (f_1(x^\star), f_2(x^\star), \ldots, f_k(x^\star))^T \mid x^\star \in PS. \tag{4}$$

All the potential solutions for the MOP can be found in PS. Furthermore, the MOP can be defined as a search for a set of non-dominated vectors that can approximate the PF.

Figure 3 shows the Near-optimal Pareto solutions generated by a Multi-Objective Particle Swarm Optimization Algorithm (MOPSO).

5 Importance of Multi-objective Clustering

As data dimensions increase and data items distribute sparsely in the data space, the results of traditional clustering methods become weak. The main problem is that considering a single criterion for clustering cannot fit all clusters in the entire feature space. There may be some clusters of mysterious shapes hidden in sub-spaces of the original feature space [27]. Thus, reliance on a single criterion may result in clusters of similar shapes and have a close data density. Obviously, a good clustering method should carefully choose the appropriate optimization criterion for clustering.

There are several valid clustering criteria for a dataset, many of which are conflicting proprieties. Furthermore, data distribution is usually unknown at the beginning of the clustering process. For these reasons, the simultaneous optimization of multiple complementary objectives will solve most of the clustering problems and obtain more reliable solutions. A multi-objective clustering algorithm applies the Pareto optimization which considers multiple clustering qualities simultaneously. The algorithm focuses on optimizing a group of clustering objectives and creating a Pareto front set consisting of a set of trade-off solutions.

6 Framework of Evolutionary Multiple Objectives Clustering

To build a general evolutionary multiple objectives clustering framework, there is a need to review some aspects of evolutionary single-objectives clustering. To design an evolutionary single-objectives clustering, there are three aspects to consider [50]:

- The encoding scheme of the solution. This describes the representation of a clustering solution using a chromosome.
- The selection of appropriate evolutionary operators to manage the update of solutions at each evolutionary cycle.
- Choose the fitness function that guides the reproductive mechanism by assessing the closeness of the candidate solution is to the optimal level.

In the case of evolutionary multi-objective clustering, there are some additional issues to consider:

- First, there is more than one fitness function that must be optimized simultaneously. The fitness functions should be carefully defined in a manner that takes into account each specific clustering validity index.
- Second, there must be a specific tracking strategy to store the non-dominated solutions (PF).
- Third, there should be an appropriate strategy for determining the clustering solution from the non-dominated Pareto near-optimal solutions.

All the aforementioned points must be taken into account when designing a new MOEA. Algorithm 1 is the Pseudocode for an evolutionary multi-objective clustering in general.

Algorithm 1 Pseudocode for a general evolutionary multi-objective clustering

```
Input: Dataset, the parameters of the algorithm.
Output: Clustering results of dataset
i = 0;
Pi=Initialize Population(Dataset,Algorithm parameters);
Fiti=Evaluate Population(Pi);
  while Stopping condition is not met do
    i = i + 1;
    Mati=Select(Pi-1, Fiti-1);
    Crossi=Crossover(Mati);
    Muti=Mutation(Crossi);
    Fiti=Evaluate Population(Muti);
    Pi=Update Population(Pi-1, Fiti-1,Muti, Fiti);
    C=Select From Pareto Front(Pi);
  end while
  return C
```

7 Overview of Multi-objective Clustering Techniques

This section provides a detailed information about the major multi-objective evolutionary methods. These methods were suggested in the literature to solve the clustering problem. An overview of the key studies will be provided based on the points discussed in Sect. 6.

7.1 Encoding Schemes of Evolutionary Clustering Solution

This subsection summarizes some of the different encoding schemes that have been adopted in the literature to represent the evolutionary algorithm solution to the clustering problem. In general, there are two common ways to represent a clustering solution: prototype-based method and point-based method. In the prototype-based method, the clustering solution encodes some points representing the cluster like medoids, centroids, and modes. On the other hand, the point-based method encodes the complete clustering solution by specifying for each point in the data space its associated cluster. Table 1 summarizes these methods based on their encoding scheme, advantages, and disadvantages and provides some references to the studies that have adopted each method.

7.2 Objective Functions of MOEA for Clustering Problem

Because the clustering problem is multi-objective, there is a set of objective functions that have to be optimized simultaneously. In the literature, different combinations

Table 1 Summary of the encoding schemes for the MOEA clustering solution

Encoding scheme		Strategy	Pros	Cons	References
Prototype-based	Centroid-based	The chromosome has real values that represent the coordinates of the cluster centers. If K is the number of cluster centers and d is the number of dimensions then the length of the chromosome will be $d \times K$	1. The lengths of chromosomes are small 2. Less time to apply genetic operators 3. The algorithm converges quickly 4. Determine the overlapping clusters	1. Cannot capture the clusters with inconsistent shapes 2. Variable lengths of chromosomes will make the design of the evolutionary operators very complex 3. Unsuitable for high dimensional data because this will increase the chromosome length	[12, 13, 43, 44, 46, 47]
	Medoid-based	Encode in the chromosomes the cluster medoids or the indices of the points representing the cluster medoids. The optimization process tries to minimize the sum of distances of the cluster points to the medoid. The chromosome length equals the number of clusters K	1. Medoids differs from centroids in that they do not suffer from the presence of outliers 2. Medoids can be computed based on a distance matrix without the information of the coordinates of the points 3. Used with categorical dataset	Finding the medoid is computationally expensive when the number of points is large	[48, 61] (not commonly used in MOEA clustering)
	Mode-based	Cluster modes are used to represent the clusters. In chromosome representation, K represents the most frequent value for that feature (cluster modes). The length of a chromosome will be $K \times d$	1. Has advantages for the datasets with categorical datatype because the centroid cannot be computed 2. The cluster modes are computed much faster than cluster medoids	Cluster modes may have noise or outliers	[49, 51]
Point-based	Cluster Label-based	All the data points are encoded instead of only the representatives/prototypes of the clusters. In this representation, the length of a chromosome is equal to the number of points in the dataset and each position denotes the cluster label of the associated point. If position i of the chromosome is assigned a value k, then the i_{th} data point is put in cluster k. Thus, a chromosome can be composed of only integer values $1, 2, \ldots, K$, where K is the maximum number of clusters	1. Easy decoding step because the data points are already given the cluster labels in the encoding 2. Can capture clusters with arbitrary shape 3. Suitable for any data types such as continuous and categorical 4. The lengths of the chromosomes are fixed and because they do not rely on the number of clusters encoded in the chromosome	1. The chromosome length equals to the number of points. Thus, it becomes slower for large datasets 2. It produces a large number of redundant solutions 3. Applying crossover/mutation may cause the clustering structure to get damaged 4. Does not support fuzzy clustering since each data point is assigned to exactly one cluster so that overlapping clusters can not be detected	[21, 29, 33, 54, 55, 57, 57] (Most common technique)
	Locus-based Adjacency Graph	In this representation, a chromosome has n positions for the number of data items. If position i has an integer value j, it means that i and j are linked. These data items are added to the same cluster in the clustering solution. Thus, each chromosome is represented as a graph, the data items are the vertices and the links between two data points are the edges in the graph	1. The number of clusters is encoded in the chromosome automatically 2. The number of redundant chromosomes is small 3. Can catch clusters with arbitrarily shaped	1. The chromosome length is equal to the number of data items so it is not practical with large datasets 2. Clustering process does not support fuzziness	[14, 30, 31]

of clustering validity indices were used to formulate the objective functions of the evolutionary clustering algorithm. The clustering validity indices represent different clustering criteria such as compactness, separation, and cluster density. These indices are either related to cluster prototypes or cluster labels. This subsection presents some of the validity indices used in MOEA to solve the clustering problem. Table 2 summarizes some of the MOEA objective functions used in the literature for the clustering problem.

7.3 Design of Evolutionary Operators

An important aspect to design a multi-objective evolutionary framework is to pay attention to the evolutionary operators such as selection, crossover, and mutation and how to refine them to work in the multi-objective space.

Selection operators are used in the evolutionary algorithm to determine which chromosomes will be more reproduced to generate new children for the next generation. In literature, there are many strategies for applying selection operators in multi-objective space. The main purpose of the selection strategy is to obtain different possible trade-offs among the fitness functions to get a variety of solutions in the non-dominated set.

A region-based selection strategy was used to divide the objective space into a set of hyperboles and each solution is put in a certain hyper box according to its fitness values [17]. Thus the selection is done for a hyper box instead of an individual solution. The main goal is to choose the hyper box with a fewer number of solutions.

Another selection strategy used in the literature is a crowding distance-based binary tournament selection strategy [20]. A population and the child population are combined in this strategy then a non-dominated sorting is used to give a rank for every solution. The P top solutions are then selected (P is the population size) using a crowding factor. The crowding factor represents the density of the area in which the solution locates in the multi-objective space. For two solutions, the one that will be selected is the one with a higher rank. However, if the two solutions have the same rank then the one that will be selected is the one from a less crowded area. Crowding distance is computed by determining the position of the solution in the objective space and drawing a hyper box around this solution which includes other succeeding and preceding solutions. The volume of the hyper box determines its crowding distance. The larger crowding distance means the solution is from a less crowded region. The idea is to have a higher diversity of solutions in the Pareto front.

Crossover is another evolutionary operator that exchanges the genes between the chromosomes to generate new children to be used in the next generation. For the multi-objective clustering algorithms, the crossover operator may be a single-point or two points operation. In prototype-based chromosomes, the crossover point is selected to fall between two cluster centroids/modes to avoid sudden disturbances of the cluster centroids/modes [12, 49]. Two points crossover is used to deal with

Table 2 MOEA objective functions for clustering problem

Validity index	Objective function	Description	Mathematical formula	References
Based on cluster prototype	J_m index	The global fuzzy cluster variance	$J_m = \sum_{k=1}^{K} \sum_{i=1}^{n} u_{ki}^m D^2(z_k, x_i)$ (5) where $D(z_k, x_i)$ is the distance of the i_{th} data point x_i from the k_{th} cluster center z_k. n is the number of data points. K is the number of clusters and u_{ki}^m is the membership degree of the i_{th} data point to the k_{th} cluster	[33]
	Overall cluster deviation	The crisp version of J_m index	$Dev(C) = \sum_{C_k \in C} \sum_{x_i \in C_k} D(z_k, x_i)$ (6)	[19]
	Davies-Bouldin Index	Represents a function of the ratio of the sum of within-cluster scatter to between-cluster separation	$DB = \frac{1}{K} \sum_{i=1}^{k} R_i$ (7) $R_i = max_{j, j \neq i} \left\{ \frac{S_i + S_j}{d_{\{ij\}}} \right\}$ (8) where S_i and S_j is the scatter within the i_{th}, j_{th} clusters, respectively. d_{ij} is defined as the distance between the centers of the two clusters	[71]
	Xie-Beni Index	Defined as a function of the ratio of the total fuzzy cluster variance σ to the minimum separation sep of the clusters	$\sigma = \sum_{k=1}^{K} \sum_{i=1}^{n} u_{ki}^2 . D^2(z_k, x_i)$ (9) $sep = min_{k \neq l} \left\{ D^2(z_k, z_l) \right\}$ (10)	[62]
	Intra-Cluster Entropy	Used to measure the average purity of clusters without using the class labels of data points	$H = \sum_{i=1}^{K} [(1 - H(C_i).g(z_i))]^{\frac{1}{K}}$ (11) where $H(C_i) = -[g(z_i) \log_2 g(z_i) + (1 - g(z_i)) log_2 (1 - g(z_i)))]$ (12) $g(z_i)$ is the average similarity between a cluster center z_i and the data points belonging to cluster i that is defined in terms of cosine distance	[39]
	Cluster Separation	Cluster Separation Sep or Inter-cluster distances is the average distance among the cluster centers	$Sep(C) = \frac{2}{K(K-1)} \sum_{i=1}^{K} \sum_{j \neq i} D^2(z_i, z_j)$ (13) where C is the set of all clusters	[42]
	Average Between Group Sum of Squares (ABGSS)	Similar to the cluster separation measure that is computed as the average distance of the cluster centers from the centroid of the dataset	$ABGSS = \sum_{i=1}^{K} n_i . D^2(z_i, \bar{z})$ (14) where C is the set of all clusters	[56]
Based on cluster labels	Dunn Index	Computes the ratio of the distance between two clusters $\delta(C_i, C_j)$ to the maximum diameter of the clusters $max_{1 \leq k \leq K} \Delta(C_k)$	$DN = min_{1 \leq i \leq K} \left\{ min_{1 \leq j \leq K, j \neq i} \left\{ \frac{\delta(C_i, C_j)}{max_{1 \leq k \leq K} \Delta(C_k)} \right\} \right\}$ (15) where $\delta(C_i, C_j) = min_{x_i \in C_i, x_j \in C_j} D(x_i, x_j)$ (16) and $\Delta(C_i) = max_{x_i, x_k \in C_i} D(x_i, x_k)$ (17)	[23]
	Cluster Connectedness	Describes the connectivity level of clusters that can be measured using the connectedness index between the data point and its neighboring points	$Conn(C) = \sum_{i=1}^{n} \left(\sum_{j=1}^{L} k_{i, nn_{i,j}} \right)$ (18) where $k_{i, nn_{ij}} = \frac{1}{j}$ if $\nexists C_k : i \in C_k \wedge nn_{i,j} \in C_k$ and $k_{i, nn_{ij}} = 0$ otherwise. nn_{ij} is the $j_t h$ nearest neighbor of data point i. L is a user defined parameter that decides the number of neighbors and impacts the connectedness measure	[33]

(continued)

Table 2 (continued)

Validity index	Objective function	Description	Mathematical formula	References
	Edge Index	Measures the distance between boundaries of the clusters	$Edge(C) = -\sum_{i=1}^{n} \sum_{j \in IF_i} \xi_{ij}$ (19) where $\xi_{ij} = D(x_i, x_j)$	[65]
	Silhouette Index	It is the ratio between a_i that represents the average distance of a point x_i from the other points in same cluster, and b_i is the minimum of the average distances of x_i from the points of the other clusters	The silhouette width s_i can be formulated as $s_i = \frac{b_i - a_i}{max\{a_i, b_i\}}$ (20) and Silhouette index S is defined as follows: $S = \frac{1}{n} \sum_{i=1}^{n} S_i$ (21) where $-1 \leq S \leq 1$	[63]
	Min-Max Cut	It is the ratio of the distances among the points within a cluster and the distances of the points from different clusters	$MinMaxCut(C) = \sum_{i=1}^{k} \frac{\sum_{j \in C_i} \sum_{k \notin C_i} D(x_j, x_k)}{\sum_{j \in C_i} \sum_{k \in C_i} D(x_j, x_k)}$ (22)	[21]
	Total Within Cluster Variance	Describes the variance between points of the same cluster	$TWCV = \sum_{i=1}^{n} \sum_{j=1}^{d} x_{ij}^2 - \sum_{k=1}^{K} \frac{1}{n_k} \left(\sum_{x_i \in C_k} x_{ij} \right)$ (23) x_{ij} denotes the j_{th} feature value of the i_{th} data point, and n_k denotes the number of points in cluster C_k	[41]

the variable lengths of chromosomes when the centroid is encoded in chromosomes [43].

The mutation is an operator that causes small changes in the chromosomes to enforce diversity in the population. There are many strategies for applying mutation on the multi-objective clustering algorithm. These strategies can be classified based on the encoding methods of the solution.

An example of applying mutation is generating a real value between 0 and 1 and from a uniform distribution and use it to shift the position of a randomly selected centroid [49]. In general, for prototype-based encoding such as medoid-based, mode-based and centroid based, the mutation operators need to change only in few positions because the length of the chromosomes is small so that so small changes may heavily change the cluster structure. For point-based encoding, mutation operation takes a longer time because the chromosome has a larger length and the mutation is done in one or two positions in the chromosomes to update the clustering structure. Low mutation probability does not result in much diversity in the population resulting in a low convergence rate. To mitigate this problem, a mutation in one position results in mutations in neighboring points also [33].

7.4 *Maintaining Non-dominated Solutions*

One important issue that the MOEA has to take into consideration is the preservation of the non-dominated solutions. Storing the group of current non-dominated solutions can be done by two ways: either by storing them in the population itself or keeping them outside the population in a separate archive.

7.4.1 Storing Non-dominated Solutions Within Population

In this method, a non-dominated sorting is done for the current population and the generated child population to give different ranks for the solutions. Elitism is used then to select half of the population for the next iteration. NSGA-II-based [20] is a multi-objective clustering algorithm that stores the non-dominated solution in the same population. This method was used by many MOEAs including [12, 16, 61].

The advantage of this method is that the multi-objective algorithm and the single-objective algorithm are similar. Furthermore, there is no need for a separate archive to maintain the non-dominated front which requires additional overhead to update the external archive in each iteration. The disadvantage of this method is that the crowding distance criterion may be used to truncate the non-dominated solutions when they are going beyond the population size. This means that this method cannot be applied in some cases that require generating a large number of trade-off solutions. For example, when the dataset is complex or the number of clusters is large.

7.4.2 Storing Non-dominated Solutions in Archive

Archive is another method for storing non-dominated solutions. PESA-II is an example of this method which has two populations: internal and external. The non-dominated solutions in the internal population enter the external population (archive) when they are non-dominated by any solution in the archive. All the other dominated solutions are truncated using a crowding factor, when the non-dominated solutions are added to the archive. Examples on MOEAs that adopted this method to store the non-dominated solutions include [33, 60, 65].

The advantage of this method is that the evolutionary operators can be applied to the internal population only. Because the archive does not include the application of evolutionary operators, population size can be increased without increasing the complexity of time. The disadvantage of this method is the requirement for maintaining the archive and updating it at each iteration.

7.5 Deciding the Final Clustering Solution

The outcome of an MOEA for clustering is a set of trade-off solutions, that cannot be optimized more. This is because the improvement of a solution for one objective may lead to degrading another objective. Thus, there should be a specific method to be applied by the MOEA clustering algorithm to obtain the final clustering solution from the resulting non-nominated solutions. In literature, there are three main methods for doing this based on an independent objective, finding the knee of the non-dominated front, and using the ensemble of the non-dominated solutions.

In the first method, the final clustering solution is obtained from the set of non-dominated front independently based on a specific validity index which was not considered in the objective function of the MOEA [12, 46]. This is a simple method that requires a little time complexity. However, it may be biased to the selected validity index. In the second method, the interesting final solution is one for which the change in one objective value results in maximizing the values of the other objectives. This solution is very important for the non-nominated Pareto front [14, 30].

The first and second methods are based on some metrics or proprieties of the Pareto optimal front. The major problem is that selecting one solution may deny important clustering information that was gathered by other discarded non-dominated solutions. To address this problem, a different ensemble method has been proposed to combine all the non-dominated solutions in one final solution [44, 47]. The idea is to integrate a set of MOEAs in one clustering framework and use a supervised classifier to select among the solutions in the non-dominated set. The instances that are usually clustered together by the majority voting of different MOEAs are most likely to be clustered correctly. These instances are used to train the supervised classifier to cluster the remaining instances that are not yet clustered. The positive of this method is that it treats all the non-nominated solutions equally and gives the same importance.

8 Applications of Multi-objective Evolutionary Clustering

The MOEAs have been used to solve the clustering problem and find optimized solutions in several domains. This section discusses some major real-life applications that have used MOEAs for clustering purposes including image processing, bioinformatics, social networks, and more.

8.1 *Image Applications*

Image segmentation is a major image application that has extensively used MOEAs to divide the image into several disjointed regions with the same features. Image segmentation serves various fields which concern image understanding and pattern recognition.

One of the image segmentation methods is the image clustering. In 2008, a Multi-Objective Evolutionary Clustering Ensemble Algorithm (MECEA) was proposed for texture image segmentation [60]. The image segmentation was applied in two phases where MECEA was applied in the first phase to trade-off compactness within a cluster and connectedness between clusters. The second phase used the MCLA to select the last segmentation solution from the group of Pareto solutions obtained from the first phase.

One year later, a new method for evolutionary multi-objective clustering image segmentation was proposed. Two objectives were optimized simultaneously: the overall deviation and edge value. They have a significant role in image segmentation.

A new multi-objective nature-inspired clustering (MoNiC) algorithm was proposed for image segmentation [70]. The study explained identified multiple objectives associated with the image segmentation problem.

In [73], a Multi-Objective Evolutionary Fuzzy Clustering (MOEFC) algorithm was proposed. It used the decomposition strategy to divide the problem into subproblems where each sub-problem incorporates local image information and represents a fuzzy clustering problem. Opposition-based learning was used to improve the search process. The results showed that the proposed clustering algorithm was able to achieve a trade-off between two necessary complementary objectives for image segmentation which are preserving image details and removing noise.

Later on, a Multi-Objective Evolutionary Intuitionistic Fuzzy Clustering Algorithm (NR-MOEIFC) was proposed for noisy image segmentation [74]. The Intuitionistic Fuzzy Set (NR-IFS) was able to overcome the noise in the corrupted image. NR-IFS considered the membership and non-membership degree to describe the blurred features in the contaminated image. The results revealed that NR-MOEIFC improved noise removal and segmentation performance, and at the same time reduced the time.

For more information about this multi-objective clustering with meta-heuristic in image segmentation, a reader can refer to a survey [15].

8.2 *Bioinformatics*

MOEAs clustering algorithms have been used in wide applications in bioinformatics. These applications include grouping coexpressed genes according to their expressions because genes that have close patterns for their expressions may have similar functionality [51]. Furthermore, microarray technology was used to monitor the levels of a large number of gene expression [48]. A recent application of MOEA in bioinformatics is the identification of complex protein such as the clustering protein interaction networks [52].

In [48], the main objective was to search for a suitable set of cluster centers to optimize the validity index of the cluster. For achieving this purpose, fuzzy partitioning and Genetic algorithm were utilized to perform the clustering. Simultaneous optimization of several validity indices was considered. The performance of the algorithm was compared with partitional and hierarchical methods on Yeast Sporulation and Human Fibroblasts Serum datasets.

In [68], a Differential Evolution (DE) algorithm was used to apply fuzzy clustering Multi-Objective Optimization (MO). The target was to simultaneously optimize two fuzzy validity indices. The user can select from the resulting Pareto optimal set from the most promising solutions based on the problem at hand. The multi-objective DE-variants were compared with two common MO clustering, algorithms namely the Non-Dominated Sorting Genetic Algorithm (NSGAII) and Multi-Objective Clustering with an unknown number of Clusters K (MOCK). The results proved that DE is a promising algorithm for MO clustering schemes.

Recently, an interactive version of MOEA has been developed to interact with human decision-makers and periodically adjust the objective functions in an interactive way to cope with new gene expression data.

8.3 *Social Networks*

The social network is a graph that represents the interaction between individuals. There is much useful information that can be obtained from social networks related to social behavior, marketing, and web characteristics. Clustering plays an important role in social networks where each cluster shares common characteristics. In [38], proposed the integration of Dynamic and Multi-Objective Evolutionary Algorithms (DMOEA) for online social network clustering in dynamic environments.

In [16], a new dynamic clustering multi-objective evolutionary algorithm was proposed. The proposed algorithm was different from traditional static clustering methods in that the length of a chromosome was variable which allowed the algorithm to search for both optimal cluster center positions and cluster numbers. The cluster number was optimized during dynamically run time instead of being prespecified in advance. NSGA-II algorithm was adopted to simultaneously optimize the compactness and connectedness.

The multi-objective optimization was used to discover the evolving communities in dynamic [26]. The proposed algorithm optimized two main objectives related to the accuracy of the resulting clusters and the difference between successive steps. Experiments on synthetic datasets showed improved performance of the method.

Authors in [11] proposed the multi-objective optimization to solve the community detection problem based on a combination of the Harmony Search Algorithm (HSA) and Chaotic Local Search (CLS). The external repository was used to save non-dominated solutions found during the search process and a fuzzy clustering method was used to control the size of the repository. The results from applying the proposed algorithm on synthetic and real networks showed that more accurate community structures were discovered.

The community detection problem was redefined in [34] using the multi-objective model to get the intra- and inter-community structures simultaneously. MOEA/D and the community detection model were applied in complex networks to detect the overlapped community sets.

8.4 Other Applications

MOEAs were used in various other applications such as web recommender systems, software module clustering, security vulnerability assessment, and classification of actions of 3D human models.

Web recommender systems is an application used to cluster data that represents the user sessions [22]. Sequence clustering is used to draw a weighted, undirected graph that shows the visited webpages by a user sequentially. Similar pages are connected via an edge. Hence, the clustering problem is actually becoming a graph partitioning problem which is considered as NP-hard problem and needs some kinds of MOEA to be solved. The primary objective of the paper was to determine the most efficient MOEA to cluster sequence data by conducting multiple comparisons with other methods in the literature on sample datasets and the best method is selected.

In [57], two novel multi-objective algorithms were proposed to automatically solve the software module clustering problem. The main target was to improve the program structure by utilizing an MOEA to organize the software units into modules.

The MOEA task is to identify the module boundaries by search procedure using an objective function. The proposed objective function worked to optimize two conflicting objectives, namely, cohesion and coupling. The required objectives were to increase cohesion and decrease coupling. This paper introduces the software module clustering problem using two new multi-objective formulations, in which several different objectives are represented independently. The experiments were performed on 17 real-world module clustering problems. The results showed that a multi-objective approach produced more promising solutions than the single-objective approach.

A new MOEA approach was proposed in [18] to cluster data of security assessments. The clustered devices share their vulnerabilities. The results showed that the quality of the obtained clusters by the proposed MOEA was better than those obtained by single-objective method.

An automatic clustering algorithm Multi-objective Immunized Particle Swarm Optimization (MOIMPSO) [53]. The algorithm provides a Pareto optimal archive for clustering aggressive actions of 3D human models. It was shown that the MOIMPSO algorithm achieved a ratio of 100% classification accuracy for the physical actions of human models.

9 Conclusion

This chapter sheds light on two main problems: the clustering problem and the multi-objective problem. Furthermore, it reviews several multi-objective evolutionary clustering algorithms proposed in the literature to provide optimized solutions to the problem of multi-objective clustering problem. The lack of systematic reviews on this subject has motivated us to undertake this effort to prepare this work.

The chapter discusses the clustering problem and identifies it as an optimization problem. Furthermore, it focuses on the basic concepts of multi-objective optimization and its importance to solve the clustering problem. Subsequently, it presents a framework for evolutionary multi-objective clustering. A detailed overview of multi-objective clustering techniques is provided with a focus on five subjects: chromosome encoding methods, objective functions for optimization, evolutionary operators, storing current non-dominated solutions, and selecting the final solution from the non-dominated front.

Finally, a discussion is presented on the use of multi-objective clustering algorithms in different applications including image segmentation, bioinformatics, social networks, etc.

For the future, there is plenty of researches that are planned in the area of multi-objective evolutionary clustering. One of the main challenges is the scalability problem. This problem requires the development of efficient evolutionary algorithms that are able to cope with large scale clustering problem. These algorithms are scalable variants for the classical multi-objective clustering algorithms that can simultaneously handle many conflicting validity indices and generate near-optimal Pareto front in a large multi-objective space.

References

1. Abu Khurma, Ruba, Ibrahim Aljarah, and Ahmad Sharieh. 2020. An efficient moth flame optimization algorithm using chaotic maps for feature selection in the medical applications. In *Proceedings of the 9th international conference on pattern recognition applications and methods—volume 1: ICPRAM*, 175–182. INSTICC, SciTePress.

2. Abu Khurma, Ruba, Ibrahim Aljarah, Ahmad Sharieh, and Seyedali Mirjalili. 2020. EvoloPy-FS: An open-source nature-inspired optimization framework in Python for feature selection. In *Evolutionary machine learning techniques*, 131–173. Springer.
3. Al-Madi, Nailah, Ibrahim Aljarah, and Simone A. Ludwig. 2014. Parallel glowworm swarm optimization clustering algorithm based on mapreduce. In *2014 IEEE symposium on swarm intelligence*, 1–8. IEEE.
4. Al Shorman, Amaal, Hossam Faris, and Ibrahim Aljarah. 2020. Unsupervised intelligent system based on one class support vector machine and grey wolf optimization for IoT botnet detection. *Journal of Ambient Intelligence and Humanized Computing* 11 (7): 2809–2825.
5. Aljarah, Ibrahim, and Simone A. Ludwig. 2012. Parallel particle swarm optimization clustering algorithm based on mapreduce methodology. In *2012 fourth world congress on nature and biologically inspired computing (NaBIC)*, 104–111. IEEE.
6. Aljarah, Ibrahim, and Simone A. Ludwig. 2013. Mapreduce intrusion detection system based on a particle swarm optimization clustering algorithm. In *2013 IEEE congress on evolutionary computation*, 955–962. IEEE.
7. Aljarah, Ibrahim, and Simone A. Ludwig. 2013. A new clustering approach based on glowworm swarm optimization. In *2013 IEEE congress on evolutionary computation*, 2642–2649. IEEE.
8. Aljarah, Ibrahim, and Simone A. Ludwig. 2013. Towards a scalable intrusion detection system based on parallel PSO clustering using mapreduce. In *Proceedings of the 15th annual conference companion on Genetic and evolutionary computation*, 169–170.
9. Aljarah, Ibrahim, Majdi Mafarja, Ali Asghar Heidari, Hossam Faris, and Seyedali Mirjalili. 2020. Clustering analysis using a novel locality-informed grey wolf-inspired clustering approach. *Knowledge and Information Systems* 62 (2): 507–539.
10. Aljarah, Ibrahim, Majdi Mafarja, Ali Asghar Heidari, Hossam Faris, and Seyedali Mirjalili. 2020. Multi-verse optimizer: Theory, literature review, and application in data clustering. In *Nature-inspired optimizers*, 123–141. Springer.
11. Amiri, Babak, Liaquat Hossain, and John Crowford. 2012. A multiobjective hybrid evolutionary algorithm for clustering in social networks. In *Proceedings of the 14th annual conference companion on Genetic and evolutionary computation*, 1445–1446.
12. Bandyopadhyay, Sanghamitra, Ujjwal Maulik, and Anirban Mukhopadhyay. 2007. Multiobjective genetic clustering for pixel classification in remote sensing imagery. *IEEE Transactions on Geoscience and Remote Sensing* 45 (5): 1506–1511.
13. Bandyopadhyay, Sanghamitra, Anirban Mukhopadhyay, and Ujjwal Maulik. 2007. An improved algorithm for clustering gene expression data. *Bioinformatics* 23 (21): 2859–2865.
14. Bezdek, James C. 1981. Objective function clustering. In *Pattern recognition with fuzzy objective function algorithms*, 43–93. Springer.
15. Bong, C.W., and M. Rajeswari. 2012. Multiobjective clustering with metaheuristic: Current trends and methods in image segmentation. *IET Image Processing* 6 (1): 1–10.
16. Chen, Enhong, and Feng Wang. 2005. Dynamic clustering using multi-objective evolutionary algorithm. In *International conference on computational and information science*, 73–80. Springer.
17. Corne, David W., Nick R. Jerram, Joshua D. Knowles, and Martin J. Oates. 2001. Pesa-II: Region-based selection in evolutionary multiobjective optimization. In *Proceedings of the 3rd annual conference on genetic and evolutionary computation*, 283–290.
18. Corral, Guiomar, Alvaro Garcia-Piquer, Albert Orriols-Puig, Albert Fornells, and Elisabet Golobardes. 2009. Multiobjective evolutionary clustering approach to security vulnerability assesments. In *International conference on hybrid artificial intelligence systems*, 597–604. Springer.
19. Davies, David L., and Donald W. Bouldin. 1979. A cluster separation measure. *IEEE Transactions on Pattern Analysis and Machine Intelligence* 2: 224–227.
20. Deb, Kalyanmoy, Amrit Pratap, Sameer Agarwal, and Tamt Meyarivan. 2002. A fast and elitist multiobjective genetic algorithm: NSGA-II. *IEEE Transactions on Evolutionary Computation* 6 (2): 182–197.

21. Demir, G.Nildem, A. Şima Uyar, and Şule Gündüz-Öğüdücü. 2010. Multiobjective evolutionary clustering of web user sessions: A case study in web page recommendation. *Soft Computing* 14 (6): 579–597.
22. Demir, Gül Nildem, A. Sima Uyar, and Sule Gündüz Ögüdücü. 2007. Graph-based sequence clustering through multiobjective evolutionary algorithms for web recommender systems. In *Proceedings of the 9th annual conference on genetic and evolutionary computation*, 1943–1950.
23. Dunn, Joseph C. 1974. Well-separated clusters and optimal fuzzy partitions. *Journal of Cybernetics* 4 (1): 95–104.
24. Faris, Hossam, Ibrahim Aljarah, and Ja'far Alqatawna. 2015. Optimizing feedforward neural networks using Krill Herd algorithm for e-mail spam detection. In *2015 IEEE Jordan conference on applied electrical engineering and computing technologies (AEECT)*, 1–5. IEEE.
25. Faris, Hossam, Ibrahim Aljarah, Seyedali Mirjalili, Pedro A. Castillo, and Juan Julián Merelo Guervós. 2016. EvoloPy: An open-source nature-inspired optimization framework in Python. In *IJCCI (ECTA)*, 171–177.
26. Folino, Francesco, and Clara Pizzuti. 2010. A multiobjective and evolutionary clustering method for dynamic networks. In *2010 international conference on advances in social networks analysis and mining*, 256–263. IEEE.
27. Gzara, Mariem, and Abdelbasset Essabri. 2011. Balanced explore-exploit clustering based distributed evolutionary algorithm for multi-objective optimisation. *Studies in Informatics and Control* 20 (2): 97–106.
28. Han, Jiawei, Jian Pei, and Micheline Kamber. 2011. *Data mining: Concepts and techniques*. Elsevier.
29. Handl, Julia, and Joshua Knowles. 2004. Evolutionary multiobjective clustering. In *International conference on parallel problem solving from nature*, 1081–1091. Springer.
30. Handl, Julia, and Joshua Knowles. 2005. Exploiting the trade-off—the benefits of multiple objectives in data clustering. In *International conference on evolutionary multi-criterion optimization*, 547–560. Springer.
31. Handl, Julia, and Joshua Knowles. 2005. Multiobjective clustering around medoids. In *2005 IEEE congress on evolutionary computation*, vol. 1, 632–639. IEEE.
32. Handl, Julia, and Joshua Knowles. 2006. Multi-objective clustering and cluster validation. In *Multi-objective machine learning*, 21–47. Springer.
33. Handl, Julia, and Joshua Knowles. 2007. An evolutionary approach to multiobjective clustering. *IEEE Transactions on Evolutionary Computation* 11 (1): 56–76.
34. Hariz, Wisam A., Mayyadah F. Abdulhalim, et al. 2016. Improving the performance of evolutionary multi-objective co-clustering models for community detection in complex social networks. *Swarm and Evolutionary Computation* 26: 137–156.
35. Hastie, Trevor, Robert Tibshirani, and Jerome Friedman. 2009. Unsupervised learning. In *The elements of statistical learning*, 485–585. Springer.
36. Hruschka, Eduardo Raul, Ricardo J.G.B. Campello, Alex A. Freitas, et al. 2009. A survey of evolutionary algorithms for clustering. *IEEE Transactions on Systems, Man, and Cybernetics, Part C (Applications and Reviews)* 39 (2): 133–155.
37. Kanungo, Tapas, David M. Mount, Nathan S. Netanyahu, Christine D. Piatko, Ruth Silverman, and YWu Angela. 2002. An efficient k-means clustering algorithm: Analysis and implementation. *IEEE Transactions on Pattern Analysis and Machine Intelligence* 24 (7): 881–892.
38. Kim, Keehyung, Robert Ian McKay, and Byung-Ro Moon. 2010. Multiobjective evolutionary algorithms for dynamic social network clustering. In *Proceedings of the 12th annual conference on genetic and evolutionary computation*, 1179–1186.
39. Kirkland, Oliver, Victor J. Rayward-Smith, and Beatriz de la Iglesia. 2011. A novel multi-objective genetic algorithm for clustering. In *International conference on intelligent data engineering and automated learning*, 317–326. Springer.
40. Kodinariya, Trupti M., and Prashant R. Makwana. 2013. Review on determining number of cluster in k-means clustering. *International Journal* 1 (6): 90–95.

41. Liu, Yimin, Tansel Özyer, Reda Alhajj, and Ken Barker. 2005. Integrating multi-objective genetic algorithm and validity analysis for locating and ranking alternative clustering. *Informatica* 29 (1).
42. Maulik, Ujjwal, and Sanghamitra Bandyopadhyay. 2002. Performance evaluation of some clustering algorithms and validity indices. *IEEE Transactions on Pattern Analysis and Machine Intelligence* 24 (12): 1650–1654.
43. Maulik, Ujjwal, Sanghamitra Bandyopadhyay, and Anirban Mukhopadhyay. 2011. Combining pareto-optimal clusters using supervised learning. In *Multiobjective genetic algorithms for clustering*, 123–145. Springer.
44. Maulik, Ujjwal, Anirban Mukhopadhyay, and Sanghamitra Bandyopadhyay. 2009. Combining pareto-optimal clusters using supervised learning for identifying co-expressed genes. *BMC Bioinformatics* 10 (1): 27.
45. Mirjalili, Seyedeh Zahra, Seyedali Mirjalili, Shahrzad Saremi, Hossam Faris, and Ibrahim Aljarah. 2018. Grasshopper optimization algorithm for multi-objective optimization problems. *Applied Intelligence* 48 (4): 805–820.
46. Mukhopadhyay, Anirban, Sanghamitra Bandyopadhyay, and Ujjwal Maulik. 2006. Clustering using multi-objective genetic algorithm and its application to image segmentation. In *2006 IEEE international conference on systems, man and cybernetics*, vol. 3, 2678–2683. IEEE.
47. Mukhopadhyay, Anirban, Sanghamitra Bandyopadhyay, and Ujjwal Maulik. 2009. Analysis of microarray data using multiobjective variable string length genetic fuzzy clustering. In *2009 IEEE congress on evolutionary computation*, 1313–1319. IEEE.
48. Mukhopadhyay, Anirban, and Ujjwal Maulik. 2007. Multiobjective approach to categorical data clustering. In *2007 IEEE congress on evolutionary computation*, 1296–1303. IEEE.
49. Mukhopadhyay, Anirban, Ujjwal Maulik, and Sanghamitra Bandyopadhyay. 2009. Multiobjective genetic algorithm-based fuzzy clustering of categorical attributes. *IEEE Transactions on Evolutionary Computation* 13 (5): 991–1005.
50. Mukhopadhyay, Anirban, Ujjwal Maulik, and Sanghamitra Bandyopadhyay. 2015. A survey of multiobjective evolutionary clustering. *ACM Computing Surveys (CSUR)* 47 (4): 1–46.
51. Mukhopadhyay, Anirban, Ujjwal Maulik, Sanghamitra Bandyopadhyay, and Carlos A. Coello Coello. 2013. Survey of multiobjective evolutionary algorithms for data mining: Part II. *IEEE Transactions on Evolutionary Computation* 18 (1): 20–35.
52. Mukhopadhyay, Anirban, Sumanta Ray, and Moumita De. 2012. Detecting protein complexes in a PPI network: a gene ontology based multi-objective evolutionary approach. *Molecular BioSystems* 8 (11): 3036–3048.
53. Nanda, Satyasai Jagannath, and Ganapati Panda. 2013. Automatic clustering algorithm based on multi-objective immunized PSO to classify actions of 3D human models. *Engineering Applications of Artificial Intelligence* 26 (5–6): 1429–1441.
54. Özyer, Tansel, Yimin Liu, Reda Alhajj, and Ken Barker. 2004. Multi-objective genetic algorithm based clustering approach and its application to gene expression data. In *International conference on advances in information systems*, 451–461. Springer.
55. Özyer, Tansel, Ming Zhang, and Reda Alhajj. 2011. Integrating multi-objective genetic algorithm based clustering and data partitioning for skyline computation. *Applied Intelligence* 35 (1): 110–122.
56. Pakhira, Malay K., Sanghamitra Bandyopadhyay, and Ujjwal Maulik. 2004. Validity index for crisp and fuzzy clusters. *Pattern Recognition* 37 (3): 487–501.
57. Praditwong, Kata, Mark Harman, and Xin Yao. 2010. Software module clustering as a multi-objective search problem. *IEEE Transactions on Software Engineering* 37 (2): 264–282.
58. Qaddoura, Raneem, Hossam Faris, and Ibrahim Aljarah. 2020. An efficient clustering algorithm based on the k-nearest neighbors with an indexing ratio. *International Journal of Machine Learning and Cybernetics* 11 (3): 675–714.
59. Qaddoura, Raneem, Hossam Faris, Ibrahim Aljarah, and Pedro A. Castillo. 2020. Evocluster: An open-source nature-inspired optimization clustering framework in Python. In *International conference on the applications of evolutionary computation (part of EvoStar)*, 20–36. Springer.

60. Qian, Xiaoxue, Xiangrong Zhang, Licheng Jiao, and Wenping Ma. 2008. Unsupervised texture image segmentation using multiobjective evolutionary clustering ensemble algorithm. In *2008 IEEE congress on evolutionary computation (IEEE world congress on computational intelligence)*, 3561–3567. IEEE.
61. Ripon, Kazi Shah Nawaz, and Mia Nazmul Haque Siddique. 2009. Evolutionary multi-objective clustering for overlapping clusters detection. In *2009 IEEE congress on evolutionary computation*, 976–982. IEEE.
62. Ripon, K.S. Nawaz, Chi-Ho Tsang, Sam Kwong, and Man-Ki Ip. 2006. Multi-objective evolutionary clustering using variable-length real jumping genes genetic algorithm. In *18th international conference on pattern recognition (ICPR'06)*, vol. 1, 1200–1203. IEEE.
63. Rousseeuw, J.P.J. 1989. A graphical aid to the interpretation and validation of cluster analysis. *Journal of Computational Application Mathematics*.
64. Sheikh, Rahila H., Mukesh M. Raghuwanshi, and Anil N. Jaiswal. 2008. Genetic algorithm based clustering: A survey. In *2008 first international conference on emerging trends in engineering and technology*, 314–319. IEEE.
65. Shirakawa, Shinichi, and Tomoharu Nagao. 2009. Evolutionary image segmentation based on multiobjective clustering. In *2009 IEEE congress on evolutionary computation*, 2466–2473. IEEE.
66. Shukri, Sarah, Hossam Faris, Ibrahim Aljarah, Seyedali Mirjalili, and Ajith Abraham. 2018. Evolutionary static and dynamic clustering algorithms based on multi-verse optimizer. *Engineering Applications of Artificial Intelligence* 72: 54–66.
67. Steinbach, Michael, Levent Ertöz, and Vipin Kumar. 2004. The challenges of clustering high dimensional data. In *New directions in statistical physics*, 273–309. Springer.
68. Suresh, Kaushik, Debarati Kundu, Sayan Ghosh, Swagatam Das, Ajith Abraham, and Sang Yong Han. 2009. Multi-objective differential evolution for automatic clustering with application to micro-array data analysis. *Sensors* 9 (5): 3981–4004.
69. Wang, Rui, Wu Shiming Lai, Lining Xing Guohua, Ling Wang, and Hisao Ishibuchi. 2018. Multi-clustering via evolutionary multi-objective optimization. *Information Sciences* 450: 128–140.
70. Wei, Bong Chin, and Rajeswari Mandava. 2010. Multi-objective nature-inspired clustering techniques for image segmentation. In *2010 IEEE conference on cybernetics and intelligent systems*, 150–155. IEEE.
71. Xie, Xuanli Lisa, and Gerardo Beni. 1991. A validity measure for fuzzy clustering. *IEEE Transactions on Pattern Analysis and Machine Intelligence* 13 (8): 841–847.
72. Zhang, Hu, Shenmin Song, Aimin Zhou, and Xiao-Zhi Gao. 2014. A clustering based multi-objective evolutionary algorithm. In *2014 IEEE congress on evolutionary computation (CEC)*, 723–730. IEEE.
73. Zhang, Mengxuan, Licheng Jiao, Wenping Ma, Jingjing Ma, and Maoguo Gong. 2016. Multi-objective evolutionary fuzzy clustering for image segmentation with MOEA/D. *Applied Soft Computing* 48: 621–637.
74. Zhao, Feng, Jiulun Fan, Hanqiang Liu, Rong Lan, and Chang Wen Chen. 2018. Noise robust multiobjective evolutionary clustering image segmentation motivated by the intuitionistic fuzzy information. *IEEE Transactions on Fuzzy Systems* 27 (2): 387–401.

A Review of Evolutionary Data Clustering Algorithms for Image Segmentation

Laila Al-Qaisi, Mohammad A. Hassonah, Mahmoud M. Al-Zoubi, and Ala' M. Al-Zoubi

Abstract Evolutionary algorithms have demonstrated their ability to solve complex optimization problems throughout the past years. Recent studies took advantage of such algorithms to enhance the performance of clustering algorithms in finding the best values for necessary parameters. Image segmentation is one of the most recognized fields in clustering algorithms that extensively applied evolutionary algorithms for different problems. This chapter presents a review on integrating several evolutionary algorithms with clustering techniques to perform image segmentation. We choose some of the most common applications on the topic, which are Medical, Multi-objective, and Multilevel Thresholding image segmentation techniques. Then, other applications in the field are also reviewed.

Keywords Data clustering · Image segmentation · Optimization · Evolutionary computation · Nature-inspired algorithms · Swarm intelligence · Metaheuristics · Unsupervised learning

---

L. Al-Qaisi
Information Systems and Networks Department, Faculty of Information Technology, The World Islamic Sciences and Education University, Amman, Jordan
e-mail: Laila.qaisi@wise.edu.jo

M. A. Hassonah
King Abdullah II School for Information Technology, The University of Jordan, Amman, Jordan
e-mail: mohammad.a.hassonah@gmail.com

M. M. Al-Zoubi
Department of Commutation and Information Technology, Yarmouk Water Company, Irbid 21110, Jordan
e-mail: mzoubi12@yahoo.com

A. M. Al-Zoubi (✉)
School of Science, Technology and Engineering, University of Granada, Granada, Spain
e-mail: alaah14@gmail.com; alzoubi@correo.ugr.es

I. Aljarah et al. (eds.), *Evolutionary Data Clustering: Algorithms and Applications*, Algorithms for Intelligent Systems,
https://doi.org/10.1007/978-981-33-4191-3_9

1 Introduction

As a result of the rapid growth of technology utilization all around our daily routine, massive amount of data exists today [48]. Which needs to be recognized, categorized, and classified in order to get the needed knowledge [4]. Therefore, techniques and concepts of acquiring, storing and discovering are required to accomplish such tasks. Data clustering is defined as concept which partition data into smaller several parts called clusters. This increases the ability of user understanding for a dataset structure. Therefore, a cluster is a set of objects that are internally similar but less or not similar to objects belonging to other surrounding clusters. Both similarity and dissimilarity are defined then assessed in terms of object attributes being examined [41].

Data clustering techniques can be applied to images, or image datasets in a process called image segmentation. It is defined as partitioning an image into several sets of pixels that are homogeneous in a certain way. Some attributes like intensity and color are used to determine homogeneity of image segments. The concept of image segmentation itself is considered as a preprocessing step for image processing, so it has been applied through the literature on various fields [10].

Image segmentation techniques can be sorted as traditional and metaheuristic. Metaheuristic techniques are described as flexible, more reliable and capable of solving image segmentation problems more than traditional ones. Various advantages such as the ability of human thinking simulation and working with their ownership functions along with having cost-effective, high performance and steadfast solutions made metaheuristics more suitable for image segmentation applications. In the literature, different algorithms and different versions of algorithms were used to solve image segmentation problem such as Genetic Algorithm (GA) [2], Particle Swarm Optimization (PSO), Grey Wolf Optimization (GWO), and Grasshopper Optimization Algorithm (GOA). [10, 46].

The rest of this chapter is organized as follows: Preliminary definitions for some of the used terminologies including Clustering, Metaheuristics, and Image Segmentation in Sect. 2. Section 3 reviews the work conducted on image segmentation using metaheuristic clustering techniques. Finally, Sect. 4 concludes and summarizes the work presented in this chapter.

2 Preliminaries

2.1 Clustering

Clustering or Cluster Analysis is the process of finding natural groupings of data points without previous knowledge of their labels [51]. Clustering is known to be applied and utilized in various areas, such as image processing, market segmentation, document classification, and many other fields.

Data clustering Algorithms are categorized into four main categories: Partition-based clustering algorithms, Fuzzy clustering, Model-based Clustering Algorithms, and Density-based clustering algorithms [35].

Partition-based clustering algorithms are established on a general idea which is An object belongs to exactly one group and each cluster has at least one object. In this case, number of clusters is set by the user making the process a semi-supervised process. While fuzzy clustering algorithms specify a set of membership coefficients to each object which determines the degree of membership to a cluster.

Model-based clustering assumes that the given data is generated from various multiple probability distributions such as Gaussian or multinomial, then Expectation Maximization Algorithm is used to estimate their parameters mean and covariance matrix. Moreover, the process of selecting optimal number of clusters may be performed by Bayesian information or Akaike information criteria.

Finally, Density-based clustering algorithms make use of the fact that describes the cluster as a connected dense component. This component may expand and grow in any direction until it reaches a threshold. As a result, an automatic negligence of outliers and identification of well-divided clusters of random shapes will occur.

2.2 *Metaheuristic Algorithms*

Nature-inspired metaheuristic algorithms have been used for optimization purposes in various problems. They were described as powerful techniques for solving complex optimization problems. The Main characteristic of metaheuristic algorithms is their independence of the addressed problem. This made metaheuristic algorithms the suitable option for optimizing discontinuous, multimodal, non-smooth, and non-convex functions. They are used for the purpose of finding global optimal solutions with limited computational effort [1].

Many different metaheuristic optimization techniques have been presented and successfully applied to different problems. Examples include Genetic Algorithms (GAs), Particle Swarm Optimization (PSO), Whale Optimization Algorithm (WOA) [3], Salp Swarm Algorithm (SSA) [14, 15, 43], Multi-Verse Optimizer (MVO) [13, 19, 43], Water Cycle, Mine Blast and Ant Lion Optimizer (ALO), Firefly Algorithm, etc.

2.3 *Image Segmentation*

Image segmentation is known as the process of decomposing an image into multiple different parts, groups of pixels. These pixels in each part comparatively have similar visual attributes that differ from the other portions. The purpose of segmentation is to make the image representation simple and easy for analysis or in order to alter the representation of the image itself. Segmentation has a wide range of applications, such as in medical implementations (i.e., recognize the anatomical structure), satellite imaging (i.e., identify civil areas), and so on. Image segmentation can be used for any problem where the image needs further analysis alongside removing the noise without the loss of important information from an image. Besides, there are five techniques usually applied when performing the image segmentation, including, model-based, threshold-based, region-based, feature-based clustering, and edge-based. Each technique has its negative and positive characteristics when used for a specific scenario, application, or problem.

3 Applications of Clustering with Metaheuristic Algorithms on Image Segmentation

3.1 Medical Image Segmentation

Medical image segmentation is one of the most common implementations, as noticed through written works. Magnetic Resonance Imaging (MRI), Computed Tomography (CT), X-Ray, Ultrasound, and many other types of medical images present the sources, from which, better diagnostic insights can be obtained, hence the popularity.

Magnetic Resonance Imaging (MRI) is one of the most reliable techniques to detect brain tumors, which are rare shaping of cells in the cerebrum, and can either be benign (non-cancerous) or malignant (cancerous). Therefore, several image segmentation methods were proposed to detect suspicious areas in MRI. Of these recent methods is the metaheuristic clustering approach. For instance, [18] proposed an intelligent system in 2010 to process MR images to diagnose brain tumors using Fuzzy C-Means (FCM) alongside Genetic Algorithm (GA) and Particle Swarm Optimization (PSO). While [16, 25] utilized Ant-Colony Optimizer (ACO) in the same manner as an initial step in identifying the suspicious region in a brain MRI, and [21] integrated Chaotic Firefly algorithm with FCM for brain tissue segmentation.

Another metaheuristic algorithm called Glowworm Swarm Optimization (GSO) algorithm used by [23] to initialize the cluster centers for FCM algorithm and change them through an iterative manner as well. The results showed improvements over conventional FCM in terms of time to reach the optimal result.

On the other hand, [40] overcame the problems of being trapped into local minima while selecting the cluster centers, and sensitivity to noise through proposing an improved PSO integrated with kernelized fuzzy entropy clustering with spatial information and bias correction. Their work was tested on simulated and real MRI brain images and compared against 5 other well-recognized methods. The results showed more effectiveness of the proposed method than the other methods. Similarly, Fuzzy Firefly algorithm was used in [36] to increase the chance of finding the optimal cluster centers in a shorter period of time by reducing the number of iterations. The method was applied on Brain MRI and superior results were obtained.

Further work was conducted to segment five brain MRI images using K-means with Galactic Swarm Optimization algorithm (GSOA) in [34]. Otsu's thresholding method was used as a fitness function for GSOA. Although the proposed required extra time for processing the images due to the number of iterations for GSOA, it showed more accuracy in segmenting the images than other benchmark algorithms in different measures.

Recently, [8] proposed a clustering method using Flower Pollination Algorithm (FPA) and Crow Search Algorithm (CSA) to process MRI images for brain tumors and breast cancer. Their work demonstrated the ability to reach global optimal solutions with the least error rates among traditional clustering methods including K-means and FCM, and metaheuristic clustering algorithms including GA and PSO. It is also worth mentioning that all metaheuristic clustering methods performed better than the traditional methods.

Apart from MRI, Computed Tomography (CT) scan is also common in medical imaging, which is often used to find irregularities in different parts of the body, including lungs and liver. Recent metaheuristic clustering methods were applied on CT images. In particular, Firefly algorithm was implemented in [37] to initialize centroids for FCM algorithm to segment lung nodules in real-time CT images. The results showed better results than in normal FCM and K-means. Ali et al. [5] applied Grey Wolf Optimizer (GWO) with FCM for Abdominal CT Liver Parenchyma image segmentation, where GWO is used to try to find the optimal cluster centers. In addition, a second type of application using Ant Bee Colony (ABC) approach with Region Growing technique for CT liver segmentation was employed as a clustering technique itself in the same study. The best similarity index was obtained compared to other well-known techniques.

Firefly, GA, and Teaching Learning-Based Optimization (TLBO) algorithms were tested in [47] to initialize and update cluster centers in images produced by lung CT scans. Nodular planes images are segmented into pixels represented by features, which are cluster center candidates and selected or not by the metaheuristic algorithm according to the corresponding solution.

A density-based algorithm for clustering called Density Peaks Clustering (DPC) algorithm was applied in combination with Fruit fly Optimization Algorithm (FOA) by [56]. The FOA was used to randomly determine the parameters of DPC such as cluster centers. Image entropy was used a fitness measurement for FOA. FOA-DPC performed competitively against robust methods such as k-means and GA-DPC.

Other medical issues require different forms of imaging. For example, X-ray images are used to detect bone fracture or other abnormalities. However, in a number of cases, the detection is delicate and requires accurate diagnosis. Therefore, many studies applied image segmentation on X-ray as well. Das [11] used Bat Algorithm with K-means algorithm in separate phases to enhance the x-ray image and then produce final image, respectively.

Ultrasound image represents another source of medical images that is mainly important for detecting a number of types of cancer. Samundeeswari et al. [45] segmented ultrasound images to detect breast cancer. They used K-means with regularization technique. ACO algorithm was implemented for cluster center initialization. The results showed improvements over normal K-means algorithm.

Some of the most known types of imaging in medical field were presented in this section. We explored recent works on image segmentation using MRI, CT, X-ray, and Ultrasound images. It is found that metaheuristic algorithms are mostly used for cluster center initialization, while some of them are utilized for setting further parameters. Only a few studies tried to operate metaheuristics as clustering mechanism themselves.

3.2 Multiobjective Image Segmentation

Solving multi-tasks or multi-mathematical optimization problems can be carried out using a single-objective function; however, in the single-objective, the priority is usually given to one solution more than the others. Therefore, in the last decade, multiobjective optimization algorithms have been applied more often to overcome this issue [32, 38, 50]. Multiobjective optimization is a non-dominant criterion to handle the multiple optimization problems, simultaneously. These problems can range between maximizing one problem while minimizing the other without affecting each other. Multiobjective optimization has been utilized in different fields such as engineering [30], economics [17, 33], science [12], and in our case, image segmentation.

Lately, multiobjective optimization for image segmentation attracted more attention in the literature, especially combined with evolutionary algorithms for cluster problems. Fuzzy clustering is one of the most popular methods for image segmentation in multiobjective cases. The concept of fuzzy clustering (soft clustering) comes from the probability of having data points that belong to more than one cluster, unlike the non-fuzzy clustering, where each point belongs to distinct clusters [28]. The fuzzy clustering can be a useful tool in image segmentation and multiobjective applications. For example, the work in [44] presented an improved multiobjective differential evolution with fuzzy clustering to optimize different validity measures at the same time. Their work generated a set of solutions with a non-dominated characteristic for image segmentation that can easily be selected according to specific requirements. Similar to the previous work, fuzzy clustering was also applied with a multiobjective evolutionary algorithm and multiple images spatial to overcome the noise in images using image segmentation [54]. The proposed approach consists of two components, the first is to get the fuzzy information from an image, while the second to apply the evolutionary multiobjective optimization. Furthermore, the intuitionistic fuzzy fitness function and evaluation are both adopted in order to optimize and select the best solution, respectively. In a recent work from [53], the authors proposed a fuzzy clustering and evolutionary multiobjective algorithm to preserve significant details from images while removing the noise using decomposition techniques for image segmentation. The decomposition strategy is designed to divide the main problem into a set of sub-problems where each sub-problem depicts one fuzzy clustering problem that corresponds to local information on the image. The outcome of their proposed approach showed a trade-off between removing noise and preserving details from the images for image segmentation.

In 2011, [52] presented an Artificial Immune System (AIS) for image segmentation of the Synthetic Aperture Radar (SAR). The proposed approach consists of the following phases; first, they combined the AIS and multiobjective optimization algorithm in order to rank the clones and maintenance diversity. Secondly, the fuzzy clustering validity integrated and then optimized. Moreover, the texture discrimination and representation feature generated and researched for an optimal solution. The SAR with synthetic texture images was compared with several clustering algorithms namely, Single-Objective Genetic Algorithm (SOGA), Spectral Clustering Ensemble (SCE), Wavelet-Domain Hidden Markov models (HMTseg), Self-Organizing Map (SOM), and Fuzzy C-Means (FCM). Experimental results for the proposed method achieved better performance than the other algorithms in image segmentation. Another aspect of using the fuzzy clustering and the evolutionary multiobjective optimization in image segmentation was presented in [55]. They proposed a Kriging model to reduce the computational cost while applying their approach (KRV-MRSFC) in removing noise from images. Furthermore, the angle penalized distance is utilized in order to enhance the Kriging model accuracy rate by selecting the optimal individuals in the exploration and exploitation phase. The results proved that their algorithm achieved acceptable segmentation with low time cost.

On the other hand, many researchers in the literature adapted and solved image segmentation problems apart from fuzzy clustering as their main approach. For instance, Multiobjective Emperor Penguin Optimizer (MEPO) was used to automatically identify the best number of clusters in image segmentation [29]. The proposed MEPO has two operations, calculate the cluster centroid and setting the threshold to tune these centroids. Besides, the MEPO compared against other well-known approaches, nevertheless, the proposed method outperforms all the other approaches. Another work was conducted by [49] on image recognition using multiobjective evolutionary clustering and image segmentation. In this work, two objectives were utilized, edge value and overall deviation, and both of them were optimized using an evolutionary multiobjective algorithm to find several solutions. They tested their method on various image segmentation problems and the performance showed good results regarding the image segmentation technique. In [6], the authors provided a detailed review of image segmentation problems using multiobjective clustering. They classified the approaches into three folds, types of optimization, formulation problems, and the kinds of datasets used. Another review proposed by [7], depicts the state-of-the-art for image segmentation clustering through metaheuristic algorithm. Their study discusses the non-evolutionary and evolutionary multiobjective clustering methods alongside their requirements and challenges for image segmentation.

However, in this brief review, we covered and focused on the recent works that utilized the evolutionary multiobjective clustering in image segmentation, unlike the mentioned reviews.

3.3 Multilevel Thresholding Image Segmentation

Image thresholding is referred to dividing images into several class-based segments. Bi-level thresholding is used to segment images into two classes: objects and background. Nevertheless, when images are complex and may contain several levels, bilevel thresholding cannot segment them. In this case, multilevel thresholding technique is required. The main problem here is increased computational complexity because of the increased levels within an image. Several studies used different metaheuristic algorithms to overcome drawbacks of increased levels in the problem of multilevel thresholding [9, 22, 24, 26, 27, 31].

The study conducted by [27] used Grey Wolf Optimizer (GWO) for solving the drawback of multilevel thresholding in image segmentation. The computation complexity increases when number of thresholds increases. Two major functions were used for the experiments, namely Kapur's and Otsu's entropy between class variance functions. The proposed algorithm and functions were tested using set of benchmark images. Afterwards, set of performance comparisons were conducted with PSO and Bacterial Foraging Optimization (BFO). Experimental results showed that GWO outperformed both PSO and BFO-based methods. Nevertheless, the proposed GWO was slower than PSO, however, faster than BFO-based algorithms.

Moth-Flame Optimization (MFO) was used to overcome the problem of multilevel thresholding in image segmentation by [26]. The threshold chosen selection criteria were Kapur's entropy and Otsu's between class variance functions. The proposed algorithm was tested and evaluated using benchmark images such as Lena, Peppers, Baboon, Man, and Airplane. Segmentation analysis measure was Peak Signal-to-Noise Ratio (PSNR) besides fitness function. Results proved that MFO can be used as a solution to multilevel thresholding problem using both functions. The proposed MFO was compared with both PSO and BFO-based methods by statistical analysis using Anova test. It was observed that the proposed MFO outperformed both PSO and BFO-based methods with increased stability and higher solution quality. Moreover, multilevel thresholding computation cost was greatly reduced.

Another solution to multilevel thresholding problem was proposed by [31] using Gravitational Search Algorithm (GSA). Basically, multilevel thresholding divides an image into several non-overlapping regions. A novel (2D) histogram-based segmentation method is used to enhance multilevel thresholding image segmentation. In the proposed method, Kbest GSA was used to reach optimal thresholds. Furthermore, a 2D Rényi entropy was redefined for the optimization purposes. The proposed methodology was tested on Berkeley Segmentation Dataset and Benchmark (BSDS300) in terms of both subjective and objective evaluations. The experimental results showed that the proposed methodology outperformed other traditional (2D) histogram-based segmentation methods.

Minimum Cross-Entropy Thresholding (MCET) was the area of study in [9]. The problem was targeting the computation cost of MCET in image segmentation. A novel Improved Particle Swarm Optimization (IPSO) algorithm was presented and deployed to conduct multilevel thresholding image segmentation. The main idea was that PSO generally has a problem of premature convergence. This was solved by IPSO which decomposed high-dimensional swarm into multiple one-dimensional swarms, then removed premature convergence from each one-dimensional swarm. The proposed technique was tested on grey-scale images, and results showed significantly better performance. It resulted in better MCET threshold values along with fast and high convergence rate. Experimental results of the proposed approach were compared with other state-of-the-art methods such as modified ABC, Cuckoo search, Firefly, PSO, and GA algorithms. It was concluded that the proposed IPSO approach performed better in terms of quantitative measures such as fitness function value, CPU time and qualitative measures such as effective misclassification error, peak signal-to-noise ratio, feature similarity index measurement, complex wavelet structural similarity index measurement values.

A new multilevel thresholding method for color images based on Water Cycle Algorithm (WCA) along with Masi entropy was suggested by [24] and the term Masi-WCA was used for it. Other exploited optimization algorithms were Bat, PSO, Wind-Driven Optimization (WDO), Monarch Butterfly Optimization (MBO), and GOA. Two fitness functions were used along with the chosen algorithms, Tsallis and Masi entropy methods for comparison purposes. Experiments tested the chosen algorithm and proposed Masi-WCA in terms of quality, and statistical analysis was provided in this regard. WCA registered a better performance in terms of quality and fast convergence rate. Masi entropy was recommended as a fitness function to be used for color image segmentation with WCA rather than Tsallis.

Hybrid GOA and Minimum Cross-Entropy (MCE) was proposed by [22] for multilevel threshold color image segmentation. The proposed model combined GOA with self-adaptive differential evolution (jDE), and called GOA–jDE. This was done to improve the efficiency of search and preserve diversity of population specifically in last iterations. The proposed multilevel was compared with standard color image thresholding methods along with advanced image thresholding techniques in different criteria. Berkeley segmentation dataset was used for validating robustness of proposed GOA–jDE. Both Friedman test and Wilcoxon's rank sum test were conducted for assessment purposes. The proposed GOA–jDE outperformed other techniques in terms of average fitness function value, Peak Signal-to-Noise Ratio, Structural Similarity Index, feature similarity index, Standard Deviation, convergence performance, and computation time.

Overall, it can be seen that previous studies tried to solve the main drawbacks of multilevel thresholding, specifically cost in addition of accuracy of the thresholds with levels nearly greater than three. Some of the proposed and suggested solutions by using different metaheuristic algorithms in recent studies are also listed, explained, and summarized. Most of them had significantly improved performance.

3.4 Other Applications on Image Segmentation

Through the literature, several image segmentation methods were applied to various systems to solve the problem of locating objects in different images such as Traffic Control Systems, Self-Driving Cars, and Locating objects in satellite images.

Rodrigues et al. [42] conducted a study on image segmentation using Firefly Algorithm (FFA). They applied five different functions for evaluating the proposed mechanism: Cross-Entropy, Exponential Cross-Entropy, Modified Cross-Entropy, Shannon and Tsallis entropy measurements. The selected images were taken from previously conducted researches on image processing and synthetic data observed from 2-, 3-, 4-, and 5-modal Gaussians. Conducting experiments on real images showed that FFA is a great alternative, where Tsallis entropies outperformed the rest tested functions as a kernel and proved to be better than cross-entropy which was showed ability to have the best results in previous research conducted by [20].

A study was performed by [39] using K-means. They introduced a criterion to decrease number of clusters while selecting the best segments; in order to overcome the problem of depending on the number of clusters. The difference between intra-class and inter-class was used to overcome the shortcoming of initialization to form the optimal criterion. The proposed criterion was tested and evaluated using eight digital images. Results showed that segmented images using the proposed criterion had stability and robustness. Also, the proposed criterion was found to be outperforming the traditional K-means methods with increased consistency.

Eventually, this section demonstrates the broadness of image segmentation applications by presenting a number of studies that applied metaheuristic algorithms on cluster analysis. The experiments were conducted on benchmark images. This indicates that the same algorithms and techniques are open to be utilized in more fields such as Traffic Control Systems, Self-Driving Cars, and Locating objects in satellite images.

4 Conclusion and Future Directions

This chapter reviewed a number of studies for performing clustering mechanism on image segmentation with integration of metaheuristic algorithms. Studies for Medical, Multiobjective, and Multilevel Thresholding applications were addressed, as well as studies of other applications were generally viewed.

It is well-observed that utilizing metaheuristic algorithms helped in overcoming the many issues faced by clustering algorithms, primarily to find global optimal solutions for computationally hard problems. The integration point where metaheuristic algorithms were exploited for the sake of clustering was mostly observed in cluster center initialization, for example, for K-means and Fuzzy C-means algorithms. Other studies were found to use metaheuristic algorithms to mimic the clustering mechanism itself.

References

1. Abdel-Basset, Mohamed, Laila Abdel-Fatah, and Arun Kumar Sangaiah. 2018. Metaheuristic algorithms: A comprehensive review. In *Computational intelligence for multimedia big data on the cloud with engineering applications*, 185–231. Elsevier.
2. Al-Qudah, Dana A., Al-Zoubi, Ala' M., Pedro A. Castillo-Valdivieso, and Hossam Faris. 2020. Sentiment Analysis for e-Payment Service Providers Using Evolutionary eXtreme Gradient Boosting. *IEEE Access*, 8:189930–189944.
3. Al-Zoubi, Ala' M., Hossam Faris, Ja'far Alqatawna, and Mohammad A. Hassonah. 2018. Evolving support vector machines using whale optimization algorithm for spam profiles detection on online social networks in different lingual contexts. *Knowledge-Based Systems* 153: 91–104.
4. Al-Zoubi, Ala' M., Ali Rodan, and Azmi Alazzam. 2018. Classification model for credit data. In *2018 fifth HCT information technology trends (ITT)*, 132–137. IEEE.
5. Ali, Ahmed Fouad, Abdalla Mostafa, Gehad Ismail Sayed, Mohamed Abd Elfattah, and Aboul Ella Hassanien. 2016. Nature inspired optimization algorithms for CT liver segmentation. In *Medical imaging in clinical applications*, 431–460. Springer.
6. Bong, Chin-Wei, and Mandava Rajeswari. 2011. Multi-objective nature-inspired clustering and classification techniques for image segmentation. *Applied Soft Computing* 11 (4): 3271–3282.
7. Bong, C.W., and M. Rajeswari. 2012. Multiobjective clustering with metaheuristic: Current trends and methods in image segmentation. *IET Image Processing* 6 (1): 1–10.
8. Braik, Malik, Alaa Sheta, and Sultan Aljahdali. 2019. Diagnosis of brain tumors in mr images using metaheuristic optimization algorithms. In *International conference Europe Middle East & North Africa information systems and technologies to support learning*, 603–614. Springer.
9. Chakraborty, Rupak, Rama Sushil, and M.L. Garg. 2019. An improved PSO-based multilevel image segmentation technique using minimum cross-entropy thresholding. *Arabian Journal for Science and Engineering* 44 (4): 3005–3020.
10. Chouhan, Siddharth Singh, Ajay Kaul, and Uday Pratap Singh. 2018. Soft computing approaches for image segmentation: A survey. *Multimedia Tools and Applications* 77 (21): 28483–28537.
11. Das, Goutam. 2013. Bat algorithm based softcomputing approach to perceive hairline bone fracture in medical x-ray images. *International Journal of Computer Science & Engineering Technology (IJCSET)* 4 (04).
12. Di Barba, Paolo, Ivo Dolezel, Maria Evelina Mognaschi, Antonio Savini, and Pavel Karban. 2014. Non-linear multi-physics analysis and multi-objective optimization in electroheating applications. *IEEE Transactions on Magnetics* 50 (2): 673–676.
13. Faris, Hossam, Mohammad A. Hassonah, Ala' M. Al-Zoubi, Seyedali Mirjalili, and Ibrahim Aljarah. 2018. A multi-verse optimizer approach for feature selection and optimizing SVM parameters based on a robust system architecture. *Neural Computing and Applications* 30 (8): 2355–2369.
14. Faris, Hossam, Ali Asghar Heidari, Ala' M. Al-Zoubi, Majdi Mafarja, Ibrahim Aljarah, Mohammed Eshtay, and Seyedali Mirjalili. 2020. Time-varying hierarchical chains of salps with random weight networks for feature selection. *Expert Systems with Applications* 140: 112898.
15. Faris, Hossam, Majdi M. Mafarja, Ali Asghar Heidari, Ibrahim Aljarah, Ala' M. Al-Zoubi, Seyedali Mirjalili, and Hamido Fujita. 2018. An efficient binary salp swarm algorithm with crossover scheme for feature selection problems. *Knowledge-Based Systems* 154: 43–67.
16. Gahukar, Sayali D., and S.S. Salankar. 2014. Segmentation of MRI brain image using fuzzy C means for brain tumor diagnosis. *International Journal of Engineering Research and Applications* 4 (4): 107–111.
17. Gharavi, H., M.M. Ardehali, and S. Ghanbari-Tichi. 2015. Imperial competitive algorithm optimization of fuzzy multi-objective design of a hybrid green power system with considerations for economics, reliability, and environmental emissions. *Renewable Energy* 78: 427–437.

18. Gopal, N. Nandha, and M. Karnan. 2010. Diagnose brain tumor through MRI using image processing clustering algorithms such as fuzzy C means along with intelligent optimization techniques. In *2010 IEEE international conference on computational intelligence and computing research*, 1–4. IEEE.
19. Hassonah, Mohammad A., Rizik Al-Sayyed, Ali Rodan, Ala' M. Al-Zoubi, Ibrahim Aljarah, and Hossam Faris. 2019. An efficient hybrid filter and evolutionary wrapper approach for sentiment analysis of various topics on Twitter. *Knowledge-Based Systems* 105353
20. Horng, Ming-Huwi, and Ren-Jean Liou. 2011. Multilevel minimum cross entropy threshold selection based on the firefly algorithm. *Expert Systems with Applications* 38 (12): 14805–14811.
21. Jansi, S., and P. Subashini. 2015. A novel fuzzy clustering based modified firefly algorithm with chaotic map for MRI brain tissue segmentation. *MAGNT Research Report* 3 (1): 52–58.
22. Jia, Heming, Chunbo Lang, Diego Oliva, Wenlong Song, and Xiaoxu Peng. 2019. Hybrid grasshopper optimization algorithm and differential evolution for multilevel satellite image segmentation. *Remote Sensing* 11 (9): 1134.
23. Kalaiselvi, T., P. Nagaraja, and Z. Abdul Basith. Fuzzy C-means clustering with GSO based centroid initialization for brain tissue segmentation in MRI head scans. In *Computational methods, communication techniques and informatics*, 53.
24. Kandhway, Pankaj, and Ashish Kumar Bhandari. 2019. A water cycle algorithm-based multilevel thresholding system for color image segmentation using Masi entropy. *Circuits, Systems, and Signal Processing* 38 (7): 3058–3106.
25. Karnan, M., and T. Logheshwari. 2010. Improved implementation of brain MRI image segmentation using ant colony system. In *2010 IEEE international conference on computational intelligence and computing research*, 1–4. IEEE.
26. Khairuzzaman, Abdul Kayom Md, and Saurabh Chaudhury. 2017. Moth-flame optimization algorithm based multilevel thresholding for image segmentation. *International Journal of Applied Metaheuristic Computing (IJAMC)* 8 (4): 58–83.
27. Khairuzzaman, Abdul Kayom Md, and Saurabh Chaudhury. 2017. Multilevel thresholding using grey wolf optimizer for image segmentation. *Expert Systems with Applications* 86: 64–76.
28. Klawonn, Frank, and Frank Höppner. 2003. What is fuzzy about fuzzy clustering? Understanding and improving the concept of the fuzzifier. In *International symposium on intelligent data analysis* 254–264. Springer.
29. Kumar, Dinesh, Vijay Kumar, and Rajani Kumari. 2019. Automatic clustering using quantum-based multi-objective emperor penguin optimizer and its applications to image segmentation. *Modern Physics Letters A* 1950193.
30. Mirjalili, Seyedali, Pradeep Jangir, and Shahrzad Saremi. 2017. Multi-objective ant lion optimizer: A multi-objective optimization algorithm for solving engineering problems. *Applied Intelligence* 46 (1): 79–95.
31. Mittal, Himanshu, and Mukesh Saraswat. 2018. An optimum multi-level image thresholding segmentation using non-local means 2D histogram and exponential Kbest gravitational search algorithm. *Engineering Applications of Artificial Intelligence* 71: 226–235.
32. Monsef, H., M. Naghashzadegan, A. Jamali, and R. Farmani. 2019. Comparison of evolutionary multi objective optimization algorithms in optimum design of water distribution network. *Ain Shams Engineering Journal* 10 (1): 103–111.
33. Mousavi-Avval, Seyed Hashem, Shahin Rafiee, Mohammad Sharifi, Soleiman Hosseinpour, Bruno Notarnicola, Giuseppe Tassielli, and Pietro A. Renzulli. 2017. Application of multi-objective genetic algorithms for optimization of energy, economics and environmental life cycle assessment in oilseed production. *Journal of Cleaner Production* 140: 804–815.
34. Nanda, Satyasai Jagannath, Ishank Gulati, Rajat Chauhan, Rahul Modi, and Uttam Dhaked. 2019. A K-means-galactic swarm optimization-based clustering algorithm with Otsu's entropy for brain tumor detection. *Applied Artificial Intelligence* 33 (2): 152–170.
35. Nerurkar, Pranav, Archana Shirke, Madhav Chandane, and Sunil Bhirud. 2018. Empirical analysis of data clustering algorithms. *Procedia Computer Science* 125: 770–779.

36. Pant, Anurag, Sai Srujan Chinta, and Balakrushna Tripathy. 2018. Kernelised clustering algorithms fused with firefly and fuzzy firefly algorithms for image segmentation. In *International conference on soft computing systems*, 125–132. Springer.
37. Parveen, S. Shaik, and C. Kavitha. 2015. Segmentation of CT lung nodules using FCM with firefly search algorithm. In *2015 international conference on innovations in information, embedded and communication systems (ICIIECS)*, 1–6. IEEE.
38. Pasquier, Philippe, Angelo Zarrella, and Denis Marcotte. 2019. A multi-objective optimization strategy to reduce correlation and uncertainty for thermal response test analysis. *Geothermics* 79: 176–187.
39. Pei, Jialun, Long Zhao, Xiangjun Dong, and Xue Dong. 2017. Effective algorithm for determining the number of clusters and its application in image segmentation. *Cluster Computing* 20 (4): 2845–2854.
40. Pham, Thuy Xuan, Patrick Siarry, and Hamouche Oulhadj. 2018. Integrating fuzzy entropy clustering with an improved PSO for MRI brain image segmentation. *Applied Soft Computing* 65: 230–242.
41. Ramadas, Meera, and Ajith Abraham. 2019. *Metaheuristics for data clustering and image segmentation*. Springer.
42. Rodrigues, Paulo S., Guilherme A. Wachs-Lopes, Horst R. Erdmann, Monael P. Ribeiro, and Gilson A. Giraldi. 2017. Improving a firefly meta-heuristic for multilevel image segmentation using Tsallis entropy. *Pattern Analysis and Applications* 20 (1): 1–20.
43. Sadiq, Ali Safa, Hossam Faris, Ala' M. Al-Zoubi, Seyedali Mirjalili, and Kayhan Zrar Ghafoor. 2019. Fraud detection model based on multi-verse features extraction approach for smart city applications. In *Smart cities cybersecurity and privacy*, 241–251. Elsevier.
44. Saha, Indrajit, Ujjwal Maulik, and Sanghamitra Bandyopadhyay. 2009. An improved multi-objective technique for fuzzy clustering with application to IRS image segmentation. In *Workshops on applications of evolutionary computation*, 426–431. Springer.
45. Samundeeswari, E.S., P.K. Saranya, and R. Manavalan. 2016. Segmentation of breast ultrasound image using regularized K-means (ReKM) clustering. In *2016 international conference on wireless communications, signal processing and networking (WiSPNET)*, 1379–1383. IEEE.
46. Saxena, Varun, Deeksha Goel, and Tarun Kumar Rawat. 2018. Image segmentation using meta-heuristic algorithms. In *2018 international conference on computing, power and communication technologies (GUCON)*, 661–666. IEEE.
47. Shakibapour, Elham, António Cunha, Guilherme Aresta, Ana Maria Mendonça, and Aurélio Campilho. 2019. An unsupervised metaheuristic search approach for segmentation and volume measurement of pulmonary nodules in lung CT scans. *Expert Systems with Applications* 119: 415–428.
48. Sheta, Alaa, Hossam Faris, Ali Rodan, Elvira Kovač-Andrič, and Al-Zoubi, Ala' M. 2018. Cycle reservoir with regular jumps for forecasting ozone concentrations: two real cases from the east of Croatia. *Air Quality, Atmosphere & Health*, 11 (5):559–569.
49. Shirakawa, Shinichi, and Tomoharu Nagao. 2009. Evolutionary image segmentation based on multiobjective clustering. In *2009 IEEE congress on evolutionary computation*, 2466–2473. IEEE.
50. Stupar, Andrija, Timothy McRae, Nenad Vukadinovic, Aleksandar Prodic, and Josh A. Taylor. 2019. Multi-objective optimization of multi-level DC-DC converters using geometric programming. *IEEE Transactions on Power Electronics*.
51. Xu, Rui, and Don Wunsch. 2008. *Clustering*, vol. 10. Wiley.
52. Yang, Dongdong, Licheng Jiao, Maoguo Gong, and Fang Liu. 2011. Artificial immune multi-objective SAR image segmentation with fused complementary features. *Information Sciences* 181 (13): 2797–2812.
53. Zhang, Mengxuan, Licheng Jiao, Wenping Ma, Jingjing Ma, and Maoguo Gong. 2016. Multi-objective evolutionary fuzzy clustering for image segmentation with MOEA/D. *Applied Soft Computing* 48: 621–637.
54. Zhao, Feng, Hanqiang Liu, Jiulun Fan, Chang Wen Chen, Rong Lan, and Na Li. 2018. Intuitionistic fuzzy set approach to multi-objective evolutionary clustering with multiple spatial information for image segmentation. *Neurocomputing* 312: 296–309.

55. Zhao, Feng, Zhe Zeng, Han Qiang Liu, and Jiu Lun Fan. 2019. A Kriging-assisted reference vector guided multi-objective evolutionary fuzzy clustering algorithm for image segmentation. *IEEE Access* 7: 21465–21481.
56. Zhu, Hong, Hanzhi He, Jinhui Xu, Qianhao Fang, and Wei Wang. 2018. Medical image segmentation using fruit fly optimization and density peaks clustering. *Computational and Mathematical Methods in Medicine* 2018.

Pavement Infrastructure Asset Management Using Clustering-Based Ant Colony Optimization

Saqib Gulzar and Hasnain Ali

Abstract Pavement asset management comprises of all the activities involved in the planning stage and design stage followed by execution involving inventory, resource allocation, and construction activities of a pavement structure. It also comprises of maintenance activities of the pavement infrastructural system during the design life supported by continuous rehabilitation of pavement sections whenever and wherever necessary. To achieve the above goals efficiently and effectively, management strategies and tools are required which help the planners and decision-makers choose the optimum combination of activities and resources, based on defined multi-objective setups. Ant colony optimization (ACO) technique offers one such option to improve the efficiency of decision-making and clustering analysis in pavement asset management. It is inspired by the natural behavioral system of ants and its probabilistic formulations which lead to the selection of optimum strategies. It extends the scope of planning through a feedback mechanism in the management of pavement systems. In this book chapter, the need for the management of pavement assets is deliberated, followed by the detailed theoretical background of ant colony optimization and its possible application in the pavement asset management is deliberated. A framework is proposed for the application of ACO in pavement management system. Further, an ACO based clustering approach is deliberated to choose optimum bids for asphalt pavement construction projects based on time of completion, total cost, and quality of construction.

Keywords Ant colony · Data clustering · Optimization · Evolutionary computation · Nature-inspired algorithms · Metaheuristics · Unsupervised learning · Machine learning · Data mining

---

S. Gulzar (✉) · H. Ali
Formerly, Indian Institute of Technology Delhi, New Delhi, India
e-mail: saqibgulzarbhat@gmail.com

H. Ali
e-mail: hasnain.mit14@gmail.com

I. Aljarah et al. (eds.), *Evolutionary Data Clustering: Algorithms and Applications*, Algorithms for Intelligent Systems,
https://doi.org/10.1007/978-981-33-4191-3_10

1 Introduction

The importance of pavement management system can be realized from the fact that the current road system has been deficient in meeting the traffic demand over the years owing to the improper and staggered planning coupled with the inherent deficiencies in the geometric design of roadways, construction methodologies employed which are very inefficiently executed and ill-planned maintenance activities add to the overall chaos [1]. Further, the need for proper use of resources subject to the constraints of feasibility, geographic and climatic conditions, have made it imperative to establish a code of practice for managing the pavement assets efficiently. Even though every year, road maintenance activities are carried out, there are multiple premature pavement failures suggesting that the pavements are not being designed, constructed, and maintained properly; the drainage system is not actively synchronized with pavement construction leading to poor performance of pavement structures and the frequent crushing, cutting, and filling of pavements for laying service lines indicate the mismanagement, lack of coordination in the implementation of developmental activities, and improper planning in the overall constructional development setup. The difficulties of not being able to predict the deterioration of pavements and design correct maintenance strategies have led to poor pavement performance, both structurally and functionally, leading to the increase in vehicle operating costs, accidents, and competing demands [2, 3]. Many a times the use of advanced materials helps in improving pavement performance [4, 5]. A pavement management system indicates the structure needed to prevent acceptable deterioration level during its design life and permits the prediction of structural and functional assessment of the deterioration condition. It outlines what potential projects are essential and where these should be done, and the type of improvements needed to prevent premature failure of pavements and also suggests means and alternate ways of extending the pavement life by operational and maintenance synchronization across multiple project levels.

The pavement management system (PMS) focusses on optimizing (minimizing) the overall cost considering the road user cost, pavement life cycle cost including the road condition and road administrator costs to reach a minimum cost which satisfies all the basic tenets of the stakeholders and efficient use of resources as shown in Fig. 1.

The overview of the components of a pavement management system is shown in Fig. 2. The data about the condition of the road and its inventory along with the pavement information which consists of surface profile, structural design, frictional and surface texture characteristics, etc., taken together are fed into the database. This database contains the present, as well as historical data, which is used for evaluating the pavement performance, optimal resources required, road user demands, road surface requirements, structural requirements, pavement layer requirements, material requirements, maintenance projects with schedules and also enables the selection of approaches for pavement monitoring, decisions for resurfacing, repair or rehabilitation, best management strategy for pavement construction, maintenance, and rehabilitation. Thus, a complete pavement management system comprises of a

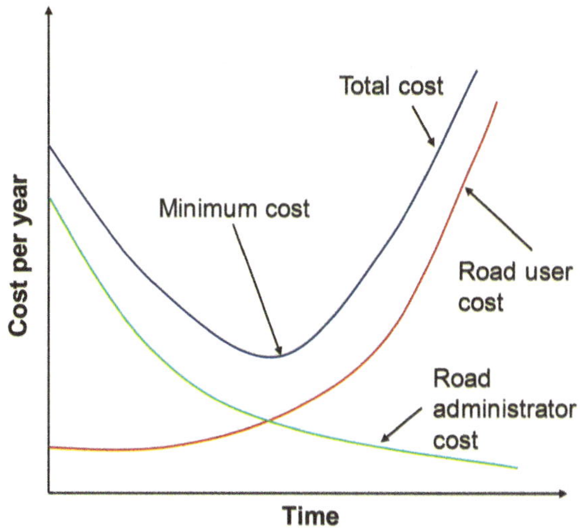

Fig. 1 Optimizing overall cost

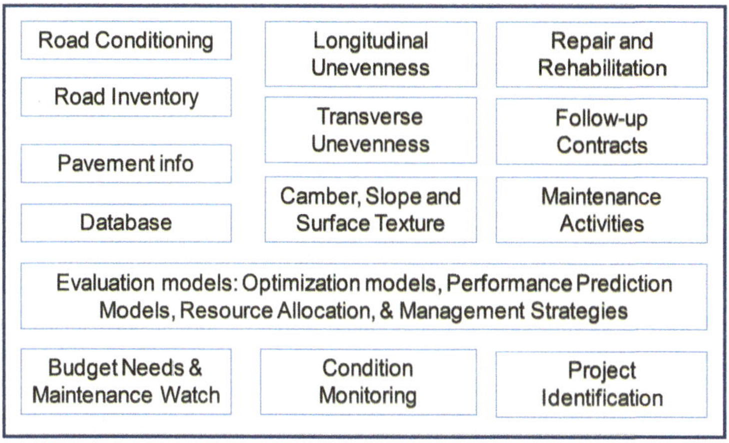

Fig. 2 Overview of PMS components

coordinated system of activities, all focused towards accomplishing the best value possible for the available public funds in constructing and maintaining safe, smooth, and economical road pavement networks.

The operation levels within a PMS are shown in Fig. 3, which can be divided under network level and project level activities, all linked to the database which forms the core for any management system. It should be noted that engineering analysis and research to come up with better evaluation models [6], optimization techniques [7] and management strategies [8] should also be funded to enable efficient utilization of resources which keep on diminishing with each passing year [9].

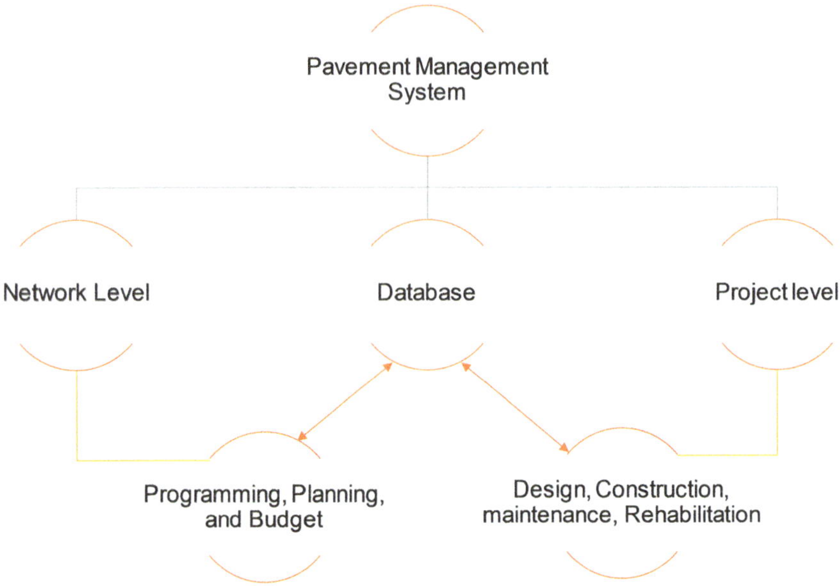

Fig. 3 Overview of PMS at network and project levels and their inter-relationships

2 Theoretical Background

2.1 Optimization Techniques

Optimization techniques form as a component of the subject of operations research which deals with the implementation of mathematical techniques to decision-making scenarios and problems in order to establish the best or optimal solutions based on the given objective function subject to the existing or future constraints. The optimization techniques vary based on the problems which are classified on the basis of constraint existence, typical nature of variables of design, structure of the overall problem, type of equations involved, allowable values of design variables, deterministic or probabilistic nature of all variables, decoupling nature, and total number of objective functions. These optimization methods may be purely mathematical, stochastic or statistical in nature. With the advancement of understanding of biological systems, many unconventional optimization techniques have been developed such as algorithms based on genetics (GA), simulated annealing (SA), particle swarm optimization (PSO), ant colony optimization (ACO), neural networks, fuzzy optimization, etc. In this paper, the ant colony optimization technique is evaluated for possible use in pavement management systems and discussed in detail in subsequent sections.

2.2 *Ant Colony Optimization (ACO)*

Ant colony optimization is based on the real coordinative behavioral traits of ants which are able to converge and locate the shortest path to the source of food from their homes. The preceding ants who successfully find food leave a volatile substance called pheromone behind them on their path which guides the following ants to reach the food source through the successful shortest path. This forms the communication basis among ants and enables them to solve complex problems, moving around a series of obstacles each time finding an optimum pathway leading them to the food source. Various experimental studies have verified this fact that an ant group perform unanimously and find the shortest path from their home to the food source spatially through their communication methodology involving pheromone deposition and evaporation, allowing for path update as when the pheromone evaporates [10, 11]. The optimization algorithms taking inspiration from ant's behavior in finding the shortest paths by using only information based on the chemical substance called pheromone, initially developed in 1990s, have been used for various discrete type optimization problems and have performed successfully [12–15].

2.3 *ACO Behavioral Process*

The ACO process can be understood by qualitatively representing the complete discrete optimization problem like a set of multilayered, multistep series of alternatives as shown in Fig. 4. The total number of design variables involved in the optimization problems is set equal to the number of layers or steps in the graphical representation while the count of nodes in each step represents the permissible value of each variable of design subject to the constraints and equal the total count of discrete values allowed for the respective design variable. Thus, each individual node is associated with an allowable discrete value of the design variable of the optimization problem.

Assuming that an ant colony consists of N ants starting from their home node and travel towards finding a food destination. The ants travel through a number of layered framework from the first layer to the last layer, and in each iteration cycle move towards their food but end up at a destination node in the way. At each node, each ant has a choice of selecting multiple paths but selects only one based on the pheromone deposition. Each ant makes a choice at each node and that forms a local solution, however, this continues at each node and with each iteration, the global path selected based on the pheromone concentration represents the optimal path as ant progress towards their food source [16]. This optimal path is shown as a bold line shown in Fig. 4. The probability of choosing the next node is called the ant searching transition rule and the trail followed by the following ants is governed by the local retracing and updating rule as explained in the subsequent sections.

At the start of the optimization process in the first iteration, all paths are allotted an equal concentration of pheromone, thereby, in the first iteration, all ants begin from

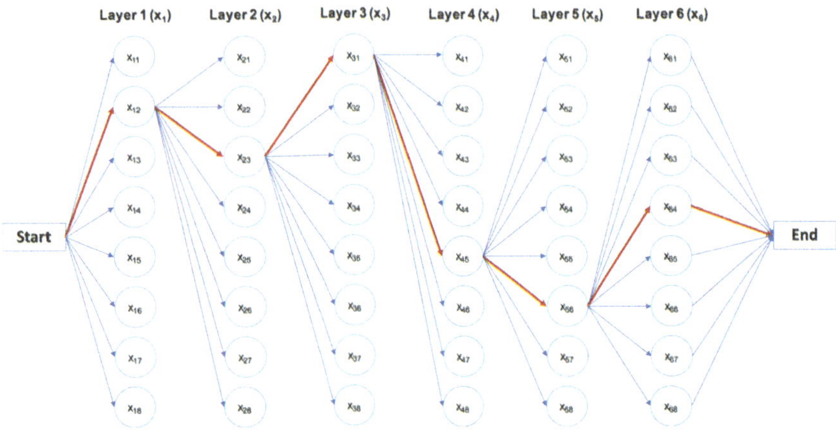

Fig. 4 ACO behavioral process

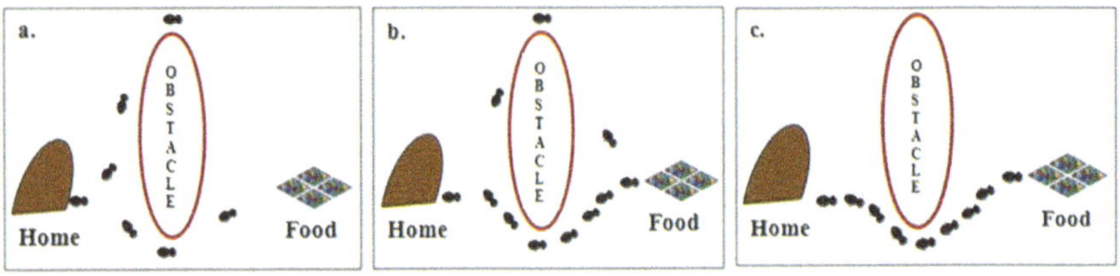

Fig. 5 **a** Ants exploring all paths with equal probability; **b** As time progresses, the shorter path witnesses more ants; **c** All ants converging on the shorter path

home or nest node and reach food or destination nodes by randomly or haphazardly selecting any node in each step. However, with the next iteration, the paths are updated based on the updating rule and finally the optimization process ends up if the maximum number of iterations specified is reached or the single solution emerges out within the specified number of iterations. The optimal solution consists of all those nodes representing each design variable, which lie on the path containing the highest amount of pheromone concentration. Thus, at the optimum, all ants choose the same best path converged after a number of iterations, which is the shortest and most efficient way of traveling from home node to food destination mode.

The ant colony optimization process has been illustrated in Fig. 5, wherein ants travel from home to food initially through all routes but with each iteration converge to an optimum path represented by the shortest path between the home node and the food destination node [17].

2.4 Ant Searching Transition Rule

An ant r, positioned at an ith node needs to choose the next jth node based on the pheromone deposition between the two nodes equal to ψ_{ij}^{δ}. This probability of choosing jth node from the preceding ith node based on the pheromone deposition ψ_{ij}^{δ} is called an ant searching transition rule and is given below

$$P_{ij}^{(r)} = \frac{\psi_{ij}^{\delta}}{\sum\limits_{j \in N_i^{(r)}} \psi_{ij}^{\delta}}, \quad \forall j \in N_i^{(r)} \tag{1}$$

$$P_{ij}^{(r)} = 0, \quad \forall j \notin N_i^{(r)} \tag{2}$$

where δ denotes the degree of pheromone importance and $N_i^{(r)}$ denotes all the nodes around node i when rth ant is position at an ith node. It should be noted that $N_i^{(r)}$ contains all nodes surrounding ith node except the preceding nodes which has been already visited before. This helps in preventing an ant from going back to the same node which has already been visited before node i. This formulation is applied at each node until an ant traveling from home node reaches food destination nodes and subsequently applied to each ant each iteration coupled with updating rule as given in the next section.

2.5 Local Retracing and Updating Rule

Before coming back to the home node, the rth ant deposits $\Delta\psi^{(r)}$ of pheromone on the path segments it has gone through. The pheromone value ψ_{ij} on the path segment (i, j) traveled is updated as per the below local retracing and updating rule

$$\psi_{ij} \leftarrow \psi_{ij} + \Delta\psi^{(r)} \tag{3}$$

This updating causes an increase in the pheromone deposit on the path segment (i, j), thus the probability of this path segment being selected from the following acts will increase, as they'll be choosing the subsequent nodes as per the ant searching transition rules which is based on the amount of pheromone deposition along each segment connecting the nodes across different layers.

2.6 Evaporation of Pheromone Deposits

One of the important aspects of ant colony optimization is the evaporation of pheromone deposits. The pheromone evaporation helps in the exploration of alternate paths during the ant searching transition process and avoids pre-timed convergence to the local sub-optimal solutions, thereby, eliminating the poor selection of segments in the process of path selection. It also helps in limiting the maximum value of the pheromone deposit attainable on any path segment.

When an rth ant travels to the next node, the pheromone evaporates from all the path segments (i, j) in accordance to the pheromone evaporation rules as given below

$$\psi_{ij} \leftarrow (1-\rho)\psi_{ij}, \quad \forall (i, j) \in S \tag{4}$$

where $\rho \in (0, 1]$ is a parameter representing the rate of evaporation and is called as pheromone decay factor and S denotes all the segments traveled by an ant r in its path from home nest to food destination.

One complete iteration consists of a complete cycle of ant's searching transition or movement, evaporation of pheromone, and deposition of pheromone through updating. Once all ants have returned from the food source to the home node, the pheromone deposit information is updated according to the updating relation, but considering the contribution of the best ant to the path segments which is inversely proportional to the length of the path traveled by the rth ant. The global updating principle at the end of each iteration is given by the below relation.

$$\begin{aligned} \psi_{ij} = (1-\rho)\psi_{ij} + \sum_{r=1}^{n} \Delta\psi_{ij}^{(r)}; \\ \text{where, } \Delta\psi_{ij}^{(r)} = \frac{B}{L_r} \\ B = \text{An arbitrary constant} \\ L_r = \text{Path length traveled by } r\text{th ant} \end{aligned} \tag{5}$$

The best ant pheromone deposition equation $\Delta\psi_{ij}^{(r)} = \frac{B}{L_r}$ is implemented as per the below scheme

$$\Delta\psi_{ij}^{(r)} = \begin{cases} \frac{\alpha\xi_{best}}{\xi_{worst}}, & \text{if (i, j)} \in \text{global optimum path} \\ 0, & \text{otherwise} \end{cases} \tag{6}$$

where ξ_{best} is the best value while ξ_{worst} is the worst value of the governing objective function among all the paths traveled by a total of N ants and α is a parameter to adjust the scale of the updating of pheromone at a global level. A higher value of α indicates that higher amount of pheromone has been deposited on the global best path, where more ants will exploit this path in the forthcoming iterations. This helps in assigning greater pheromone levels to the paths which results in better objective function values.

2.7 Global Implementation Algorithm

The algorithm for implementing ant colony optimization technique for a minimization problem such as pavement management system which requires to achieve best possible use of resources and services at the minimum cost is given in Table 1. It involves four steps consisting series of sub-schemes starting with basic assumptions, followed by computation of probability of path segment selection and evaluation of objective function for each set of design variables for each ant and determination of best and worst values, following which test for convergence is done and subsequently pheromone evaporation and updating is done for the next iteration until we get the optimum path or the maximum iterations number is achieved.

3 ACO in Pavement Management System

3.1 Prospects

The construction of a roadway involves numerous resources, labor, and machinery amounting to crores of rupees and once construction is completed, the next most important task is the management of pavement infrastructure, maintenance activities which have to be periodic as these help in extending the service life of pavements, improve their functional, as well as structural performance, and ensure sufficient drainage systems for the pavement structure. The road infrastructure contributes a lot to the economy of a nation, thus with increasing awareness about the importance of an efficient road infrastructure at the policy level, more funds are being allocated to the national highway development and maintenance activities across all countries, especially in the developing world. Therefore, it is very important that all the construction, as well as maintenance work, should be well planned, ensuring the best possible utilization of all the resources subject to the constraint of resources and with the climate change happening, more focus has been on the efficient use of natural resources involved in the construction of highways, as well as reduction in the emission during asphalt laying and vehicle operations, with well-maintained highways reducing the vehicle operating costs and emissions. Further, it is very important to select the optimum strategy for repair, rehabilitation, and maintenance works, as well as to cater the demands of these works, depend on how well the planning has been done. Therefore, it becomes inevitably essential to carry out the optimization of various alternatives and select the best to avoid any losses and reduce the complexity

Table 1 Global ant colony optimization algorithm

Step 1: Select Hyper-parameters

Assume the following

- N = Number of ants in the colony
- Permissible discreet values of n design variables
- Add equal amounts of pheromone ψ_{ij}^{1} initially along all segments
- Set iteration number l = 1
- Pheromone decay factor, $\rho = 0.5$ to 0.8

Step 2: Compute path selection probability

2.1 Compute the probability of selecting segment between (i, j)

$$P_{ij} = \frac{\psi_{ij}^{(1)}}{\sum_{m=1}^{p} \psi_{im}^{(1)}}; \quad (7)$$
$$i = 1, 2, \ldots, m;$$
$$j = 1, 2, \ldots, p; \alpha = 1 \text{ or large}$$

2.2 Find cumulative probability ranges associated with different paths based on probabilities given in step 2.1

Step 3: Evaluation of different paths

3.1 Find the specific path chosen by ant r by generating N random numbers in the range of (0, 1), one for each ant and determining the cumulative range previously found in step 2.2 for any ant r for a particular variable i

3.2 Repeat step 3.1 for all design variable $i = 1, 2, 3, \ldots, n$

3.3 Evaluate the objective function corresponding to complete paths ∀ ants $(r = 1, 2, 3, \ldots, N)$ and design variables $(i = 1, 2, 3, \ldots, n)$; $\xi_r = \xi(r, n)$

3.4 Determine the best and worst paths among N paths chosen by different ants

$$\xi_{best} = Min.\xi(r, n) \ \forall k = 1, 2, 3, \ldots, N \quad (8)$$
$$\xi_{worst} = Max.\xi(r, n) \ \forall k = 1, 2, 3, \ldots, N$$

Step 4: Obtain best path upon convergence of algorithm

4.1 Test for convergence of the whole ant colony optimization process. It is assumed that the process has converged if N ants take the best path

4.2 If convergence is not achieved, assume all ants return home and start again in search of food. The new iteration number is set $l = l + 1$ and updating rule is applied to the pheromones on different path segments as

$$\psi_{ij}^{l} = \psi_{ij}^{old} + \sum_{r=1}^{N} \Delta\psi_{ij}^{(r)} \quad (9)$$
$$\text{where } \psi_{ij}^{old} = (1 - \rho)\psi_{ij}^{l}$$

4.3 With new value of ψ_{ij}^{l}, go to step 2 and repeat the whole algorithm, until the process converges

of the problem. Ant colony optimization offers one such technique to help resolve engineering-cum-managerial optimization problems in pavement asset management system.

3.2 Application in Pavement Management Systems

Serdel and Serin proposed a metaheuristic method based on an ant colony optimization algorithm for planning the economic evaluation and maintenance works on pavements considering the rehabilitation, routine, and major maintenance options for flexible pavements [18] governed by the availability of funds, historical accounts, and network or project level considerations based on organizational and political factors. In the proposed model, six constraints such as requirement of production, availability of labor, budget, equipment, material, and rehabilitation have been considered using the methodological framework of ant colony optimization algorithm implemented through visual basic software. The authors were able to compute optimum maximum work load for maintenance work priority wise and found that it decreased with increasing the number of ant in the colony. Their work has been based on the formulations proposed by Fwa et al. [19], which provide an integration optimization programming model tool to prioritize routine maintenance works in highway pavements. Further, Fwa et al. [20] suggested that pavement maintenance is a multi-objective optimization problem as such agencies need to employ advanced network level optimization techniques comprising of dynamic route searching features and inherent capability to handle multiple solutions must be adopted. They suggested the use of a genetic algorithm-based optimization technique and illustrated its use through a numerical example performed for two- and three-objective optimization, respectively. Ant colony optimization is also a similar optimization technique which can serve as a tool to handle multi-objective optimization problems encountered in pavement management system. Chikezie et al. [21] also suggested the use of multi-objective optimization in pavement maintenance and rehabilitation works. Gao et al. [22] has illustrated the network level selection feature of ant colony optimization technique and proposed a model based on the methodological framework of ant colony optimization. Sivagaminathan and Ramakrishnan [23] have proposed a hybrid approach for optimum network level selection using neural networks and ant colony optimization. It helps in the selection of input variables which help in the prediction of an optimum route with an acceptable level of accuracy. Christodoulou [24] has presented a methodology for scheduling resource-constrained construction projects using ant colony optimization technique. The author has used the ACO process methodology and efficiently computed optimum solution in less than 50 iterations. Further, the author has compared the ACO technique with other traditional and modern algorithms and concluded that ACO is relatively simple than a genetic algorithm and particle swarm optimization, and representatively similar to traditional CPM but more efficient and robust. Fileccia Scimemi [25] have used ant colony optimization technique coupled with Levy perturbation for back-calculating the elastic moduli and thickness of multiple layers of airport pavements using surface deflection data obtained from nondestructive testing. They showed that the ant colony optimization technique can offer a viable technique to obtain the above parameters correctly while considering the inherent variation of inputs and calculated parameters. The ACO algorithm removes the anomalies of local optima through its global searching rule and gives accurate results over other optimization methods. Zhitao et al. [26] have given an algorithmic framework for scheduling of construction works based on ant colony optimization. These construction works involve building, as well as pavement infrastructure, and may be optimized well using the ACO process. In road construction, there is an additional delay caused by the ongoing traffic due to the construction activities leading to a loss in speed and increase in travel time which result in huge economic losses, as such the optimum scheduling is even more essential in such case and ACO offers a viable method to achieve this objective. The authors have integrated the inputs based on expert knowledge into the metaheuristic ACO algorithm called as CoANT to achieve an optimum scheduling scheme and have implemented it on realistic transportation networks. CoANT is seen to work faster, better, and is able to produce competitive schedules which aid the agencies in decision-making. In our previous work [2], the applicability of ant colony optimization in pavement management system was showcased and the ACO algorithm was used to explore the resource allocation and time scheduling of pavement construction at a project level. Further, the use of ACO algorithm in clustering analysis for project allocation and efficient bidding process in pavement construction projects was also deliberated.

Julia and Meyer [27] reviewed the two network level optimization clustering techniques based on ant colony optimization and concluded that the algorithms based on ant colony optimization are very effective for multi-objective, decentralized, and self-organizing problems and can offer a robust solution alternative to compute the optimum route, whether based on priority, budgetary constraints or resources constraint. It implicitly implies that, optimization problems in pavement management systems can be effectively handled using ant colony algorithms and may prove beneficial than any other conventional optimization technique. Based on the use of ant colony optimization in pavement management system, a general framework as shown in is proposed. This framework uses the existing ant colony optimization algorithms and provides a way to use it in handling pavement construction management problems at a project and network level.

4 Ant Colony Optimization-Based Clustering Analysis in PMS

As described in Sect. 2, the pavement management system incorporates multiple processes and activities. All such activities revolve around the construction of a road infrastructure. A road construction project involves a range of activities and many a times, there are different subcontractors for different activities. So, a project given out by an agency is let by a major contracting company, which in turn, sublets it to different subcontractors to handle a different set of activities. As stated before, both the agency and the contractor want to get the project done at minimum cost and duration without compromising the quality of the project. The whole process

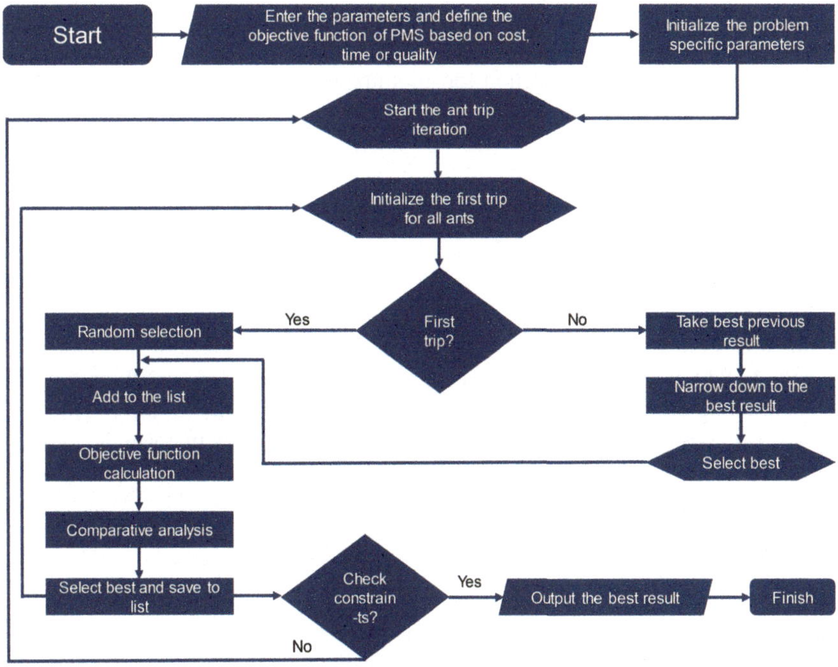

Fig. 6 Ant colony optimization framework in pavement management system

can be treated as a multi-objective optimization problem and we will now proceed to formulate the objective functions and constraints of such a problem. In order to reach an optimal or feasible solution, an ant colony optimization-based clustering algorithm is used, and it is shown that such a protocol proves to be very effective in such a scenario [28] (Fig. 6).

In an asphalt pavement construction project, the consideration of Indirect cost (I_C) per time period over the entire duration of the project (D_p) plays an important role in letting the project to contractors or subcontractors. Every road project is assigned a due date (D_d). Depending on the actual time of completion, the contractor may get incentive, say α cost for the time saved as a reward or may be charged a penalty, say δ for the overtime taken to finish the project. For any activity l, let us suppose there are a total of S_l available subcontractors, such that each subcontractor m bids a cost c_{lm} with a proposed duration of d_{lm} with an expected quality of q_{lm}. In order to complete all the project activities, we need to select a near-optimal group of subcontractors based so as to minimize the total duration (D_p) and cost (C_p) of the project, maximize the quality of the project ($Q_{p\tau}$). The objective function for minimizing the total duration of the project can be written as below

$$\text{Objective 1 : min } D_p = \max \sum_{l \in W_p} \sum_{m=1}^{n} d_{lm} k_{lm} \tag{10}$$

where k_{lm} is a binary variable (0, 1) for each activity l, and W_p is the activity sequence of pth path.

Similarly, the objective function for minimizing the total cost of the project for all the activities of the project (N), can be written as below

$$\text{Objective 2}: \min C_P = \sum_{l=1}^{N}\sum_{m=1}^{S_l} c_{lm}k_{lm} + I_C D_p + \delta(D_p - D_d) + \alpha(D_d - D_p) \tag{11}$$

The final objective function is to maximize the quality of the project. It should be noted that there is a minimum threshold, Q_{min} for all the possible alternatives of an activity which is measured against the average quality, Q_{avg} for all the possible subcontracting alternatives eligible for the project. The deviation between the two, say τ sets the desirability of each parameter by putting the relative weights on each and determines the overall quality of the project. So, the objective function may be written as below

$$\text{Objective 3}: \max Q_{p\tau} = \tau Q_{min} + (1 - \tau)Q_{avg} \tag{12}$$

The constraints are set in the quality and also on the way the contracting process is done. The constraints on the quality of the project are as below

$$Q_{min} = min\{q_{lm} : k_{lm} = 1\},\ 1 \le l \le N,\ 1 \le m \le S_l \tag{13}$$

$$Q_{min} \le \{q_{lm} \times k_{lm} + H(1 - k_{lm})\},\ 1 \le l \le N,\ 1 \le m \le S_l \tag{14}$$

$$Q_{avg} = \frac{\sum_{l=1}^{N}\sum_{m=1}^{S_l} q_{lm}k_{lm}}{N},\ 1 \le l \le N,\ 1 \le m \le S_l \tag{15}$$

It should be noted that H is an arbitrarily large number and Q_{min} can never exceed H. This is done to keep the constraint nonbinding. The indicators that are used to check the quality of an asphalt pavement include total air voids in the mix, asphalt binder content, in-situ air void content, ride quality, etc. Similarly, the constraint on the time to ensure that project duration is within the due date can be written as

$$D_p \le D_d \tag{16}$$

Also, to ensure that each activity is done successively in the project network, a constraint on the starting time of a scheduled activity, t_i is applied as given below

Table 2 An illustrative alternative matrix showing different combinations of cost, duration, and quality of different project activities proposed by different project contractors

Project activity	Cost (C_p)—Duration (D_p)—Quality ($Q_{p\tau}$) alternatives								
	Alternative 1			Alternative 2 …			Alternative 10		
	C_p	D_p	$Q_{p\tau}$	C_p	D_p	$Q_{p\tau}$	C_p	D_p	$Q_{p\tau}$
1	2000	10	100	2500	12	99	2000	13	98
2	4000	12	98	4500	14	100	35000	13	97
3	6000	12	95	5000	13	97	4500	10	100
4	4500	9	97	4200	11	98	6000	15	100
5	3000	8	99	6000	9	95	6000	12	100
6	5000	12	95	3500	8	98	3750	11	99
7	8000	13	100	9000	15	96	8500	14	98
8	800	5	95	550	6	96	750	8	96
9	500	3	97	950	7	98	900	9	96
10	600	7	98	320	8	100	450	5	97
11	25000	25	99	29000	23	100	35000	29	97
12	30000	29	100	35000	35	95	29500	32	98
13	30000	24	100	20000	20	98	25000	28	99
14	2000	11	98	2500	9	99	2410	10	97
15	9000	9	97	7000	11	99	6000	15	96
16	6000	10	95	3000	12	98	5550	13	95
17	3000	9	96	4000	8	97	3500	9	98
18	4000	10	95	5000	7	96	3590	11	99
19	3000	13	98	3000	10	95	2500	15	99
20	5000	8	99	4580	5	100	7000	3	100

$$t_P \geq t_l + \sum_{m=1}^{S_l} q_{lm} k_{lm},\ 1 \leq l \leq N \tag{17}$$

To ensure that each activity is assigned only one subcontracting alternative, the constraint can be written as

$$\sum_{m=1}^{S_l} k_{lm} = 1,\ 1 \leq l \leq N \tag{18}$$

Now, in order to solve this multi-objective optimization problem, the steps highlighted in Table 3 are systematically executed and the flow of the algorithm is as per Fig. 7. It should be noted that clustering analysis is done along with ACO by considering the groups of subcontractors based on the bid time, cost and quality of the project activity. A list of activities grouped by the available subcontractors is evaluated and a near-optimal cluster is selected. It should be noted that there is always a trade-off involved in multi optimization problems such as this and achieving an exact solution is not always feasible. A brief illustration on synthetic data for a flexible pavement construction project based on the data from Feng et al. [29] is given in Table 2.

Consider a flexible pavement construction project involving a set of 20 (say) activities. Let us suppose the bidding process receives quotations from 10 contractors proposing a different combination of time, cost, and quality for each activity. The options matrix will look similar to Table 2. It shows the different combinations of cost, duration, and quality of different project activities proposed by different project contractors. The cost is in dollars, the duration is in days, and the quality is rated on a scale of 1 to 100. Other parameters such as the proposed date of completion is 220 days, reward and penalty costs are chosen to be $250 and $150 per day, etc. With 20 activities in the project and 10 contractors, there are over 4 billion combinations for scheduling and executing the project. In order to obtain the optimal combination of contracting alternatives such that the project time and cost are minimized, while the quality is maximized, the ant colony optimization algorithm searches the subspace and with each successive iteration leaves out the nonoptimal solutions. In order to initialize the implementation algorithm, several trials were conducted and finally a suitable value of initial parameters is obtained. After running the algorithm for several thousand iterations, a trade-off matrix is obtained and a near-optimal solution is chosen. An example solutions trade-off table is shown in Table 3.

It can be seen that there are different solutions with different optimal conditions. For example, solution 26 offers lowest cost but project time and quality are not highest. Similarly, solution 24 offers lowest project duration, but is not necessarily cheapest or of best quality. Also, solution 25 offers the best quality but is quite costly and takes a longer time. So, it is essential to select a solution by trading-off on the selection criteria and selecting a near-optimal solution. The performance for all the seven sample solutions is compared and contrasts are drawn as shown in Figs. 7, 8, and 9 for cost, time or duration, and quality of the project. Finally, the trade-off plot for all three selection criteria is shown in Fig. 10. The trade-off analysis is done for each pair of selection criteria and the set of project alternatives that give the near-optimal solution is selected. Further, in order to evaluate the efficiency of the ant colony-based algorithm, a performance analysis is conducted using the metrics such as degree and speed of convergence, sparsity, and number of local optima, etc. The variance of solutions, the computation time, and the root mean square distance are often used as well.

The activities that result in a similar selection criterion are clustered together to achieve higher efficiency and quicker convergence at sufficiently low computation time. Daniel [30] reports that the solution obtained using ant colony optimization algorithm is highly dependent on the correct implementation of the pheromone metric. For small changes in the pheromone metric, there were huge changes in the optimal solution, highlighting the sensitivity of ACO to pheromone trail. As discussed by Christodoulou [24], the algorithm provides an efficient methodology in constructing solutions in network level optimization problems. There is a lot of scope for the improvement in ACO algorithms by incorporating forward coupling, as well

Table 3 Solution trade-off table

Solution	Project activities	Selection criteria		
		Cost ($)	Time (days)	Quality (%)
21	[1, 1, 2, 1, 3, 1, 5, 4, 1, 8, 6, 1, 2, 1, 6, 5, 4, 3, 8, 9]	187900	275	83.526
22	[2, 1, 5, 4, 2, 6, 5, 1, 2, 3, 9, 8, 2, 4, 4, 5, 1, 2, 4, 5]	190065	265	85.996
23	[3, 3, 3, 1, 1, 1, 1, 5, 1, 6, 4, 1, 8, 1, 6, 2, 9, 1, 5, 2]	191565	289	84.153
24	[2, 1, 5, 4, 1, 8, 3, 2, 9, 1, 5, 2, 8, 4, 1, 5, 2, 1, 5, 1]	220513	255	87.225
25	[2, 3, 2, 2, 2, 3, 1, 2, 5, 1, 4, 5, 1, 6, 5, 2, 8, 7, 4, 5]	251200	295	89.551
26	[5, 1, 5, 4, 2, 5, 4, 5, 6, 2, 3, 1, 8, 1, 1, 1, 2, 6, 5, 1]	151559	288	81.322
27	[8, 1, 9, 7, 9, 1, 9, 8, 1, 6, 5, 9, 1, 1, 5, 6, 1, 1, 1, 5]	187900	277	79.854

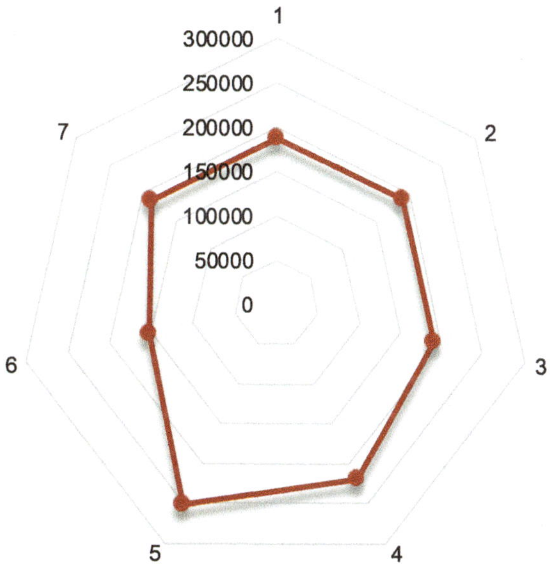

Fig. 7 Cost performance of the sample solutions

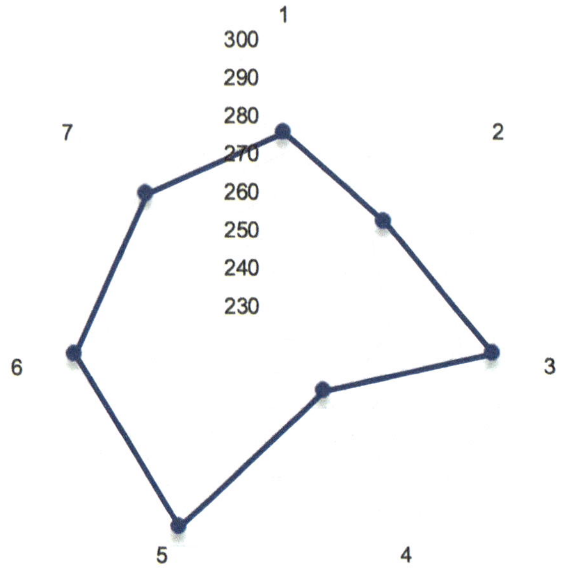

Fig. 8 Time performance of the sample solutions

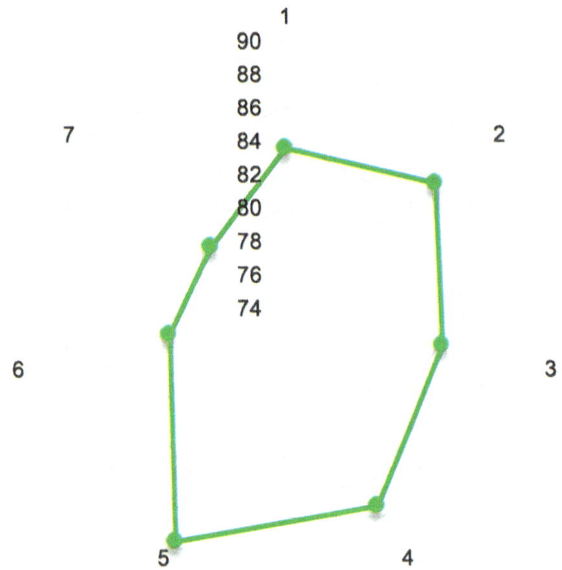

Fig. 9 Quality performance of the sample solutions

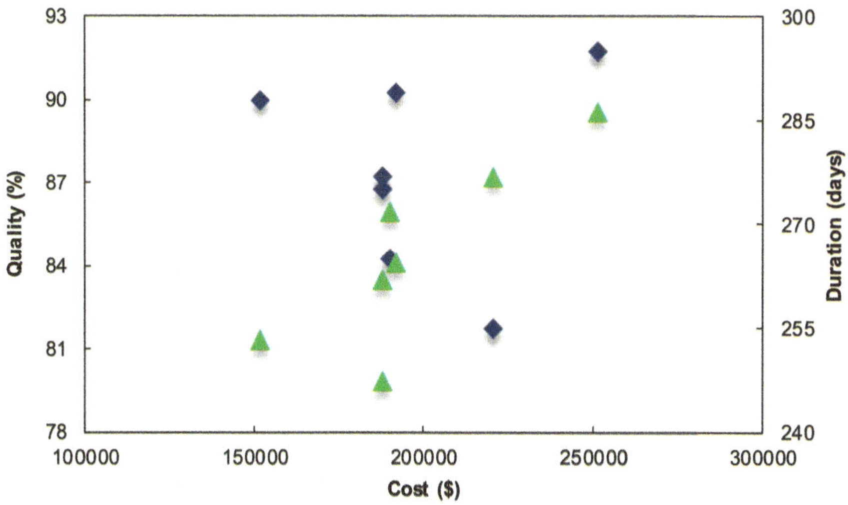

Fig. 10 Trade-off plot of selection criteria

as backward indentation mechanism at the project, as well as network levels. There is also prospective scope in design optimization as well as rehabilitation prioritization problems and scheduling of maintenance activities. The overall computation efficiency and varied applicability of the ACO algorithms for handling complex problems comprising of huge number of activities, resources, and manpower needs to be evaluated using numerical simulations and programming tools.

5 Conclusion

Pavement management system involves multiple optimization and clustering problems both at project level, as well as at network level. In order to improve efficiency in pavement construction, management, and maintenance, it is important to use evolutionary algorithms such as ant colony optimization to reduce losses as showcased in this chapter. An introduction to pavement infrastructure was given, followed by the description of ant colony optimization technique. The steps involved in the implementation of ACO in pavement management system were highlighted. Ant colony optimization proves to offer the advantage of having a robust search algorithm at a global level with good accuracy. The objective functions for an asphalt pavement construction project along with the constraints were defined in a multi-objective optimization problem and the steps to use of ACO to obtain near-optimal solution based on clustering analysis involving project duration, cost, and quality of each activity involved were presented. The implementation algorithm presented in this chapter can be used for different types of pavement management scenarios and it is hoped that a full case study showing the complete application of the ant colony optimization-based clustering analysis in PMS will be presented in future.

References

1. Vajjarapu, Harsha, Ashish Verma, and Saqib Gulzar. 2019. Adaptation policy framework for climate change impacts on transportation sector in developing countries. *Transportation in Developing Economies* 5 (1): 3.
2. Gulzar, Saqib, Hasnain Ali, and Chetan Doddamani. 2017. Ant colony optimization in pavement asset management. In *Proceedings of ASCE India conference 2017*. Indian Institute of Technology Delhi.
3. Gulzar, Saqib, Hasnain Ali, and Chetan Doddamani. 2017. A conceptual framework for introducing 'mobility as a service' in India: Opportunities & challenges. In *Proceedings of ASCE India conference 2017*. Indian Institute of Technology Delhi.
4. Gulzar, Saqib, and Shane Underwood. 2019. Use of polymer nanocomposites in asphalt binder modification. In *Advanced functional textiles and polymers: Fabrication, processing and applications*, 405.
5. Gulzar, Saqib, and Hazratullah Paktin. 2017. Scope of using nanomaterials in pavement engineeering. In *Proceedings of nano India 2017*. Indian Institute of Technology Delhi.
6. Ali, Hasnain, Yash Guleria, Sameer Alam, and Michael Schultz. 2019. A passenger-centric model for reducing missed connections at low cost airports with gates reassignment. *IEEE Access* 7: 179429–179444.
7. Ali, Hasnain, Yash Guleria, Sameer Alam, Vu N. Duong, and Michael Schultz. 2004. Impact of stochastic delays, turnaround time and connection time on missed connections at low cost airports. In *Proceedings of the 13th USA/Europe air traffic management. R&D Seminar*.
8. Ali, Hasnain, Raphael Delair, Duc-Thinh Pham, Sameer Alam, and Michael Schultz. Dynamic hot spot prediction by learning spatial-temporal utilization of taxiway intersections. In *International conference on artificial intelligence and data analytics in air transportation (AIDA-AT 2020)* (in press).
9. Haas, Ralph, and W. Ronald Hudson. 2015. *Pavement asset management*. Wiley.
10. Dorigo, Marco. 1992. Optimization, learning and natural algorithms. PhD thesis, Politecnico di Milano.
11. Heidari, Ali Asghar, Hossam Faris, Seyedali Mirjalili, Ibrahim Aljarah, and Majdi Mafarja. 2020. Ant lion optimizer: Theory, literature review, and application in multi-layer perceptron neural networks. In *Nature-inspired optimizers*, 23–46. Springer.
12. Mirjalili, Seyedali, Jin Song Dong, and Andrew Lewis. 2020. Ant colony optimizer: Theory, literature review, and application in auv path planning. In *Nature-inspired optimizers*, 7–21. Springer.
13. Mirjalili, Seyedali. 2015. The ant lion optimizer. *Advances in Engineering Software* 83: 80–98.
14. Mirjalili, Seyedali, Pradeep Jangir, and Shahrzad Saremi. 2017. Multi-objective ant lion optimizer: A multi-objective optimization algorithm for solving engineering problems. *Applied Intelligence* 46 (1): 79–95.
15. Mafarja, Majdi, Derar Eleyan, Salwani Abdullah, and Seyedali Mirjalili. 2017. S-shaped vs. v-shaped transfer functions for ant lion optimization algorithm in feature selection problem. In *Proceedings of the international conference on future networks and distributed systems*, 21. ACM.
16. Rao, Singiresu S. 2019. *Engineering optimization: Theory and practice*. Wiley.
17. Ali, Hasnain, and Arpan Kumar Kar. 2018. Discriminant analysis using ant colony optimization-an intra-algorithm exploration. *Procedia Computer Science* 132: 880–889.
18. Terzi, Serdal, and Sercan Serin. 2014. Planning maintenance works on pavements through ant colony optimization. *Neural Computing and Applications* 25 (1): 143–153.

19. Fwa, Tien Fang, Kumares C. Sinha, and John D.N. Riverson. 1988. Highway routine maintenance programming at network level. *Journal of Transportation Engineering* 114 (5): 539–554.
20. Fwa, T.F., W.T. Chan, and K.Z. Hoque. 2000. Multiobjective optimization for pavement maintenance programming. *Journal of Transportation Engineering* 126 (5): 367–374.
21. Chikezie, Clarkson Uka, Adekunle Taiwo Olowosulu, and Olugbenga Samuel Abejide. 2013. Multiobjective optimization for pavement maintenance and rehabilitation programming using genetic algorithms. *Archives of Applied Science Research* 5 (4): 76–83.
22. Gao, Hai-Hua, Hui-Hua Yang, and Xing-Yu Wang. 2005. Ant colony optimization based network intrusion feature selection and detection. In *2005 international conference on machine learning and cybernetics*, vol. 6, 3871–3875. IEEE.
23. Sivagaminathan, Rahul Karthik, and Sreeram Ramakrishnan. 2007. A hybrid approach for feature subset selection using neural networks and ant colony optimization. *Expert Systems with Applications* 33 (1): 49–60.
24. Christodoulou, Symeon. 2009. Scheduling resource-constrained projects with ant colony optimization artificial agents. *Journal of Computing in Civil Engineering* 24 (1): 45–55.
25. Fileccia Scimemi, G., T. Turetta, and C. Celauro. 2016. Backcalculation of airport pavement moduli and thickness using the lévy ant colony optimization algorithm. *Construction and Building Materials* 119: 288–295.
26. Xiong, Zhitao, David Rey, Vinayak V. Dixit, and S. Travis Waller. 2014. An algorithmic framework for the scheduling of construction projects based on ant colony optimization and expert knowledge. In *17th international IEEE conference on intelligent transportation systems (ITSC)*, 2446–2452. IEEE.
27. Handl, Julia, and Bernd Meyer. 2007. Ant-based and swarm-based clustering. *Swarm Intelligence* 1 (2): 95–113.
28. Mungle, Santosh, Lyes Benyoucef, Young-Jun Son, and M.K. Tiwari. 2013. A fuzzy clustering-based genetic algorithm approach for time-cost-quality trade-off problems: A case study of highway construction project. *Engineering Applications of Artificial Intelligence* 26 (8): 1953–1966.
29. Feng, Chung-Wei, Liang Liu, and Scott A. Burns. 1997. Using genetic algorithms to solve construction time-cost trade-off problems. *Journal of Computing in Civil Engineering* 11 (3): 184–189.
30. Angus, D. 2005, January. Solving a unique shortest path problem using ant colony optimisation. *Communicated by T. Baeck*, 1–26.

A Classification Approach Based on Evolutionary Clustering and Its Application for Ransomware Detection

Raneem Qaddoura, Ibrahim Aljarah, Hossam Faris, and Iman Almomani

Abstract Ransomware analysis and detection has been recently applied using supervised and unsupervised machine learning approaches. Combining both approaches simplifies the complexity of the data and builds an expert classifier for each predicted cluster. In this paper, a hybrid approach for detecting ransomware is proposed. It combines evolutionary clustering approach using Grey Wolf Optimizer (GWO) with an ensemble of Support Vector Machine (SVM) classification on a dataset collected from Android benign and ransomware applications. Experiments are applied to identify the best number of clusters for the proposed approach on the selected dataset and to compare the results with those obtained from applying the SVM single classification approach. Results show that applying $SVM\text{-}GWO_{k=2}$ approach outperforms the corresponding SVM classifier in terms of accuracy.

Keywords Clustering-based classification · Grey Wolf Optimizer · Support Vector Machine classifier · Ransomware detection · Evolutionary computation · EvoCluster

R. Qaddoura
Information Technology, Philadelphia University, Amman, Jordan
e-mail: rqaddoura@philadelphia.edu.jo

I. Aljarah (✉) · H. Faris · I. Almomani
King Abdullah II School for Information Technology, The University of Jordan, Amman, Jordan
e-mail: i.aljarah@ju.edu.jo

H. Faris
e-mail: hossam.faris@ju.edu.jo

I. Almomani
e-mail: imomani@psu.edu.sa; i.momani@ju.edu.jo

I. Almomani
Security Engineering Lab, Prince Sultan University, Riyadh, Saudi Arabia

I. Aljarah et al. (eds.), *Evolutionary Data Clustering: Algorithms and Applications*, Algorithms for Intelligent Systems,
https://doi.org/10.1007/978-981-33-4191-3_11

1 Introduction

Data has become an important asset with the emergence of cloud and IoT services and their consequences of big data acquisition and storage [5, 31]. This draws everyone's attention to the importance of data protection and security. Thus, ransomware detection approaches have been investigated to protect data from external threats.

Machine learning approaches have recently been applied on ransomware detection. These approaches fall under two main types: supervised learning and unsupervised learning[10, 23, 24]. Supervised learning is a classification task of constructing a training model by learning from previously labeled observations for future data predictions [2]. Unsupervised learning is a clustering task which explores the structure of unlabeled data and explores useful patterns [4, 8, 9, 11–13, 45, 49] by grouping data according to their similarities and dissimilarities [26, 47, 48]. Evolutionary clustering is another form of unsupervised learning which started to gain attention by many researchers [25, 44, 47] due to the improved performance over the traditional unsupervised clustering algorithms.

In the area of ransomware detection, both supervised and unsupervised machine learning approaches are used to generate high performance detection of threats. Combining both approaches while considering evolutionary clustering as an unsupervised learning approach is an attractive direction to investigate.

In this work, a hybrid classification approach for detecting ransomware in Android applications is proposed. The hybrid approach consists mainly of two stages: in the first stage, an evolutionary clustering approach using Grey Wolf Optimizer (*GWO*) is applied to segment the training dataset into a number of sub training datasets. Then in the second stage, an ensemble of support vector machine classifiers (SVM) is trained, where each SVM in the ensemble is trained based on a different subtraining dataset (i.e., cluster). Basically, the idea is to simplify the complexity of the data and build an expert classifier for each cluster of it. Therefore, the aim of the approach is to boost the detection accuracy in comparison with the conventional single classification approach.

The remainder of the paper is organized as follows: The next section shows the related work in ransomware detection. Section 3 presents GWO optimizer. Section 4 describes in detail the proposed algorithm. Section 5 discusses the data preparation and description. Section 6 presents the experiments conducted to evaluate and verify the efficiency of the algorithm. Section 7 concludes the main findings of this work.

2 Related Work

Ransomware is a malware which forbids users from accessing their files unless they pay attackers to gain access. Automatic analysis and detection of ransomware has been recently applied using supervised and unsupervised machine learning approaches.

Supervised classification approaches can be found extensively for ransomware detection. The work of [28] considers creating machine learning models using Random Forest, Support Vector Machine, and K-Nearest Neighbors approaches, and evaluating their performance in detecting ransomware. Two frameworks combining Reverse engineering, Natural Language Processing, and Machine Learning approaches are proposed in the work of [42, 43]. Authors of [51] applied various experiments related to network parameters and structures of the multi-layer perceptron to detect and classify ransomware threats. The work of [31] includes classifying infected files based on file entropy analysis for ransomware detection in backup systems using K-nearest neighbor, linear model, decision tree, decision tree ensemble, kernel trick, and neural network approaches. The paper [6] introduces NetConverse, which is a machine learning evaluation study to detect Windows ransomware network traffic. Other studies for detecting ransomware can be found in [18, 20, 38].

On the other hand, fewer works can be found for ransomware detection using unsupervised clustering machine learning approaches compared to those found for supervised approaches. Authors of [1] proposed a framework considering a modified version of sequential K-means clustering algorithm for detecting cloud insiders or malware perform malicious tasks on cloud customers' virtual machines. Another study [16] analyzed ransomware threats using the Cyber-genome clustering method, which are then combined with anti-malware product logs to identify distinct ransomware families. Other works [35–37] considers fuzzy hashing and a clustering method to detect ransomware threats.

However, another approach of clustering-based classification has proven in many applications that it is efficient and performs higher quality of results compared to those obtained by single approach of applying either classification or clustering. These applications include injury severity prediction of road traffic accidents [50], Superpixel Segmentation for PolSAR Image Classification [29], color histopathological image segmentation [19], hyperspectral image classification [33], indoor localization [3], breast tumor detection [52], Medical Relation detection[30], and text classification [39].

Thus, the clustering-based classification approach can be examined to detect ransomware threats, which to the best of our knowledge, could not be found in the literature. In this paper, we propose a clustering-based classification approach for ransomware detection using GWO clustering algorithm and SVM classification algorithm.

3 Grey Wolf Optimizer (GWO)

GWO is a nature-inspired optimizer inspired by the social activities performed by grey wolfs, which are the social hierarchy of leadership and the group hunting. The former activity considers a leader wolf (α), supporting wolves (β), and remaining wolves (δ and ω) [22, 34], and the later activity considers tracking, chasing, pursuing, encircling, harassing, and attacking the prey [34]. These activities are reflected mathematically in GWO algorithm.

The first activity of leadership maps the best solution to the α wolf, the other best solutions are mapped to the β and δ wolves. The remaining solutions are mapped to the ω wolves.

The second activity of hunting is reflected by predicting the prey position when converging to a better solution. This follows four processes: encircling, hunting, attacking, and searching. The encircling process is performed by two vectors calculated by Eqs. 1 and 2, and two coefficient values calculated by Eqs. 3 and 4. The hunting process is performed by a vector calculated by Eq. 5 and three values calculated by Eqs. 6, 7, and 8. The attacking process reflects exploiting the space by moving towards the prey and attacking it. The searching process reflects exploring the space by diverging from the prey to avoid local optima solutions.

$$D = |C \cdot X_p(t) - X(t)| \tag{1}$$

$$X(t+1) = X(t) - A.D \tag{2}$$

$$C = 2 \cdot r_2 \tag{3}$$

$$A = 2a \cdot r_1 - a \tag{4}$$

$$X(t+1) = (X_1 + X_2 + X_3)/3 \tag{5}$$

$$X_1 = X_\alpha - A_1 \cdot (|C_1 \cdot X_\alpha - X|) \tag{6}$$

$$X_2 = X_\beta - A_2 \cdot (|C_2 \cdot X_\beta - X|) \tag{7}$$

$$X_3 = X_\delta - A_3 \cdot (|C_3 \cdot X_\delta - X|) \tag{8}$$

where t is the iteration number, r_1 and r_2 are random vectors of values between 0 and 1, a is a vector which linearly decreases from 2 to 0, X is the position of the wolf, X_p is the position of the prey, X_α is the position vector of the alpha wolf, X_β is the position vector of the beta wolf, X_δ is the position vectors of the delta wolves. A_1 & C_1, A_2 & C_2, and A_3 & C_3 are coefficient vectors.

4 Proposed Approach

This section discusses the proposed approach, which is followed when running the experiments. The proposed approach is presented in Figure 1. First, the dataset is split into training and testing instances with a ratio of 80%:20%, respectively. The clustering process is then applied to the training instances using GWO algorithm. This generates k clusters in which each cluster has similar instances. For each cluster,

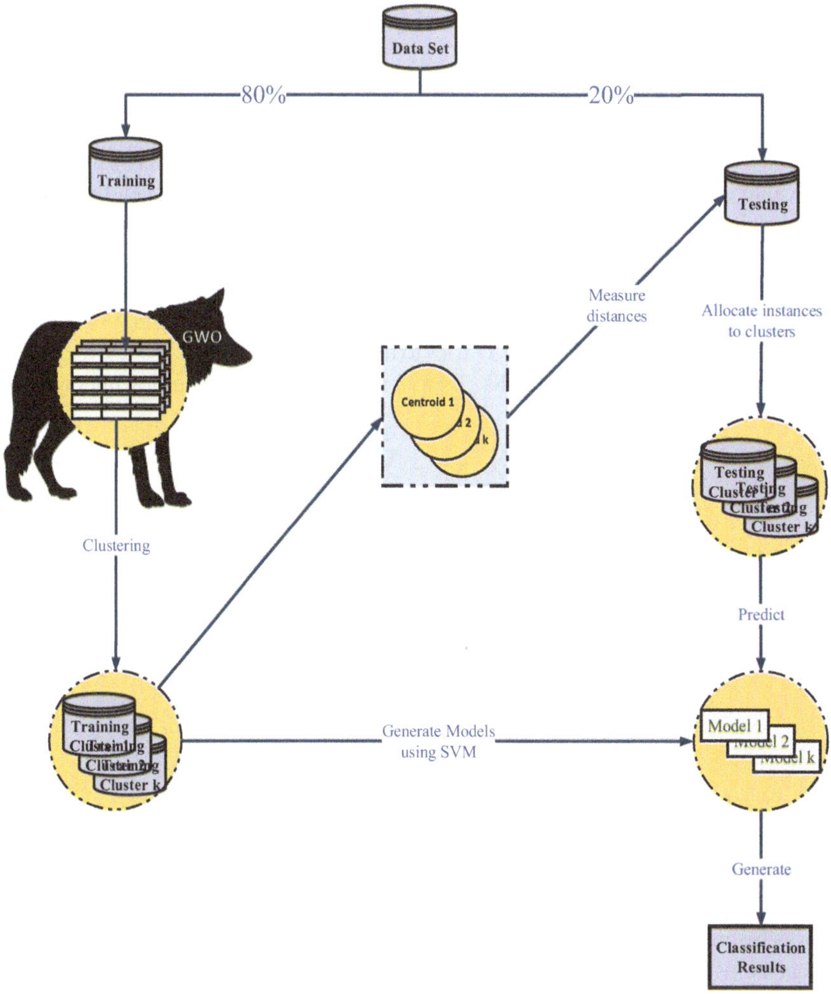

Fig. 1 Proposed approach

classification is performed by fitting a model to the training instances of the cluster using SVM algorithm. Then, the testing instances are allocated to clusters according to the euclidean distance between the instances and the centroids, by which an instance is allocated to a cluster having the minimal distance between the instance and the corresponding centroid. Finally, the labels for the instances of each cluster are predicted using the corresponding model, which forms the classification results.

The remainder of this section discusses the two main processes of the proposed approach: the clustering process of GWO algorithm using TWCV fitness function, which tests the convergence of GWO, and the classification process of the clustered instances using SVM.

4.1 Clustering with GWO

Clustering with GWO is performed by enhancing the centroids' selection of the clustering task [46]. Random selection of centroids for each individual is performed at the first iteration of the algorithm, which is enhanced in later iterations by minimizing the value of the Total Within Cluster Variance (TWCV) [41] between the centroid and the cluster instances. The best enhanced individual at the last iteration is returned as a solution.

Total Within Cluster Variance (TWCV) [21, 32, 41] evaluates the convergence of evolutionary algorithms. As the name indicates, it calculates the total within cluster variance. In the experiments, GWO tries to minimize the value of TWCV to enhance the quality of the clustering results. TWCV is calculated by the following equation:

$$TWCV = \sum_{n=1}^{N}\sum_{f=1}^{F} p_{nf}^2 - \sum_{k=1}^{K}\frac{1}{|p_k|}\sum_{f=1}^{F}\left(\sum p_{kf}\right)^2 \tag{9}$$

where F is the number of features, p_{nf} is feature f of the data point n, p_{kf} is feature f of the data point k, and $|p_k|$ is the number of data points in cluster k.

The solution obtained consists of the training instances grouped to k different training clusters. The centroids for each cluster are used to map each testing instance to the closest cluster, forming k testing clusters.

4.2 Classification with SVM

For the training instances of each cluster generated from the previous process, a linear kernel SVM classification is performed, which creates a fitted model. SVM works by creating a hyperplane in space [40]. The algorithm separates the instances of each cluster from others by finding the hyperplane with the largest distance from the training instances [27, 40].

The fitted model of the training instances of each cluster is used to predict the labels of the corresponding testing instances of the cluster. Recall that the testing instances of the same cluster are recognized by the previous clustering process. Performing this step for each cluster results in predicting the labels for all testing instances.

The power of performing the clustering process as a preparatory process to the classification process is to find a specialized model for the instances in a training cluster, which takes into consideration similar characteristics of the instances. This helps the classification algorithm generate enhanced classification results.

5 Dataset Description

Selection of Android benign and ransomware apps: The group of Android applications that were collected in [14, 15] has been used to build the dataset under study. The dataset was constructed based on 1000 Android application samples. This dataset is balanced and consists of 500 benign apps in addition to 500 ransomware apps. The benign apps were downloaded from Google play store.[1] These apps were chosen from different categories which are free and highly downloaded by users. Ransomware samples were collected from different sources including Koodous,[2] RansomProper project[17] and VirusTotal[3] to ensure gathering a unique set of ransomware that belong to different families with locking, encryption, and threatening messages features.

Reverse Engineering Process: Android apps are usually available and distributed in the format of APK files. APK is a compressed file that holds the overall app's source code that was originally written in java. To analyze the app, this APK needs to be decompiled to retrieve important files like ".xml" and ".dex" files. There are many tools available and can be used to decompile the APK through a process called "Reverse Engineering" APKtool[4] was used by this research to decompile the APK files into ".smali" files.

Code parsing and features extraction: The small files were then parsed to extract a finite set of features. The selected features were the API packages and permissions belong to Oreo Android release 27 (API 27). From this release, 199 API packages and 115 permissions were considered; with an overall 314 features in addition to the identifier of the app such as the hash code and also the category of the app whether benign or ransomware. The features extraction process counted the frequencies of API calls and permissions in each Android app to generate the following joint vector V for each one of them:

$$V = \{a_1, a_2, \ldots a_n\} \cup \{p_1, p_2, \ldots p_m\} \tag{10}$$

where a_i is the number of occurrences of the ith API call in the application, p_j is the number of occurrences of the jth permission in the application, n is the total number of API packages and m is the total number of permissions.

Data Cleansing: As part of data cleansing, duplicate vectors were removed to ensure unique records and produce a quality dataset. The duplicate vectors could have resulted from identical Android apps using different identities (different hash codes).

---

[1] https://play.google.com/store.

[2] https://koodous.com/apks.

[3] https://www.virustotal.com.

[4] https://ibotpeaches.github.io/Apktool.

6 Experimental Results

The experiments are conducted using two tools: EvoCluster framework [46] for the clustering process with GWO, which is a recent framework for clustering datasets using nature-inspired algorithms, by which GWO is one of the algorithms of the framework. The popular scikit learn tool [40] is used for the classification process with linear SVM.

The experiments are evaluated by measuring the accuracy value, which reflects the quality of the classification process. It is the ratio of the number of the true predicted values (TP + TN) to the number of all predicted values (TP + TN + FP + FN), which is observed by Eq. 11.

$$Accuracy = (TP + TN)/(TP + TN + FP + FN) \tag{11}$$

where TP is the true positive, FP is the false positive, TN is the true negative, and FN is the False negative.

The proposed approach is evaluated by running the experiments 30 times for each value of k, which are 2, 3, and 4. Then, comparing the results of the average value of the accuracy measure for each value of k against the average accuracy value of applying a single SVM classification. Table 1 shows the performance of SVM-GWO for different values of k against single SVM classification approach.

As observed from Table 1, SVM-GWO with k value of 2 outperforms itself with other values of k, and outperforms the corresponding SVM algorithm. It has the largest value of the average accuracy compared to the others on the selected dataset. It also has a recognizable highest value for the best and worst results.

The reason behind the high value of accuracy for the algorithm having $k = 2$ compared with the other values of k is due to the nature of the dataset instances where an instance is a ransomware or benign attack. This confirms that the classification of the dataset should be considered binary. In addition, the high value of accuracy for the SVM-GWO with $k = 2$ compared to SVM is due to the powerful effect of clustering as a preparatory step for the classification process.

Figure 2 shows the interquartile range, average accuracy value, best accuracy values, and worst accuracy value [7] for 30 runs of SVM, SVM-$GWO_{k=2}$, SVM-$GWO_{k=3}$, and SVM-$GWO_{k=4}$. It is observed from the figure that the SVM-GWO

Table 1 Results for using SVM with GWO with different values of k using the accuracy evaluation measure (Highest values of the average, standard deviation, best, and worst are marked with **bold**)

	Average	Std	Best	Worst
SVM	0.9517	**0.0024**	0.955	0.95
SVM-$GWO_{k=2}$	**0.9803**	0.0066	**0.99**	**0.96**
SVM-$GWO_{k=3}$	0.8608	0.0831	0.94	0.605
SVM-$GWO_{k=4}$	0.7063	0.0836	0.89	0.6

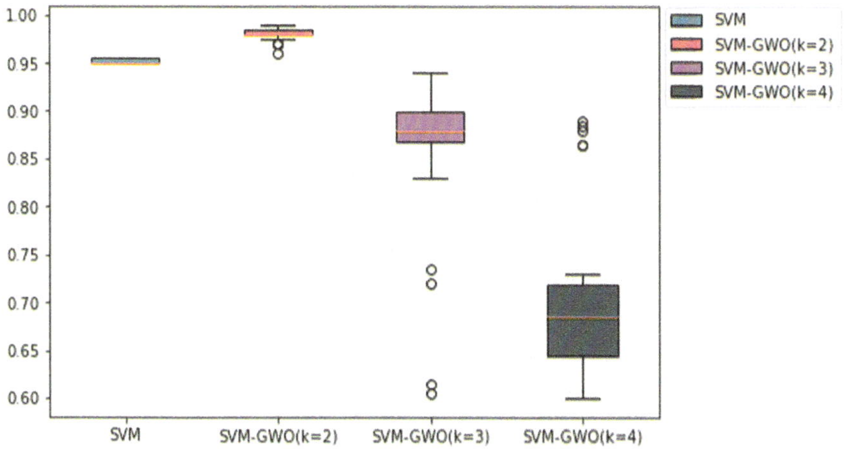

Fig. 2 Box plot for the accuracy values of SVM, $SVM\text{-}GWO_{k=2}$, $SVM\text{-}GWO_{k=3}$, and $SVM\text{-}GWO_{k=4}$

approach with $k = 2$ gives better results for all runs. The worst accuracy value of it is better than the best one for the others. In addition, the standard deviation for the proposed approach is recognized to be very minimal, which indicates a stable approach.

7 Conclusion

In this paper, ransomware detection has been applied by proposing a hybrid approach of evolutionary clustering approach using GWO with an ensemble of SVM classifier on a dataset of an Android benign and ransomware applications, to boost the detection accuracy of this approach in comparison with the conventional single classification approach. Experiments are made to detect the best number of clusters for GWO, and to compare the results achieved by the proposed approach and the corresponding single classification approach. Results show that $SVM\text{-}GWO$ with $k = 2$ outperforms the single classification approach in terms of accuracy.

For future work, this paper can be extended by enhancing the proposed approach and experimenting it with different evolutionary clustering algorithms and different classification algorithms, in which a different classifier can be identified for different clusters obtained from clustering. Further, the proposed approach could also be experimented on unbalanced ransomware dataset or any other applications.

References

1. Abdelsalam, Mahmoud, Ram Krishnan, and Ravi Sandhu. 2017. Clustering-based IaaS cloud monitoring. In *2017 IEEE 10th international conference on cloud computing (CLOUD)*, 672–679. IEEE.
2. Aggarwal, Charu C. 2015. Data classification. In *Data mining*, 285–344. Springer.
3. Akram, Beenish Ayesha, Ali Hammad Akbar, and Ki-Hyung Kim. 2018. CEnsLoc: Infrastructure-less indoor localization methodology using GMM clustering-based classification ensembles. *Mobile Information Systems* 2018.
4. Al-Madi, Nailah, Ibrahim Aljarah, and Simone A. Ludwig. 2014. Parallel glowworm swarm optimization clustering algorithm based on mapreduce. In *2014 IEEE symposium on swarm intelligence*, 1–8. IEEE.
5. Al Shorman, Amaal, Hossam Faris, and Ibrahim Aljarah. 2020. Unsupervised intelligent system based on one class support vector machine and grey wolf optimization for IoT botnet detection. *Journal of Ambient Intelligence and Humanized Computing* 11 (7): 2809–2825.
6. Alhawi, Omar M.K., James Baldwin, and Ali Dehghantanha. 2018. Leveraging machine learning techniques for windows ransomware network traffic detection. In *Cyber threat intelligence*, 93–106. Springer.
7. Aljarah, Ibrahim, Al-Zoubi Ala'M, Hossam Faris, Mohammad A. Hassonah, Seyedali Mirjalili, and Heba Saadeh. 2018. Simultaneous feature selection and support vector machine optimization using the grasshopper optimization algorithm. *Cognitive Computation* 1–18.
8. Aljarah, Ibrahim, and Simone A. Ludwig. 2012. Parallel particle swarm optimization clustering algorithm based on mapreduce methodology. In *2012 fourth world congress on nature and biologically inspired computing (NaBIC)*, 104–111. IEEE.
9. Aljarah, Ibrahim, and Simone A. Ludwig. 2013. Mapreduce intrusion detection system based on a particle swarm optimization clustering algorithm. In *2013 IEEE congress on evolutionary computation*, 955–962. IEEE.
10. Aljarah, Ibrahim, and Simone A. Ludwig. 2013. A new clustering approach based on glowworm swarm optimization. In *2013 IEEE congress on evolutionary computation*, 2642–2649. IEEE.
11. Aljarah, Ibrahim, and Simone A. Ludwig. 2013. Towards a scalable intrusion detection system based on parallel PSO clustering using mapreduce. In *Proceedings of the 15th annual conference companion on Genetic and evolutionary computation*, 169–170.
12. Aljarah, Ibrahim, Majdi Mafarja, Ali Asghar Heidari, Hossam Faris, and Seyedali Mirjalili. 2020. Clustering analysis using a novel locality-informed grey wolf-inspired clustering approach. *Knowledge and Information Systems* 62 (2): 507–539.
13. Aljarah, Ibrahim, Majdi Mafarja, Ali Asghar Heidari, Hossam Faris, and Seyedali Mirjalili. 2020. Multi-verse optimizer: Theory, literature review, and application in data clustering. In *Nature-inspired optimizers*, 123–141. Springer.
14. Alsoghyer, S., and I. Almomani. 2020. On the effectiveness of application permissions for android ransomware detection. In *2020 6th conference on data science and machine learning applications (CDMA)*, 94–99.
15. Alsoghyer, Samah, and Iman Almomani. 2019. Ransomware detection system for android applications. *Electronics* 8 (8): 868.
16. Arrott, Anthony, Arun Lakhotia, Ferenc Leitold, and Charles LeDoux. 2018. Cluster analysis for deobfuscation of malware variants during ransomware attacks. In *2018 international conference on cyber situational awareness, data analytics and assessment (Cyber SA)*, 1–9. IEEE.
17. Chen, Jing, Chiheng Wang, Ziming Zhao, Kai Chen, Ruiying Du, and Gail-Joon Ahn. 2017. Uncovering the face of android ransomware: Characterization and real-time detection. *IEEE Transactions on Information Forensics and Security* 13 (5): 1286–1300.
18. Chen, Li, Chih-Yuan Yang, Anindya Paul, and Ravi Sahita. 2018. Towards resilient machine learning for ransomware detection. arXiv:1812.09400.

19. Dhal, Krishna Gopal, Arunita Das, Swarnajit Ray, and Sanjoy Das. 2019. A clustering based classification approach based on modified cuckoo search algorithm. *Pattern Recognition and Image Analysis* 29 (3): 344–359.
20. Dion, Y., and Sarfraz N. Brohi. 2020. An experimental study to evaluate the performance of machine learning alogrithms in ransomware detection. *Journal of Engineering Science and Technology* 15 (2): 967–981.
21. Du, Jun, Emin Erkan Korkmaz, Reda Alhajj, and Ken Barker. 2005. Alternative clustering by utilizing multi-objective genetic algorithm with linked-list based chromosome encoding. In *International workshop on machine learning and data mining in pattern recognition*, 346–355. Springer.
22. Faris, Hossam, Ibrahim Aljarah, Mohammed Azmi Al-Betar, and Seyedali Mirjalili. 2018. Grey wolf optimizer: A review of recent variants and applications. *Neural Computing and Applications* 30 (2): 413–435.
23. Faris, Hossam, Ibrahim Aljarah, and Ja'far Alqatawna. 2015. Optimizing feedforward neural networks using krill herd algorithm for e-mail spam detection. In *2015 IEEE Jordan conference on applied electrical engineering and computing technologies (AEECT)*, 1–5. IEEE.
24. Faris, Hossam, Ibrahim Aljarah, Seyedali Mirjalili, Pedro A. Castillo, and Juan Julián Merelo Guervós. 2016. EvoloPy: An open-source nature-inspired optimization framework in Python. In *IJCCI (ECTA)*, 171–177.
25. Hajeer, Mustafa, and Dipankar Dasgupta. 2017. Handling big data using a data-aware HDFS and evolutionary clustering technique. *IEEE Transactions on Big Data* 5 (2): 134–147.
26. Han, Jiawei, Jian Pei, and Micheline Kamber 2011. *Data mining: Concepts and techniques*. Elsevier.
27. Hastie, Trevor, Robert Tibshirani, and Jerome Friedman. 2009. *The elements of statistical learning: Data mining, inference, and prediction*. Springer Science & Business Media.
28. Hirano, Manabu, and Ryotaro Kobayashi. 2019. Machine learning based ransomware detection using storage access patterns obtained from live-forensic hypervisor. In *2019 sixth international conference on internet of things: Systems, Management and security (IOTSMS)*, 1–6. IEEE.
29. Hou, Biao, Chen Yang, Bo Ren, and Licheng Jiao. 2018. Decomposition-feature-iterative-clustering-based superpixel segmentation for PolSAR image classification. *IEEE Geoscience and Remote Sensing Letters* 15 (8): 1239–1243.
30. Kim, Youngjun, Ellen Riloff, and Stéphane M. Meystre. 2017. Exploiting unlabeled texts with clustering-based instance selection for medical relation classification. In *AMIA annual symposium proceedings*, vol. 2017, 1060. American Medical Informatics Association.
31. Lee, Kyungroul, Sun-Young Lee, and Kangbin Yim. 2019. Machine learning based file entropy analysis for ransomware detection in backup systems. *IEEE Access* 7: 110205–110215.
32. Liu, Yimin, Tansel Özyer, Reda Alhajj, and Ken Barker. 2005. Integrating multi-objective genetic algorithm and validity analysis for locating and ranking alternative clustering. *Informatica* 29 (1).
33. Lu, Ting, Shutao Li, and Jón Atli Benediktsson. 2017. Iterative clustering based active learning for hyperspectral image classification. In *2017 IEEE international geoscience and remote sensing symposium (IGARSS)*, 3664–3667. IEEE.
34. Mirjalili, Seyedali, Seyed Mohammad Mirjalili, and Andrew Lewis. 2014. Grey wolf optimizer. *Advances in Engineering Software* 69: 46–61.
35. Naik, Nitin, Paul Jenkins, Jonathan Gillett, Haralambos Mouratidis, Kshirasagar Naik, and Jingping Song. 2019. Lockout-tagout ransomware: A detection method for ransomware using fuzzy hashing and clustering. In *2019 IEEE symposium series on computational intelligence (SSCI)*, 641–648. IEEE.
36. Naik, Nitin, Paul Jenkins, and Nick Savage. 2019. A ransomware detection method using fuzzy hashing for mitigating the risk of occlusion of information systems. In *2019 international symposium on systems engineering (ISSE)*, 1–6. IEEE.
37. Naik, Nitin, Paul Jenkins, Nick Savage, and Longzhi Yang. 2019. Cyberthreat hunting-part 2: Tracking ransomware threat actors using fuzzy hashing and fuzzy c-means clustering. In *2019 IEEE international conference on fuzzy systems (FUZZ-IEEE)*, 1–6. IEEE.

38. Niture, Nandkumar A. 2020. Machine learning and cryptographic algorithms—analysis and design in ransomware and vulnerabilities detection. *Machine Learning* 1: 19.
39. Onan, Aytug. 2017. Hybrid supervised clustering based ensemble scheme for text classification. *Kybernetes*.
40. Pedregosa, F., G. Varoquaux, A. Gramfort, V. Michel, B. Thirion, O. Grisel, M. Blondel, P. Prettenhofer, R. Weiss, V. Dubourg, J. Vanderplas, A. Passos, D. Cournapeau, M. Brucher, M. Perrot, and E. Duchesnay. 2011. Scikit-learn: Machine learning in Python. *Journal of Machine Learning Research* 12: 2825–2830.
41. Peng, Peter, Omer Addam, Mohamad Elzohbi, Sibel T. Özyer, Ahmad Elhajj, Shang Gao, Yimin Liu, Tansel Özyer, Mehmet Kaya, Mick Ridley, et al. 2014. Reporting and analyzing alternative clustering solutions by employing multi-objective genetic algorithm and conducting experiments on cancer data. *Knowledge-Based Systems* 56: 108–122.
42. Poudyal, Subash, Dipankar Dasgupta, Zahid Akhtar, and K. Gupta. 2019. A multi-level ransomware detection framework using natural language processing and machine learning. In *14th international conference on malicious and unwanted software, MALCON*.
43. Poudyal, Subash, Kul Prasad Subedi, and Dipankar Dasgupta. 2018. A framework for analyzing ransomware using machine learning. In *2018 IEEE symposium series on computational intelligence (SSCI)*, 1692–1699. IEEE.
44. Qaddoura, Raneem, Waref Al Manaseer, Mohammad A.M. Abushariah, and Mohammad Aref Alshraideh. 2020. Dental radiography segmentation using expectation-maximization clustering and grasshopper optimizer. *Multimedia Tools and Applications*.
45. Qaddoura, Raneem, Hossam Faris, and Ibrahim Aljarah. 2020. An efficient clustering algorithm based on the k-nearest neighbors with an indexing ratio. *International Journal of Machine Learning and Cybernetics* 11 (3): 675–714.
46. Qaddoura, Raneem, Hossam Faris, Ibrahim Aljarah, and Pedro A Castillo. 2020. EvoCluster: An open-source nature-inspired optimization clustering framework in Python. In *International conference on the applications of evolutionary computation (part of EvoStar)*, 20–36. Springer.
47. Qaddoura, R., H. Faris, and I. Aljarah. 2020. An efficient evolutionary algorithm with a nearest neighbor search technique for clustering analysis. *Journal of Ambient Intelligence and Humanized Computing* 1–26.
48. Qaddoura, R., H. Faris, I. Aljarah, J. Merelo, and P. Castillo. 2020. Empirical evaluation of distance measures for nearest point with indexing ratio clustering algorithm. In *Proceedings of the 12th international joint conference on computational intelligence*, vol. 1, 430–438. NCTA. ISBN 978-989-758-475-6, https://doi.org/10.5220/0010121504300438.
49. Shukri, Sarah, Hossam Faris, Ibrahim Aljarah, Seyedali Mirjalili, and Ajith Abraham. 2018. Evolutionary static and dynamic clustering algorithms based on multi-verse optimizer. *Engineering Applications of Artificial Intelligence* 72: 54–66.
50. Taamneh, Madhar, Salah Taamneh, and Sharaf Alkheder. 2017. Clustering-based classification of road traffic accidents using hierarchical clustering and artificial neural networks. *International Journal of Injury Control and Safety Promotion* 24 (3): 388–395.
51. Vinayakumar, R., K.P. Soman, K.K. Senthil Velan, and Shaunak Ganorkar. 2017. Evaluating shallow and deep networks for ransomware detection and classification. In *2017 international conference on advances in computing, communications and informatics (ICACCI)*, 259–265. IEEE.
52. Zhang, Qi, Yang Xiao, Jingfeng Suo, Yu. Jun Shi, Yi Guo Jinhua, Yuanyuan Wang, and Hairong Zheng. 2017. Sonoelastomics for breast tumor classification: A radiomics approach with clustering-based feature selection on sonoelastography. *Ultrasound in Medicine & Biology* 43 (5): 1058–1069.

The manufacturer's authorised representative in the EU is Springer Nature Customer Service Centre GmbH, Europaplatz 3, 69115 Heidelberg, Germany. If you have any concerns regarding our products, please contact ProductSafety@springernature.com

Printed and bound by CPI Group (UK) Ltd, Croydon, CR0 4YY

30/09/2026

02225295-0004